Design and Applications of Heat Exchangers

Design and Applications of Heat Exchangers

Edited by Edgar Miller

STATES
ACADEMIC PRESS
www.statesacademicpress.com

States Academic Press,
109 South 5th Street,
Brooklyn, NY 11249, USA

ISBN: 978-1-63989-149-8

Cataloging-in-Publication Data

Design and applications of heat exchangers / edited by Edgar Miller.
p. cm.
Includes bibliographical references and index.
ISBN 978-1-63989-149-8
1. Heat exchangers. 2. Heat exchangers--Design and construction.
3. Heat--Transmission. I. Miller, Edgar.
TJ263 .D47 2022
621.402 5--dc23

For information on all States Academic Press publications
visit our website at www.statesacademicpress.com

STATES
ACADEMIC PRESS

Contents

Preface

The main aim of this book is to educate learners and enhance their research focus by presenting diverse topics covering this vast field. This is an advanced book which compiles significant studies by distinguished experts in the area of analysis. This book addresses successive solutions to the challenges arising in the area of application, along with it; the book provides scope for future developments.

A system that is used to transfer heat between two or more fluids is known as a heat exchanger. It can be used for cooling as well as heating processes. The fluids may be in direct contact or may be separated by a solid wall to prevent mixing. A prominent example of heat exchanger is present within an internal combustion engine. The circulating fluid in an internal combustion engine is known as engine coolant. It flows through radiator coils and air flows past the coils. This cools the coolant and heats the incoming air. Heat exchangers are widely used in refrigeration, air conditioning, space heating, petrochemical plants and petroleum refineries. The topics included in this book on heat exchangers are of utmost significance and bound to provide incredible insights to readers. Also included herein is a detailed explanation of the various designs and applications of heat exchangers. This book is a complete source of knowledge on the present status of this important field.

It was a great honour to edit this book, though there were challenges, as it involved a lot of communication and networking between me and the editorial team. However, the end result was this all-inclusive book covering diverse themes in the field.

Finally, it is important to acknowledge the efforts of the contributors for their excellent chapters, through which a wide variety of issues have been addressed. I would also like to thank my colleagues for their valuable feedback during the making of this book.

Editor

1

Design of Heat Transfer Surfaces in Agitated Vessels

Vitor da Silva Rosa and Deovaldo de Moraes Júnior

Abstract

The project on heat transfer surfaces in agitated vessels is based on the determination of the heat exchange area, which is necessary to abide by the process conditions as mixing quality and efficiency of heat transfer. The heat transfer area is determined from the overall heat transfer coefficient (*U*). The coefficient (*U*) represents the operation quality in heat transfers being a function of conduction and convection mechanisms. The determination of *U* is held from the Nusselt's number, which is related to the dimensionless Reynolds and Prandtl's, and from the fluid's viscosity relation that is being agitated in the bulk temperature and the viscosity in the wall's temperature of heat exchange. The aim of this chapter is to present a summary for the literature concerning heat transfer in agitated vessels (equipped with jackets, helical coils, spiral coils, and vertical tube baffles) and also the many parameters of Nusselt's equation for these surfaces. It will present a numerical example for a project in an agitated vessel using vertical tube baffles and a 45° pitched blade turbine. Subsequently, the same procedure is held with a turbine radial impeller, in order to compare the heat transfer efficiencies.

Keywords: vessels, overall heat transfer coefficient, Nusselt, jackets, coils, vertical tube baffles, mechanical impeller, convection

1. Introduction

Vessels with mechanical impellers are largely used by chemical, petrochemical, food, textile, and other industries, operating as extractors, concentrators, flotators, storage vessels, and chemical reactors (producing polymers, paints, fertilizers, and resins, for instance).

The aforementioned processes need heating or cooling, these being provided by jackets, coils, and vertical tube baffles, where the heat transfer area is the target parameter for the project as a function of the overall heat transfer coefficient.

The jackets are surfaces of heat transfer characterized by encompassing the tank, given that the thermal fluid roams the space between the jacket and the tank. On these equipments, the heat transfer efficiency is low due to the heat source being in the wall, which provokes uneven heating of the fluid within the tank, besides the structural problems in big units. Nevertheless, this system presents advantages as easy cleaning and available project data provided in the current literature [1].

The helical coils are made out of tubes between the impeller center line and the tank's wall. The contact area with the fluid to be heated or cooled is much bigger around the jackets, which end up increasing the heat transfer efficiency; however, the difficulty is in the cleaning of the equipment. Most agitated vessels involving heat transfer operations are projected with helical coils due to the availability of project data [2].

The spiral coils consist of wounded tubes generally placed on the bottom of the tank. They are mostly applied to heating viscous fluids in pumping transport. Spiral coils have as disadvantage the located heat transfer, fostering uneven heating. Another downside is the lack of data in the current literature [3].

The three heating transfer surfaces mentioned above demand baffles alongside the tank's wall to avoid vortex, which is characterized by the development of a bottleneck formed by the agitated fluid around the mechanical impeller. The vortex is an indicator of inefficiency at mixing and low heat transfer, due to the circular and organized stream lines.

The vertical tube baffles are surfaces that provide significant heating transfer rate and also eradicate the development of a bottleneck, due to its spatial structure, which makes necessary to allocate extra baffles alongside the tank's wall. The drawback of this technique is the lack of data project, especially on continuous operation [4].

However, each type of heat transfer surface will present parameters for Nusselt's equation particular to each system in which they were determined, in such a way that it is possible to obtain a general model that covers all surfaces for heat transfer in agitated vessels.

For batch processes the heat transfer area is determined by transient energy balances, while for continuous processes, on the large majority, the area is obtained by energy balances for steady-state operation.

2. Fundamentals

2.1. Steady-state operation

Heat transfer in agitated vessels is carried through heat exchange surfaces, like jackets, helical coils, spiral coils, and vertical tubular baffles [5]. The surfaces of heat exchange are designed as a function of the area necessary to carry the heating or cooling, based on the overall heat transfer coefficient, which is a function of the dimensionless groups Reynolds, Prandtl, and Nusselt. The classic project equation of the heat exchange area for steady-state operation is shown in Eq. (1).

$$A = Q/ULMTD \tag{1}$$

in which the overall heat transfer coefficient (U) is a function of local coefficients of internal convection, related to the external surface area or internal and external convection coefficient (h_i and h_o) for the hot and cold fluid, respectively.

Eq. (2) shows the dependence of the overall coefficient U to the convection coefficients (internal and external), the resistance to heat transfer presented by the heat exchange surface (conduction), and the resistances by internal and external fouling in relation to the pipelines.

$$1/UA = 1/h_iA_i + Rd_i/A_i + R_c + Rd_0/A_0 + 1/h_oA_o \tag{2}$$

The thermal resistance presented by conduction can be disregarded when the thickness of the surface wall is negligible in relation to the internal and external diameter and when it is made out of a material with high thermal conductivity. Therefore, the term R_c (referring to the thermal resistance presented by heat conduction) in Eq. (2) can be disregarded [6]. The terms regarding the resistance by fouling can also be neglected if the pipelines used are new or with operational time inferior to 5 years. Eq. (2) can be written as shown in Eq. (3), being the overall heat transfer coefficient only dependent of the internal and external convection.

$$1/UA = 1/h_iA_i + 1/h_oA_0 \tag{3}$$

The internal film coefficient by definition is associated to the fluid's flow in the interior of the heat exchange surface, while the external coefficient, in the case of a vessel's surface, is related to the mixture of the fluids that will be agitated. If the mixture, which will be agitated, is the one that will be heated, the heat transfer happens on the external surface of contact of the hot source with the cold fluid; therefore, the internal convection coefficient must be corrected and placed as a function of this external area, called then as corrected internal coefficient (h_{io}), calculated from the relation between the internal and external diameter of the pipeline, as shown in Eq. (4).

$$h_{io} = h_i\frac{D_i}{D_e} \tag{4}$$

Adding Eq. (4) to Eq. (3) and indicating the areas as a function of the diameters, Eq. (5) is obtained.

$$1/U\pi D_eL = 1/(D_e/D_i)h_{io}\pi D_iL + 1/h_o\pi D_eL \tag{5}$$

It can be seen in Eq. (5) that the area of heat transfer A (πD_eL) corresponds to the external area of the tube's surface A_e. Grouping the common terms in Eq. (5), the overall coefficient of heat transfer U can be calculated as a function of the corrected internal convection coefficient and the external convection coefficient, as shown in Eq. (6).

$$1/U = 1/h_{io} + 1/h_o \tag{6}$$

The determination of the rectified internal convection coefficient (h_{io}) implies the determination of the coefficient h_i, which is a function of the fluid's physical properties (viscosity, density, specific heat, and thermal conductivity), the heat transfer surface geometry (tube diameter), the kind of flow (laminar or turbulent), the speed flow, and the temperature gradient [7].

The analytic determination of the equation, which relates the internal film coefficient to the aforementioned variables, can be done through the application of the hydrodynamic and thermal boundary layer theory on the heat transfer surface exposed to the hot fluid and the cold fluid. After using dimension analysis and some similar equations to those of continuity, momentum, and energy, it can be noticed that the Nusselt's number is a function of the Reynolds and Prandtl's number. However, this analytical method can be difficult to be solved due to the nonhomogeneous partial differential equations, which will depend of the specific boundary conditions as a function of the heat transfer surface geometry. To avoid this situation, the Buckingham's Pi theorem can be used, which relates the variables that may affect the answer variables, in this case the film coefficient, through dimensionless groups [8]. With dimension analysis the internal film coefficient can be related with the many variables aforementioned with a function in which h_i is the dependent variable, as seen in Eq. (7).

$$h_i = f(\mu, \rho, c_p, D_i, k, u) \tag{7}$$

The function expressed in Eq. (7) can be written powering the dependent variables to an exponent, adding a proportionality constant K and a dimensional constant K_h, which will be reduced to 1,0 if all the other variable dimensions, when combined, lead to thermal quantities, as illustrated in Eq. (8).

$$h_i = Ku^a \rho^b c_p{}^c D_i{}^d k^e \mu^f K''^g \tag{8}$$

Expressing all the terms in Eq. (8) as fundamental quantities and then rearranging it, the semiempiric model known as Nusselt's equation can be obtained (Eq. (9)).

$$\frac{h_i.D_i}{k} = K\left(\frac{D_i u \rho}{\mu}\right)^a \left(\frac{c_p \mu}{k}\right)^b \tag{9}$$

The term on the left side of Eq. (9) is the dimensionless Nusselt, and the terms on the right, going from left to right, are the dimensionless Reynolds and Prandtl, respectively; hence, Eq. (9) can be written in the following manner.

$$Nu = K.Re^a.Pr^b \tag{10}$$

Many researches established the exponents and constant values in Eq. (24) for forced flow on the interior of the cylindrical heat transfer surface, that is, on the interior of the pipelines.

Sieder-Tate [9] added a correction factor in Eq. (9), taking into account the viscosity effects on wider variations of temperature, a relation between the fluid's viscosity on the bulk temperature, and the fluid's viscosity on the wall's temperature, as shown in Eq. (11), which is valid for turbulent flows on the interior of cylindrical surfaces with Reynolds above 10,000 and smooth pipes.

$$Nu = 0.027Re^{0.8}Pr^{1/3}\left(\mu/\mu_w\right)^{0.14} \tag{11}$$

Eq. (11) is a simplified form to determine the internal coefficient film; however, it carries an overall error of 40%, being necessary to implement correction factors on the design of the heat transfer area. Geankoplis [10] presents an equation to determine the internal coefficient film for water with temperatures ranging from 4 to 105°C on smooth pipes' interior in fully developed turbulent flow, as shown in Eq. (12), which carries an overall error of 25%.

$$h_i = 1429\left(1 + 0.0146\overline{T}\right).u^{0.8}/D_i^{0.2} \tag{12}$$

Eq. (11) and Eq. (12) are simple to be used and present satisfactory results for the majority of heat exchanger projects, despite the high error associated with them. Gnielinski [11] presents a correlation to determine the internal convection coefficient, with an error lower than 10%, for fully developed turbulent flows, with Reynolds ranging from 3000 to 5 × 10^6, Prandtl between 0.5 and 2000, and for a relation (L/D_i) greater than 10. The correlation is presented in Eq. (13).

$$Nu = \frac{(f'/8)(Re - 1000)Pr}{1 + 12.7(f'/8)^{1/2}\left(Pr^{2/3} - 1\right)} \tag{13}$$

The friction factor (f') can be obtained from Moody's diagram or by specialized correlations. In the case of Eq. (13), which is for fully developed turbulent flow, the friction factor can be calculated through Eq. (14).

$$f' = (0.790 lnRe - 1.64)^{-2} \tag{14}$$

The external convection coefficient (h_o) is a function of the same variables of the internal convection coefficient (h_i); however, the heat transfer surface's geometry and the mechanical impeller have large influence in heat transmission. The expression to determine the h_o can be obtained through the Buckingham's Pi theorem [5], similar to the determination of h_i; hence, Eq. (9) can be rewritten by modifying the Nusselt and Reynolds' numbers for agitation (Re_a), presented in Eq. (15), and adding the viscosity correction factor proposed by Sieder-Tate, as shown in Eq. (16).

$$Re_a = \frac{ND_a^{\,2}\rho}{\mu} \tag{15}$$

$$\frac{h_o D_t}{k} = K\left(\frac{ND_a^{\,2}\rho}{\mu}\right)^a\left(\frac{C_p\mu}{k}\right)^b\left(\frac{\mu}{\mu_w}\right)^c \tag{16}$$

Differently from Eq. (9), which is generic for any type of fluid, as long as the heat transfer surface is cylindrical and has a fully developed turbulent flow, Eq. (16) has the following limitations: (i) the position and geometry of the mechanical impeller, (ii) the flow's regime, (iii) the geometry of the heat transfer surface, and (iv) the tank's geometry. These limitations do not allow to obtain the values for the K constant and the exponents generically, for all surfaces of heat transfer in agitated vessels [12].

2.2. Unsteady-state operation

The project of the heat transfer surface for agitated vessels will depend on whether the tank is operating in discontinuous mode (batch) or continuous mode.

The batch processes in agitated vessels are common in many industrial processes, specially, in chemical reactors. In batch operations, there is the occurrence of two typical situations: (1) the design parameter is the operation time for heating or cooling; hence, the surface area is unknown; (2) the area for heat exchange is known and the operation time is unknown.

These processes can be carried out on isothermal operations and with phase change (heating with steam or cooling with thermal fluid vaporization) or non-isothermal operations and fluids without phase change.

An important consideration must be highlighted regarding the U coefficient: for heat exchangers in systems operating on countercurrent or cocurrent flow, if the fluid's physical properties do not change significantly during the process, the U coefficient is almost constant over time. However, when the system has at least one of its fluids with considerable variation of physical properties, like viscosity in the case of oils heating, the U coefficient undergoes variation as function of time, making its determination complex, since it is necessary to know the pattern in which it changes over time. Similarly, this consideration can be applied to tanks' heating or cooling.

Considering a vessel with a generic heat transfer surface for heating a fluid using a isothermal source, the heat exchange area is determined from the macroscopic energy balance (Eq. (17)) with the following hypothesis:

$$\Sigma\dot{Q} + \dot{w}_e\left(h_e + \frac{v_e^2}{2} + gz_e\right) = \frac{dE_{vc}}{d\theta} + \dot{w}_s\left(h_s + \frac{v_s^2}{2} + gz_s\right) + \Sigma\dot{W} \tag{17}$$

- **a.** Few physical properties vary in the fluid to be heated; hence, the U coefficient is assumed to be constant over all the heat exchange surface.
- **b.** Hot fluid constant flow (on this isothermal heating, water steam will be considered).
- **c.** Perfect mixing tank (in such a way that the temperature will be the same at any point of the tank).
- **d.** Perfect tank insulation.
- **e.** Workflow in control volume is neglected, being only considered to be significant when the fluid has a large viscosity.

Eq. (18) presents a simplification of Eq. (17) following the aforementioned hypothesis.

$$\dot{w}(h_e - h_s) = UA(T - t_b) = Mcp_c \frac{dt_b}{d\theta} \tag{18}$$

Integrating Eq. (18) with the specific boundary condition in $\theta = 0$ with $t_b = t_{b1}$ (fluid initial temperature in the beginning of the heating process), Eq. (19) is obtained, which allows to calculate the heat exchange area with a constant U coefficient over the entire surface.

$$A = \frac{Mcp_c}{U\theta} ln \frac{T - t_{b1}}{T - tb_2} \tag{19}$$

In a similar way, Eq. (20) presents the expression to calculate heat transfer area in the case of isothermal cooling.

$$A = \frac{Mcp_c}{U\theta} .ln \frac{t_{b1} - T'}{t_{b2} - T'} \tag{20}$$

The energy balance described in Eq. (18) can be applied to nonisothermal processes, where Eq. (21) presents the heat exchange area calculation for heating and Eq. (22) presents the heat exchange area calculation for cooling, highlighting that both equations are valid for hot and cold fluid constant flows and constant inlet temperature.

$$A = \frac{Mcp_c}{U} ln \left\{ \frac{1}{1 - \left[ln \left(\frac{T - t_{b1}}{T - tb_2} \right) \frac{Mcp_c}{\dot{w}_{h} cp_h \, \theta} \right]} \right\} \tag{21}$$

$$A = \frac{Mcp_h}{U} ln \left\{ \frac{1}{1 - \left[ln \left(\frac{t_{b1} - T'}{tb_2 - T'} \right) \frac{Mcp_h}{\dot{w}_{c} cp_c \, \theta} \right]} \right\} \tag{22}$$

On continuous operations, the same hypotheses assumed for the processes in batch mode can be considered, adding only that the flow of both fluids will be constant during the whole operation with constant inlet temperatures.

Eqs. (23)–(25) show the obtainment of the heat transfer area for an isothermal process operating at continuous mode.

$$ln \frac{K_1 - K_2 t_{b2}}{K_1 - K_2 tb_1} = -\frac{K_2}{Mcp_c} \theta \tag{23}$$

$$K_1 = \dot{w}_c cp_c t_{b1} + UAT \tag{24}$$

$$K_2 = \dot{w}_c cp_c + UA \tag{25}$$

Similarly, Eqs. (26)–(28) present the heat exchange area for a nonisothermal process operated in continuous mode.

$$ln\frac{K_3 - K_4t_{b2}}{K_3 - K_4t_{b1}} = -\frac{K_4}{Mcp_c}\theta \tag{26}$$

$$K_3 = \dot{w}_c cp_c t_{b1} + \dot{w}_h cp_h\left[\left(1 + \frac{1}{e^{\frac{U.A}{\dot{w}_h cp_h}}}\right) + T_1\right] \tag{27}$$

$$K_4 = \dot{w}_c cp_c + w_h cp_h \tag{28}$$

The equations for continuous operation described in Eqs. (26)–(28) can only be used for processes operated in an unsteady state; however, for projects on tanks operating continuously, they are carried as steady state, in such a way that the design equation described by Eq. (26) is reduced to Eq. (1), as shown in item 1.

3. Expressions for Nusselt's number in several heat transfer surfaces

3.1. Jackets

Jackets are surfaces for heat exchange often used by the majority of processes of heating and cooling. Structurally, they consist of an external coating for the tank like a shirt, where the thermal fluid flows through the space between the jacket and the tank. The jackets are mainly classified as standards, spiral, and half-jackets, in which the difference is only the structure; however, these three types have the same placement inside the tank.

The heat transference through these thermal surfaces is not efficient due to its location, since the heat transfer occurs on the tank's walls; therefore, baffles are placed in order to increase the turbulence aiming to implement the heat transfer efficiency [13]. A big disadvantage of using jackets, more than just the low heat transference, is its dimension, since on tanks with large volumes, like fermentation vats with an average volume of 5000 m^3, the construction becomes structurally unfeasible due to the high cost. Industrially, jackets are used in large scale, even with the aforementioned disadvantages, because the exponents and the constant in Eq. (16) are already known for many tank's dimensions and several kinds of fluids operating in continuous or discontinuous mode.

Chilton et al. [14] carried out a pioneer work concerning the determination of the external convection coefficient, in a tank with 0.3 m of diameter for water and glycerin solution heating with axial mechanical impeller. Eq. (29) presents Nusselt's equation with constant and exponents determined by Chilton.

$$Nu = 0.36{Re_a}^{0.67}Pr^{0.33}\left(\frac{\mu}{\mu_w}\right)^{0.14} \tag{29}$$

The exponents and the constant in Eq. (29) were determined in a tank without baffles and discontinuous operation, although, as the dimension of the studied tank was quite inferior to industrial dimensions, Eq. (29) has deviations of 40% magnitude from the real values of external convection coefficients obtained experimentally in tanks of other sizes.

Uhl and Gray [15] carried experiments in a tank with 0.6 m of diameter and axial impeller, turbine type, and anchor type, with ultra-viscous fluids and four vertical baffles, being operated on discontinuous mode. However, because of the elevated viscosity, the equations are only valid if Reynolds ranges between 20 and 300. Eq. (30) presents the expression for the Nusselt determination using the axial impeller.

$$Nu = 0.415{Re_a}^{0.67}Pr^{0.33}\left(\frac{\mu}{\mu_w}\right)^{0.24} \tag{30}$$

Bourne et al. [16] determined the exponents and the constant for Eq. (16) in a tank with diameter of 0.51 m on the standard conditions for dimensions, described on Rushton et al. [17] pioneer work, conditions that are being used up until the present day. Electrolytic solutions and a turbine radial impeller with six flat blades were used; however, Bourne et al. [16] disregarded the term related to viscosity due to electrolytic solutions having a viscosity close to the one of the water. In Eq. (31), the model obtained on the aforementioned conditions is shown.

$$Nu = 0.42{Re_a}^{0.694}Pr^{0.33} \tag{31}$$

Having mathematical software's backing, the exponents and the constant of Eq. (16) have been expressed in terms of more complex functions, decreasing the error generated by the model. For instance, Karcz and Streck [18] carried out experiments for continuous operations on water heating using steam, varying the type of impeller and its height in relation to the tanks bottom. With the data obtained and arbitrating the exponents values on Prandtl's number as 0.33 and on viscosity relation as 0.14, like Mohan et al. [19], Karcz and Streck [18] developed a polynomial for the K constant determination in Eq. (16) for each mechanical impeller used. Eq. (32) presents the model determined on the experiments.

$$Nu = K{Re_a}^{0.67}Pr^{0.33}\left(\frac{\mu}{\mu_w}\right)^{0.14} \tag{32}$$

In Eq. (33), the polynomial obtained from Eq. (32) with axial mechanical impeller to determine h_o is illustrated.

$$\begin{aligned} K = 0.3119 + 0.3333\times10^{-3}.\left(\frac{F-6}{2}\right) + 1.75\times10^{-3}\left(\frac{G-0.06}{0.015}\right) - 6.8333\times10^{-3}\left(\frac{Y-0.184}{0.0694}\right) \\ + 2.3333\times10^{-3}\left(\frac{X'-0.0128}{0.0638}\right) \end{aligned} \tag{33}$$

where F, G, Y, and X are dimensional parameters, expressed in International System of Units relating to the tank and mechanical impeller.

Nassar and Mehrotra [20] present the coefficients in Eq. (16) for an isothermal heating equipped with a six-flat-blade turbine impeller, using condensed steam as heating fluid. This is shown in Eq. (34).

$$Nu = 0.44{Re_a}^{0.67}Pr^{0.33}\left(\frac{\mu}{\mu_w}\right)^{0.24} \tag{34}$$

3.2. Helical coils

The helical coils present several advantages over jackets, because they have a large heat transfer area, are compacts, and, due to their geometry, promote an efficient turbulence on the fluids to be agitated in the tank. Moreover, it is an equipment that has low cost and it is easy to build [21]. The increase on the heat transfer through this kind of surface has two disadvantages: the need for baffles to avoid the formation of vortexes and the difficulty of maintenance and cleaning when submitted to fouling fluids. The helical coils can be single ribbon and double ribbon. The single ribbon coils are more efficient than the double ribbon ones, because as they have only one tube filament, there is a small formation of stationary zones, while for double ribbon coils, there is formation of stationary zones between the two tube filaments, decreasing the turbulence on this region and, as a consequence, the heat transference.

The characteristics of an efficient mixture with helical coils are consequence of developments of secondary flows throughout the pipeline, which is known as Dean's effect [22]. The secondary flows appear due to the action of centrifugal force on the fluid unit, being the difference of axial speed on the fluid unit for the tubes cross section as a function of a force gradient. The fluid elements in the tube center are projected perpendicularly to the tubes, that is, against the tank's wall, where they undergo a speed decrease, and when returning to the tube center, they find the main flows, resulting in vortexes throughout the helical coil. These vortexes are called Dean's cells, which are responsible for the increase of heat transfer efficiency by providing intensification on local turbulence close to the coil [23].

However, this effect is significant when in laminar and transient flows, because the vortexes' stability is kept, although, in turbulent flows, the vortexes end up being negligible in relation to the turbulence generated on the system.

Similarly with jackets, several researchers provided the exponents and the constant for Eq. (16) for single ribbon helical coils, which can also be used on double ribbon coils just adding safety factors.

Cummings and West [24] determined the exponents of Eq. (16) for organic liquids heating in a 0.76 m diameter tank equipped with a six-flat-blade radial impeller without baffles, on discontinuous operation. The obtained model is expressed in Eq. (35).

$$Nu = 1.01{Re_a}^{0.62}Pr^{0.33}\left(\frac{\mu}{\mu_w}\right)^{0.14} \tag{35}$$

DeMaerteleire [25] carried out a similar study to the one done by Cummings and West [24]; however, he added four baffles to the tank and varied the turbine impeller diameter. A term was added to the obtained model, which relates the tank's diameter to the mechanical impeller diameter, thus obtaining a more realistic model that is valid for a Reynolds ranging from 26,000 to 110,000. The obtained expression is illustrated in Eq. (36).

$$Nu = 1.778 Re_a^{0.628} Pr^{0.33} \left(\frac{\mu}{\mu_w}\right)^{0.20} \left(\frac{D_t}{D_a}\right)^{0.382} \tag{36}$$

The dimensionless group (D_t/D_a) relates the tanks' diameter to the impeller's diameter; however, as demonstrated by Karcz and Streck [18], the best mixing conditions occur when the relation(D_t/D_a) = 3. For instance, if the tank's diameter is 1 m, according to the standard condition, the impeller's diameter should be 0.33 m; replacing these values with the dimensionless group and powering it to 0.382, the value 1.527 is found. This value will be fixed regardless of the chosen tank's diameter, being the only request to follow the aforementioned standard condition. The influence of an increase or decrease on the tank's diameter and consequently the mechanical impeller diameter is evident on the flow regime (Reynolds) and on the heat transference (Nusselt).

Havas et al. [26] determined the exponents and the constant in Eq. (16) for water heating in a tank with 0.4 m of diameter and radial impeller with different diameters. The same experiments were carried out in a tank with 0.8 m of diameter. The model obtained by Havas et al. shows a greater coverage compared with the ones presented in Eqs. (35) and (36), because it takes into account a wider range of tank's diameter. Eq. (37) shows the determined model based on the obtained results.

$$Nu = 0.187 Re_a^{0.688} Pr^{0.36} \left(\frac{\mu}{\mu_w}\right)^{0.11} \left(\frac{D_t}{D_a}\right)^{0.62} \tag{37}$$

Yet, regarding Eq. (16), Couper et al. [27] recommended the Prandtl's number powered to a 0.37 exponent and the c exponent, which relates the fluid viscosities, determined by an empirical expression as function of the viscosity in the average inlet and outlet temperature of the fluid to be heated, as pointed by Eq. (38).

$$c = \frac{0.714}{\overline{\mu}^{0.21}} \tag{38}$$

Dias et al. [28] determined the exponents and the constant in Eq. (16) for water heating equipped with an axial impeller with four flat blades inclined by 45° and a turbine radial impeller with six flat blades, based on Couper's assumption. The obtained models for axial and radial impellers are expressed by Eqs. (39) and (40), respectively.

$$Nu = 0.317.Re_a^{0.589} Pr^{0.37} \left(\frac{\mu}{\mu_w}\right)^{0.79} \quad (39)$$

$$Nu = 0.195 Re_a^{0.664} Pr^{0.37} \left(\frac{\mu}{\mu_w}\right)^{0.79} \quad (40)$$

3.3. Spiral coils

Spiral coils are used, primarily, in tanks for storing heavy oil, where flow is not recommended when at ambient temperatures, due to the oil's high viscosity. As a solution, these coils are placed on the tank's base, where heating steam transfers energy to the oil, which increases oil's temperatures, making the viscosity on that spot decrease, allowing pumping transportation.

As these heat transfer surfaces meet in the vessel's bottom, even with intense turbulence generated by the mechanical impeller, it is evident that the heat transfer will occur predominantly on the tank's bottom and will spread on the axial axis over time. In view of this problem, Ho and Wijeysundera [29] simulated the heat transfer on the air dehumidification process with spiral coil and numerical simulations; they determined the internal heat transfer coefficient (h_i) and, as a consequence, the heat transference performance throughout the process.

Naphon [30] determined a correlation for the h_i coefficient in a vessel without mechanical impeller for cooling of hot water by using cold water. The model was based on Prandtl's number and Dean's number; however, the expression presented is of a totally particular nature to the experiment conditions used, not being possible, and then to extrapolate it for industrial units. The unit is operated in discontinuous state and natural convection, given that the obtained model is expressed in Eq. (41).

$$Nu = 27.358 De^{0.287} Pr^{-0.949} \quad (41)$$

Rosa et al. [31] conducted a study as to predict the Nusselt's number when heating organic solutions in tanks equipped with spiral coils for two kinds of mechanical impellers, Rushton turbine (RT) with six flat blades and pitched blade turbine (PBT) with four blades at 45°. The obtained models are based in Eq. (16) and can be applied for Reynolds between 2000 and 500,000 and Prandtl around 3.8 and 140 and tanks containing baffles. The models for RT and PBT impellers are presented in Eqs. (42) and (43).

$$Nu = 0.10 Re_a^{0.83} Pr^{0.33} \left(\frac{\mu}{\mu_w}\right)^{0.14} \quad (42)$$

$$Nu = 0.81 Re_a^{0.64} Pr^{0.33} \left(\frac{\mu}{\mu_w}\right)^{0.14} \quad (43)$$

3.4. Vertical tube baffles

The vertical tubular baffles consist of tube bank connected between themselves, which are located in the interior side of the tank, in the same way of vertical baffles used in tanks with jackets and coils. The dimensions to construct tube banks follow the standard agitation configurations determined by Rushton et al. [32], which states that the tube bank's diameter should have 1/10 of the tank's diameter in order to reach the best thermal and mixing efficiency.

According to Tatterson [33], the vertical tubular baffles increases 37% the heat transfer efficiency on the turbulent region; this happens due to the large heat transfer area of tubular baffles and the intense turbulence caused by the mechanical impeller, in such a way that tanks with this kind of surface do not need additional baffles for vortex breaking, since the very thermal surface also works as a baffle.

Studies carried out by Rosa et al. [34] with water and sucrose solutions heating in tanks equipped with axial and radial impellers determined that the rotation in which the heat transfer is maximized with the lowest energy usage by mechanical impeller ranges from 200 to 300 rpm.

The external convection coefficient of heat transfer (h_o) is a function of several variables, which have already been mentioned on item 1; however, when using vertical tubular baffles, this coefficient is also reliant on the amount of bench tubes used, on their spatial placement in the tank's interior, and on the kind of operation (continuous or discontinuous) [35].

The exponents and the constant in Eq. (16) were not generically determined for any tank dimension, vertical tubular baffles, and amount of tube banks, due to the complexity presented in the previous paragraph, that is, most part of the models presented on today's literature are for discontinuous operations.

The first model determined in order to estimate the h_o coefficient using vertical tubular baffles was presented on the pioneer work carried out by Rushton et al. [17] in tanks abiding by standard dimensions. The experiments were ran in a 0.366 m diameter tank where hot water was used to heat oil through four vertical tube bank using a turbine impeller with six flat blades and a radial impeller with four flat blades. The operation was performed as a discontinuous operation.

The models obtained by Rushton et al. were found by plotting directly the experimental values of the h_o coefficient in function of the variables within the Reynolds number modified for agitated tanks, which are the impeller's diameter, the mechanical impeller's rotation, the density and the viscosity of the fluid in study.

The models for turbine radial impellers with six flat blades and with four flat blades are represented in Eqs. (44) and (45), respectively.

$$h_o = 0.00285 D^2 N \rho \mu^{-1} \tag{44}$$

$$h_o = 0.00235 D^{1.1} N^{0.7} \rho^{0.7} \mu^{-0.7} \tag{45}$$

Dunlap and Rushton [36] continued the study carried by Rushton et al. [17] with vertical tube baffles, incorporating to the obtained model the effect of the impeller's diameter in relation to the tank's diameter and the effect of the amount of tube bank (n_b) on the heat transfer from hot water to oils. Eq. (46) presents the determined model.

$$Nu = 0.09{Re_a}^{0.65}Pr^{0.33}\left(\frac{\mu}{\mu_w}\right)^{0.4}\left(\frac{D_a}{D_t}\right)^{0.33}\left(\frac{2}{n_b}\right)^{0.2} \quad (46)$$

Havas et al. [37] carried out experiments in a tank with 0.4 and 0.8 m of diameter for water and fuel oils heating with five-tube bank and radial impeller equipped with six flat blades. The operation was conducted on discontinuous mode. Based on the experimental data, the researchers concluded that the effect caused by the impellers' diameter and by the amount of tube bank is negligible in relation to the turbulence generated by the mechanical impeller. The obtained model incorporates the effects, which are directly disregarded on the Reynolds number, as presented in Eq. (47).

$$Nu = 0.208{Re_a}^{0.65}Pr^{0.33}\left(\frac{\mu}{\mu_w}\right)^{0.4} \quad (47)$$

Karcz and Strek [38] defined a model to determine the external convection coefficients using an axial mechanical impeller with three inclined blades in nonstandard conditions for the vertical tubular baffle dimensions. The model is presented in Eq. (48).

$$Nu = 0.494{Re_a}^{0.67}Pr^{0.33}\left(\frac{\mu}{\mu_w}\right)^{0.14} \quad (48)$$

Lukes [39] expanded the studies carried by Karcz and Strek [38], using an axial impeller with three inclined blades and a standard axial impeller with four blades inclined 45°, obtaining an expression similar to Eq. (48). The obtained function is presented in Eq. (49).

$$Nu = 0.542{Re_a}^{0.65}Pr^{0.33}\left(\frac{\mu}{\mu_w}\right)^{0.40} \quad (49)$$

Rosa et al. [40] carried experiments in a tank with 0.4 m diameter for heating sucrose solutions at 20% and 32% (concentration?) using hot water, through four-tube bank using an axial impeller with four blades inclined 45°; however, this is the first model presented operating in continuous operation, as illustrated in Eq. (50).

$$Nu = 17.88{Re_a}^{0.27}Pr^{0.29}\left(\frac{\mu}{\mu_w}\right)^{0.37} \quad (50)$$

Rosa et al. [41] carried out experiments under the same conditions that Rosa et al. [35], however, this time using a turbine radial impeller. The model is shown in Eq. (51).

$$Nu = 25.03{Re_a}^{0.38}Pr^{0.11}\left(\frac{\mu}{\mu_w}\right)^{0.20} \quad (51)$$

4. Project of a tank with heat exchange by vertical tubular baffles

Specify the heat exchange system model with vertical tube baffles for a tank with useful volume of 3 m^3 on continuous operation, as shown in **Figure 1**. The vertical tube baffles are made of brass composed by vertical pipelines (one baffle/three pipelines). The unit must heat an aqueous solution with 20% in mass of sucrose and a 2.0 m^3/h flow ($\dot{V}_{sol}$) from 20 to 42°C, in steady-state operation. The liquid available for heating is water at 90°C with a 10 m^3/h flow. The impeller, which has four blades, inclined 45°, works at 150 rpm. Data: (a) solutions' specific mass (ρ_{sol}) of 1074.2 kg/m^3, (b) solutions' specific heat (C_{psol}) of 3650 J/kg K, (c) water specific mass (ρ_{water}) of 1000 kg/m^3, (d) water specific heat (C_{pwater}) of 4180 J/kg K, (e) water viscosity (μ_{water}) of 0.001 kg/(m s), (f) solutions' viscosity (μ_{sol})of 0.0017 kg/(m s), (g) solutions' thermal conductivity (k_{sol}) of 0.43 W/(m °C), and (h) combined fouling factor (R_d) of 0.001 (h ft^2 °F)/Btu.

The described system has as project equation for heat transfer area, Eq. (52) (nonisothermal continuous operation) (see Eq. (1)).

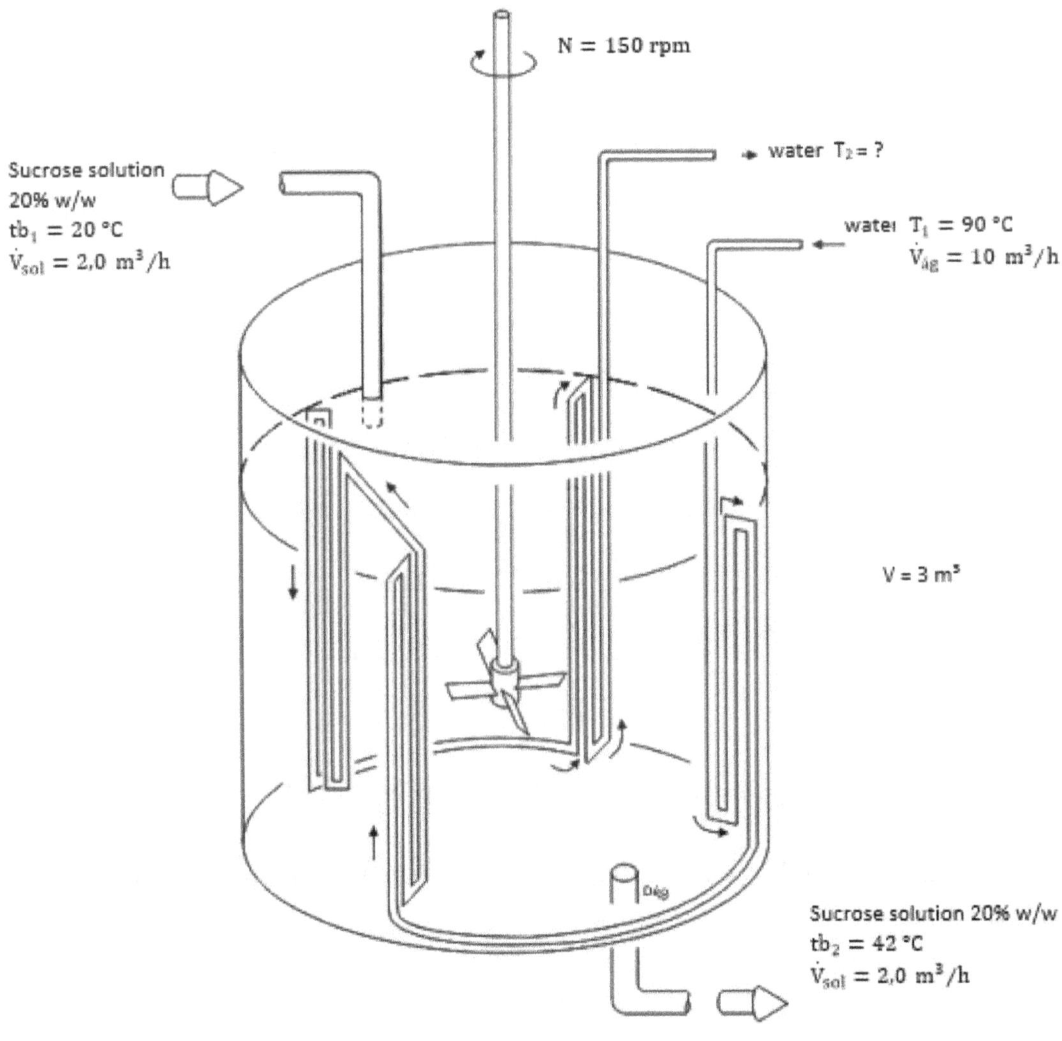

Figure 1. Tank with axial impeller and four blades inclined by 45° and system of heat exchange by vertical tubular baffles. Continuous operation [42].

$$A = Q/U.LMTD \tag{52}$$

The solution of Eq. (52) includes the determination of the overall coefficient of heat transfer. Due to the high thermal conductivity of brass, the conduction will be disregarded in this process; hence, the U_d coefficient will only depend on the convection coefficients (Eq. (53)).

$$1/U_d = 1/h_{io} + 1/h_o \tag{53}$$

The h_i coefficient will be calculated from the expression proposed by Bondy and Lippa [6] (Eq. (12) of item 2)—Eq. (54)—and corrected regarding the pipeline's external surface (Eq. (55)).

$$h_i = 1429\left(1 + 0.0146.\overline{T}\right).u^{0.8}/D_i^{0.2} \tag{54}$$

$$h_{io} = h_i \frac{D_i}{D_e} \tag{55}$$

The average speed u in the interior of the vertical tubular baffles' pipeline is calculated from the continuity equation ($\dot{V} = u.A$); however, the pipeline's diameter is not known yet. This procedure involves the determination of the vessel's internal geometry, which will be calculated from the standards recommended by Rushton et al. [32]. With 3 m^3 volume and applying the geometric relations, the following is obtained:

$$V = \left(\pi D_t^2/4\right)H = 1.56\,\text{m} \tag{56}$$

$$D_t = H = 1.56\,\text{m} \tag{57}$$

$$D_t/D_a = 3 \rightarrow D_a = 0.52\,\text{m} \tag{58}$$

$$E/D_a = 1 \rightarrow E = 0.52\,\text{m} \tag{59}$$

$$W/D_a = 1/5 \rightarrow W = 0.104\,\text{m} \tag{60}$$

$$X = W/0.707 \rightarrow X = 0.147\,\text{m} \tag{61}$$

$$J/D_t = 0.1 \rightarrow J = 0.156\,\text{m} = 6.14\,\text{in.} \cong 6\,\text{in} \tag{62}$$

Since the baffle is composed by three pipelines, their external (D_e) and internal (D_i) diameters can be specified:

$$D_e \cong 6\,\text{in.}/3 \cong 2\,\text{in.} \tag{63}$$

Considering a commercial pipeline with D_N of $1^{1/2}$ in., Sch 40S and D_i of 1.61 in., and thickness (ep) of 0.145 in. therefore

$$D_e = D_i + 2\,\mathrm{ep} = 0.04826\,\mathrm{m} \tag{64}$$

$$D_i = 0.040894\,\mathrm{m} \tag{65}$$

The average temperature $\overline{T}$ is calculated from the arithmetic average between the hot fluid's inlet and outlet (T_1 and T_2), respectively. However, on non-isothermal conditions, the temperature T_2 also varies over time. Considering that the vessel is perfectly insulated (isolated?), in such a way that all the heat given by the hot fluid (water) will be transferred to the cold fluid (solution), the temperature T_2 and the heat flow Q can be obtained by an energy balance, as shown in Eq. (66).

$$\dot{w}_h cp_h \frac{dT}{d\theta} = \dot{w}_c cp_c \frac{dt_b}{d\theta} \tag{66}$$

Integrating in $\theta = 0$ with $T = T_1$ and $t_b = t_{b1}$, the following is obtained:

$$T2 = \frac{\dot{w}_c cp_c}{\dot{w}_{h\,cph}} (t_{b2} - t_{b1}) + T_1 = 85.9°\mathrm{C} \tag{67}$$

$$Q = \dot{w}_h cp_h \frac{dT}{d\theta} = 48180\,W \tag{68}$$

Hence, the average temperature $\overline{T}$is given by Eq. (69) and the average speed u by Eq. (70).

$$\overline{T} = 87.95°\mathrm{C} \tag{69}$$

$$u = \frac{4.\dot{V}}{\pi D_i^2} = 2.11\,\mathrm{m/s} \tag{70}$$

Replacing Eqs. (65), (69), and (70) in Eq. (54), the h_i coefficient (Eq. (71)) is obtained, and replacing the value of h_i and of the Eqs. (64) and (65) in Eq. (55), the h_{io} coefficient is given (Eq. (72)).

$$h_i = 11238.20\,\mathrm{W/m^2\,°C} \tag{71}$$

$$h_{io} = 9522.90\,\mathrm{W/m^2\,°C} \tag{72}$$

The h_o coefficient for the current project example will be calculated with the expression given by Rosa et al. [40] (Eq. (73)).

$$Nu = 17.88 Re_a^{\ 0.27} Pr^{0.29} . \left(\frac{\mu}{\mu_w}\right)^{0.37} \tag{73}$$

Eqs. (74) and (75) show the calculation for the Reynolds (Re_a) and Prandtl numbers. The relation μ/μ_wwill be assumed as 1, considering that fluid temperature in the tank t_b will be equal to the wall temperature t_w.

$$Re_a = \frac{D_a^2 N \rho_{sol}}{\mu_{sol}} = 424654.5 \tag{74}$$

$$\Pr = \frac{C_p \mu_{sol}}{K_{sol}} = 14.52 \tag{75}$$

Replacing Eqs. (74) and (75) in Eq. (73), the following is given:

$$Nu = 1247.14 \tag{76}$$

$$h_0 = \frac{Nu\, K_{sol}}{D_t} = 343.76\,\mathrm{W/m^2\,°C} \tag{77}$$

$$U_C = \frac{h_{i0}\, h_0}{h_{i0} + h_0} = 331.78\,\mathrm{W/m^2\,°C} \tag{78}$$

$$U_D = 313.47\,\mathrm{W/m^2\,°C} \tag{79}$$

The LMTD will be calculated considering the agitation system operating on countercurrent. Eqs. (80) and (81) present the LMTD calculation.

$$LMTD = \frac{\Delta t_q - \Delta t_f}{\ln\left(\Delta t_q/\Delta t_f\right)} = \frac{\Delta t_f - \Delta t_q}{\ln\left(\Delta t_f/\Delta t_q\right)} \tag{80}$$

$$LMTD = 56.4°\mathrm{C} \tag{81}$$

Finally, replacing Eqs. (68), (79), and (81) in Eq. (52), the necessary heat exchange area for this project is found (Eq. (82)).

$$A = 2.72\,\mathrm{m^2} \tag{82}$$

The total pipeline length (L), the number of pipelines (N_t), and the number of tubes per baffle (N_c) are given by Eqs. (83)–(85), respectively.

$$L = \frac{A}{\pi D_e} = 17.9\,\mathrm{m} \tag{83}$$

$$N_t = \frac{L}{H} = 11.5\,\mathrm{tubes} \cong 12\,\mathrm{tubes} \tag{84}$$

$$N_b = \frac{N_t}{4\,\mathrm{baffles}} 3\, \frac{\mathrm{tubes}}{\mathrm{baffles}} \tag{85}$$

Hence, the vessel described by the given example must have four vertical tubular baffles, and each one must have three tubes. If the tank's heating were to be carried out with agitation promoted by a radial impeller, the h_o coefficient should be calculated by the equation proposed by Dunlap and Rushton [36], shown in item 3.4 (Eq. (51)), obtaining a heat transfer area of just 0.91 m^2.

5. Conclusions

Comparing the value obtained to the areas for both impellers, the agitation with radial impeller is much more efficient in terms of heat transfer when compared to the axial impeller, due to the large turbulence promoted by the radial impeller and to the fact that the radial impeller sends the fluid directly to the tank's wall, where the vertical tubular baffles are located. However, the power consumed by the mechanical impeller must also be analyzed in order to find the most economical rotation with the maximum heat exchange.

Therefore the choice of the kind of heat transfer surface suitable for the process to be projected in agitated vessels (jackets, helical coils, spiral coils, and vertical tube baffles) must be done very strictly, specially doing an analysis between the kind of impeller and its interaction with the adopted surface, because, as present on this paper, the difference between the areas obtained by each kind of surface can vary significantly.

Nomenclature

a	Exponent of Eq. (16)
A	Heat transfer area (m^2)
A_i	Heat transfer area (inside a pipe) (m^2)
A_0	Heat transfer area (outside a pipe) (m^2)
b	Exponent of Eq. (16)
c	Exponent of Eq. (16)
cp	Specific heat (kJ/kg °C)
cpc	Specific heat of cold fluid (kJ/kg °C)
cph	Specific heat of hot fluid (kJ/kg °C)
D_i	Internal diameter of tube (m)
D_e	External diameter of tube (m)
D_t	Vessel diameter (m)
D_a	Impeller diameter (m)
DE	Dean's number
e_p	Wall thickness of tube (m)
E	Distance of impeller from the bottom of the tank (m)
E_{vc}	Global energy in control volume (kJ)
f'	Friction factor of Moody

F	Constant of Eq. (33)
G	Constant of Eq. (33)
g	Acceleration of gravity (m/s^2)
h_i	Heat transfer coefficient at internal surface of tube (W/m^2 °C)
h_{io}	Heat transfer coefficient at the internal surface in relation of external surface (W/m^2 °C)
h_o	Heat transfer coefficient at the external surface
h_e	Specific enthalpy of flow inlet in control volume (kJ/kg)
h_s	Specific enthalpy of flow outlet in control volume (kJ/kg)
J	Diameter of bank of tubes (m)
k	Thermal conductivity of fluid (W/m °C)
K	Constant of Eq. (16)
K''	Constant of Eq. (8)
$LMTD$	Logarithmic mean temperature difference (°C)
L	Length of tube (m)
M	Mass of fluid in the vessel (kg)
N_b	Number of bank tube of tubular vertical baffles tubes
N_t	Total number of tubular vertical baffles tubes
N	Rotation of impeller (rpm)
Nu	Nusselt's number
Pr	Prandtl's number
Q	Heat transfer rate in control volume (W)
R_c	Thermal conduction resistance (m^2 °C/W)
Rd_i	Internal resistance by fouling (m^2 °C/W)
Rd_0	External resistance by fouling (m^2 °C/W)
Re	Reynolds number for inside tube
Re_a	Reynolds number for agitation in vessels
T_b	Bulk temperature in vessel (°C)
T_{b1}	Inlet bulk temperature in vessel (°C)
T_{b2}	Outlet bulk temperature in vessel (°C)
T	Constant temperature of hot fluid in isothermal process (°C)
T'	Constant temperature of cold fluid in isothermal process (°C)
T_1	Inlet temperature of hot fluid in nonisothermal process (°C)
T_2	Outlet temperature of hot fluid in nonisothermal process (°C)
$\overline{T}$	Mean temperature of hot fluid (°C)
u	Velocity of fluid (m/s)
U, U_c, U_d	Overall heat transfer coefficient, clean coefficient, and design coefficient (W/m^2 °C)

v_e	Inlet velocity of fluid in control volume (m/s)
v_s	Outlet velocity of fluid in control volume (m/s)
W	Height of blade impeller (m)
$\dot{W}$	Workflow in control volume (W)
X	Length of blade impeller (m)
X'	Constant of Eq. (33)
Y	Constant of Eq. (33)
z	Elevation (m)
Greek letters	
μ	Dynamic viscosity of agitated liquid at bulk temperature in the vessel (kg/m s)
μ_w	Dynamic viscosity of agitated liquid at wall temperature of thermal surface (kg/m s)
$\overline{\mu}$	Dynamic viscosity of agitated liquid at mean temperature (kg/m s)
ρ	Density of agitated liquid (kg/m^3)
Δt_q	Hot terminal temperature difference (°C)
Δt_c	Cold terminal temperature difference (°C)

Author details

Vitor da Silva Rosa* and Deovaldo
de Moraes Júnior

*Address all correspondence to: victor@unisanta.br

Chemical Engineering Department, Santa Cecília University, Santos, São Paulo, Brazil

References

[1] Karcz, J. and Cudak, M., Efficiency of the Heat Transfer Process in a Jacketed Agitated Vessel Equipped with an Excentrically Located Impeller. Chemical Papers, 56 (6), pp. 382–386, 2002.

[2] Kumpinsky, E. Heat Transfer Coefficients in Agitated Vessels. Latent Heat Models. Industrial and Engineering Chemistry Research, 35, pp. 938–942, 1996.

[3] Naphon, P. Study on the Heat Transfer and Flow Characteristics in a Spiral-Coil Tube. International Communications Heat and Mass Transfer, 38 (1), pp. 69–74, 2011.

[4] Havas, G., Deak, A. and Sawinsky, J., The Effect of the Impeller Diameter on the Heat Transfer in Agitated Vessels Provides with Vertical tube Baffles, The Chemical Engineering Journal, 27 (3), pp. 197–198, 1983

[5] Perarasu, V. T., Arivazhagan, M., Sivashanmugam, P., Heat Transfer Studies in Coiled Agitated Vessel with Varying Heat Imput. International Journal of Food Engineering, 7, 4, 2011.

[6] Bondy, F. and Lippa, S. Heat-transfer in agitated vessels. Chemical Engineering, v. 90, n. 7, p. 62–71, 1983.

[7] Goldstein, R. J., Eckert, E. R. G., Ibele, W. E., Patankar, T. W., Simon, T. H., Kuehn, P. J., Strykowski, K. K., Tamma, A., Bar-cohen, J. V. R., Heberlein, J. H., Bischof, F. A., Kulacki, U., Kortshagen, S. G., Heat Transfer – A Review of 1999 Literature. International Journal of Heat and Mass Transfer, 44, pp. 3579–3699, 2001.

[8] Kern, D. Q., Processes Heat Transfer, Singapore: McGraw-Hill, 1950.

[9] Sieder, E. N., Tate, G. E., Heat Transfer and Pressure drop of Liquids in Tubes. Industrial and Engineering Chemistry, pp. 1429–1435, 1936.

[10] Geankoplis, C. J., Transport Processes and Separations Process Principles, includes Unit Operations, 4th ed., United States of America, Prentice Hall, 2008.

[11] Gnielinski, V. New equations for heat and mass-transfer in turbulent pipe and channel flow. International chemical engineering, v. 16, n. 2, p. 359–368, 1976.

[12] Seth, K. K., Stahel, E. P., Heat Transfer from Helical Coils Immersed in Agitated Vessels. Industrial and Engineering Chemistry, 61, 6, pp. 39–49, 1969.

[13] Dhotre, M. T., Murthy, Z. V. P., Jayakumar, N. S., Modeling & Dynamic Studies of Heat Transfer Cooling of Liquid in Half-coil Jackets, Chemical Engineering Journal, 118, pp. 183–188, 2006.

[14] Chilton, T. H., Drew, T. B., Jeleens, R. H., Heat Transfer Coefficients in Agitated Vessels, Industrial and Engineering Chemistry, 36, 6, 1944.

[15] Uhl, V. W., Gray, J. B., Mixing Theory and Practice, Academic Press: New York, Vol. 1, Cap. V, 1966.

[16] Bourne, J. R., Dossenbach, O., Post, T., Local and Average Mass and Heat Transfer due to Turbine Impellers. Fifth European Conference on Mixing, Wurzburg, West Germany, Stanbury, pp. 199–207, 1985.

[17] Rushton, J. H., Lichtmann, R. S., Mahony, L. H., Heat Transfer to Vertical Tubes in a Mixing Vessel, Industrial and Engineering Chemistry, vol. 40 (6), pp. 1082–1087, 1948.

[18] Karcz, J., Strek, F., Heat Transfer in Jacketed Agitated Vessels Equipped with Non-Standard Baffles, The Chemical Engineering Journal, 58, pp. 135–143, 1995

[19] Mohan, P., Emery, A. N., Al-Hassan, T., Review of Heat Transfer to Newtonian Fluids in Mechanically Agitated Vessels, Experimental Thermal and Fluid Science, 5, pp. 861–883, 1992.

[20] Nassar, N. N., Mehrotra, A. K., Design of a Laboratory Experiment on Heat Transfer in an Agitated Vessel, Education for Chemical Engineers, 6, pp. e83–e89, 2011.

[21] Pimenta, T. A., Campos, J. B. L. M., Heat Transfer Coefficients from Newtonian and Non-Newtonian Fluids Flowing in Laminar Regime in a Helical Coil, International Journal of Heat and Mass Transfer, 58, pp. 676–690, 2013.

[22] Dean, W. R., The Stream-line Motion of Fluid in a Curved Pipe, Philosophical Magazine 5, pp. 673–695, 1928.

[23] Kalb, C. E., Seader, J. D., Fully Developed Viscous-flow Heat Transfer in Curved Circular Tubes with Uniform Wall Temperature, AICHE Journal, 2, pp. 340–346, 1974.

[24] Cummings, G. H., West, A. S., Heat Transfer Data for Kettles with Jackets and Coils, Industrial and Engineering Chemistry, 42, pp. 2303–2313, 1950.

[25] DeMaerteleire, E., Heat Transfer in Turbine Agitated Gas–Liquid Dispersions, International Symposium on Mixing, Mons, Belgium, European Federation of Chemical Engineering, pp. XC7–CX35, 1978.

[26] Havas, G., Deak, A., Sawinsky, J., Heat Transfer to Helical Coils in Agitated Vessels, The Chemical Engineering Journal, 35, pp. 61–64, 1987.

[27] Couper, J. R., et al., Chemical Process Equipment—Selection and Design, 2nd ed. revised, Burlington, MA, Gulf Professional Publishing, 2010.

[28] Dias, M. D., Moino, C. A. A., Santos, A. R., Roseno, K. T. C., Rosa, V. S., Moraes Júnior, D., Experimental Comparison of Heat Exchange in Tanks Equipped with Helical Coil Using Radial and Axial Mechanical Impellers, Science and Technology, 1, pp. 33–38, 2012.

[29] Ho, J. C., AndWijeysundera, N. E., An Unmixed-Air Flow Model of a Spiral Coil Cooling Dehumidifying Unit, Applied Thermal Engineering, 19, pp. 865–883, 1999.

[30] Naphon, P. W., An Experimental Study the in-Tube Convective Heat Transfer Coefficients in a Spiral-Coil Heat Exchanger, International Communications in Heat and Mass Transfer, 29, pp. 797–809, 2002.

[31] Rosa, V. S., Taqueda, M. E. S., Paiva, J. L., Moraes, M. S., Moraes Júnior, D., Nusselt's Correlations in Agitated Tanks Using the Spiral Coil with Rushton Turbine and PBT 45° Impeller. Comparison with Tanks Containing Vertical tube Baffles. Applied Thermal Engineering, 110, pp. 1331–1342, 2017.

[32] Rushton, J. H., E. W. Costich, H. J. Everett, Power characteristics of mixing impellers, Chemical Engineering Progress, 46(8), 395–476, 1950.

[33] Tatterson, G. B., Fluid Mixing and Gas Dispersion in Agitated Tanks, 2nd ed., McGraw Hill, North Carolina, 1991.

[34] Rosa, V. S., Santos, A. R., Rosento, K. T. C., Souza Pinto, T. C., Moino, C. A., Fernandes, F. M., Lopes, M. D., Moraes Júnior, D., Ideal Rotation in Tank with Axial Impeller and

Vertical Tubular Baffles for Heating Solutions. Tercer Congresso Argentino de Ingeniería Mecánica, 2012, Buenos Aires. III CAIM 2012. Buenos Aires: FODAMI, 2012. v. 1.

[35] Dostál, M., Petera, K., Rieger, F., Measurement of Heat Transfer Coefficients in an Agitated Vessel with Tube Baffles, Acta Polytechnica, 50, 2, pp. 46–57, 2010.

[36] Dunlap, I., Rushton, J. H., Heat Transfer Coefficients in Liquid Mixing using Vertical Tube Baffles, Chemical Engineering Progress Symposium, 19, 1953.

[37] Havas, G., Deak, A., Sawinky, J., Heat Transfer Coefficients in an Agitated Vessel using Vertical Tube Baffles, Chemical Engineering Journal, 28, pp. 161–165, 1982.

[38] Karcz, J., Strek, F., Heat Transfer in Agitated Vessel Equipped with Tubular Coil and Axial Impeller, MIESZANIE'99, pp. 135–140, 1999.

[39] Lukes, J., Mixing Equipment with Tube Baffles, Master Thesis, Czech Technical University in Prague, Czech Republic, 2000.

[40] Rosa, V. S., Souza Pinto, T. C., Santos, A. R., Moino, C. A. A., Roseno, K. T. C., Lia, L. R. B., Tambourgi, E. B., Dias, M. L., Toneli, J. T. C. L., MoraesJúnior, D., External Coefficient of Heat Transfer by Convection in Mixed Vessels Using Vertical Tube Baffles, Industrial & Engineering Chemistry Research, 52, pp. 2434–2438, 2013.

[41] Rosa, V. S., Moraes, M. S., Toneli, J. T. L., Moraes Júnior, D., External Heat Transfer Coefficient in Agitated Vessels Using a Radial Impeller and Vertical Tube Baffles. Industrial & Engineering Chemistry Research 53, pp. 13797–13803, 2014.

[42] Moraes, M. S., Moraes Júnior, D., Laboratory of Unit Operations II, Author's Edition (Deovaldo de Moraes Júnior), São Paulo, Brazil, 2012;.

2

A Multi-Period Synthesis Approach to Designing Flexible Heat-Exchanger Networks

Adeniyi Jide Isafiade and Alireza Bahadori

Abstract

This chapter presents a new synthesis method for designing flexible heat-exchanger networks. The methodology used involves a two-step approach: In the first step, a multi-period network is designed for a large number of critical operating periods using a finite set of operating points which lie within the uncertain parameter range, while considering the impact of potential fluctuations in periodic durations of each of the chosen critical points on the network. In the second step, the flexibility of the resulting multi-period network of the first step is tested using very large, randomly generated set of finite potential operating points together with their periodic durations. The key criteria used in determining the finite set of operating points that would participate in the initial multi-period network synthesis of the first step are the nominal operating points, the extreme operating points in terms of heat-load requirements as well as their length of periods. This implies that the resulting flexible network can feasibly transfer heat irrespective of possible fluctuations in periodic durations for any of the potential process-operating points. The solutions obtained using the new approach compare favourably with those in the literature.

Keywords: heat-exchanger networks, multi-period, synthesis, flexible networks, mathematical programming

1. Introduction

An efficiently designed heat-exchanger network (HEN) can be used to achieve significant reductions in energy consumption and pollutant emission into the environment by chemical plants. However, most design methods that have been presented for the synthesis of HENs have assumed fixed process-operating parameters. However, in reality, process parameters may fluctuate around some nominal operating points due to various factors such as changes in environmental conditions, plant start-ups/shutdowns, changes in product quality demand, and so on. In some other cases, the process parameters may deliberately be moved away from their set point/nominal conditions due to reasons such as planned transition from one product quality to another. For these cases, even though the set point tracking of the process parameters is required in ensuring a smooth transition to the new set of operating points, the network of heat exchangers still have to be flexible to handle these new set of operating conditions. Despite the fact that these new set of operating points lie within the possible range of variability of the process parameters, their length of duration needs to be taken into consideration while designing a flexible heat-exchanger network.

The methods that have been used for the synthesis of flexible heat-exchanger networks have been both sequential [1–3] and simultaneous in nature [4–16]. Some of the sequential methods are an automated multi-period version of the mixed integer linear programme (MILP) transshipment model [17] and the non-linear programme (NLP) minimum investment network cost model [18]. The simultaneous methods have mostly been based on a multi-period version of the simplified stage-wise superstructure (SWS) model [19]. Some of the existing design methods for flexible HENs may only be feasible to transfer heat for a finite set of process-operating parameters for which the network is designed in what is known as multi-period networks [4–8]. In the multi-period networks [4–8], each period of operation is distinct in that the process parameters, as well as the length of periods for each of the periods of operations, are known upfront. Since multi-period networks are capable of transferring heat within the specified finite set of operating periods, they can be termed flexible networks. However, their degree of flexibility may only be limited to the set of finite operating points for which the network is designed. The degree of flexibility of the SWS-based multi-period networks has been improved through the use of the timesharing mechanism [9], where heat exchangers may be shared by different stream pairs in more than one period of operation. This is unlike other SWS-based multi-period models where either the average area [4] or the maximum area [5–7] of the same pair of streams exchanging heat in the same stage of the superstructure, and in different periods of operations, is used as the representative heat-exchanger area in the objective function.

According to Jiang and Chang [9], a major shortcoming of the average area or maximum area approach, as presented in the literature, is that an exchanger may be overdesigned for some periods of operations such that when unforeseen changes in period durations occur, the multi-period network may not be feasible to transfer heat any longer, or it may require a significantly higher utility flows. Even though the timesharing approach overcomes some of the aforementioned shortcomings of the other SWS-based methods [4–7], the complexities involved in having to thoroughly clean exchangers during the process of exchanger swapping can be enormous. Furthermore, additional costs and complex controllability issues will be incurred due to excessive piping and associated instrumentations. Hence, some other synthesis methods that result in networks that have a greater degree of flexibility have been presented in the literature [10–15]. One of the methods entails [10] a three-step approach where a network is designed based on a finite set of operating points in the first step. In the second step, the resulting network from the first step is tested for flexibility using the active set strategy [20]. The third step entails using integer cuts to exclude non-qualifying networks in the flexibility tests. Chen et al. [11] extended the aforementioned method by modifying the flexibility analysis step such that area restrictions are not considered during a first step of flexibility analysis but are considered in a later step.

Chen and Hung [12] extended the method of Chen and Hung [10] with some modifications to the synthesis of flexible heat and mass exchange networks. The flexibility test of this method is carried out on a large number of randomly generated parameters within the range of uncertainty. Another method [13] in the area of flexible HEN designs used a two-stage strategy which is based on the SWS model for the synthesis of flexible and controllable networks. Li et al. [14] developed a two-step approach for flexible HENS. The first step entails the synthesis of the network structure using the nominal set of operating conditions. The flexibility of the resulting network is further improved through a structural union with the topology of the critical operating points. In the second step, the areas of the heat exchangers obtained in the first step are further optimised considering flexibility and total annual cost (TAC). It should be known that in the method of Li et al. [14], the area optimisation is only done after obtaining structures that qualify from the flexibility step. This implies that a true simultaneous optimisation may somewhat be omitted. The method has the advantage that it can synthesise flexible networks with non-convex feasible regions. Li et al. [15] used a simulated annealing and decoupling strategy to determine the flexibility index of large-scale non-convex HEN optimisation problem.

It is worth stating at this point that, apart from the use of the multi-period version of the SWS model, a common feature of most of the flexible HENS methods is that they involve a first step where a candidate single period or a multi-period network is synthesised for a minimum total annual cost (TAC) scenario, followed by a flexibility analysis step. For the first step, the candidate multi-period network has mostly been made comprising few periods of operations which may include the nominal operating conditions and the critical operating points. The critical operating points are the periods of operations that require the maximum heat load. The authors of this chapter are of the view that if these sets of operating points that are used to generate the candidate multi-period network are carefully chosen, there may not be a need for complex and mathematically intensive flexibility analysis step, especially in small- to medium-scale HENS problems. This implies that as many as possible critical operating points that lie within the overall range of potential disturbance/fluctuation should be selected for participation in the candidate multi-period network synthesis of the first step. A further criterion that needs to be considered while designing the representative candidate multi-period network, which has erstwhile been ignored by existing flexible HENS methods, is the length of periods for each of the critical operating points used to generate the representative candidate multi-period network in the first step. The existing methods [5, 9–14] used the average costs of utility usage by each period of operation present in the first-step candidate multi-period network to determine its minimum TAC and associated network structure as shown in Eqs. (1) and (2)

$$\min TAC = AF \cdot \left\{ \sum_{i\epsilon H} \sum_{j\epsilon C} \sum_{k\epsilon K} CFz_{i,j,k} + \sum_{i\epsilon H} \sum_{CU} CFz_{i,cu} + \sum_{j\epsilon C} \sum_{HU} CFz_{j,hu} \right.$$

$$\left. + \sum_{i\epsilon H} \sum_{j\epsilon C} \sum_{k\epsilon K} AC_{i,j,k} A_{i,j,k}^{AE_{i,j}} + \sum_{j\epsilon C} AC_{j,hu} A_{j,hu}^{AE_{j,hu}} + \sum_{i\epsilon H} AC_{i,cu} A_{i,cu}^{AE_{i,cu}} \right\} \tag{1}$$

$$+ \left[\sum_{p\epsilon P} \frac{DOP_p}{NOP} \sum_{j\epsilon C} CUHq_{j,hu,p} + \sum_{p\epsilon P} \frac{DOP_p}{NOP} \sum_{i\epsilon H} CUCq_{i,cu,p} \right]$$

$$\min TAC = \frac{1}{NOP+1} \left[\sum_{p\epsilon P} \sum_{i\epsilon H} CUCq_{i,cu,p} + \sum_{p\epsilon P} \sum_{j\epsilon C} CUHq_{j,hu,p} \right]$$

$$+ \sum_{i\epsilon H} \sum_{j\epsilon C} \sum_{k\epsilon K} AC_{i,j,k} A_{i,j,k}^{AE_{i,j}} + \sum_{j\epsilon C} AC_{j,hu} A_{j,hu}^{AE_{j,hu}} + \sum_{i\epsilon H} AC_{i,cu} A_{i,cu}^{AE_{i,cu}} \tag{2}$$

where AF is the annualisation factor, CF is the fixed charge for heat exchangers, AC is the area costs for heat exchangers, AE is the area cost exponent for heat exchangers, CUH and CUC are the costs of hot and cold utilities, respectively, DOP is the duration of period p, NOP is the number of periods/operating conditions, $A_{i,j,k}$ is the area of heat exchanger for hot and cold-process stream pairs i,j in interval k. $A_{j,hu}$ and $A_{i,cu}$ are the area of heat exchangers exchanging heat between hot utility and cold-process streams and cold utility and hot-process streams, respectively. H,C,HU,CU are the set of hot streams, cold streams, hot utilities and cold utilities, respectively. It should be known that the area $A_{i,j,k}$ is the representative heat exchange area which, as explained previously, are used by the same pair of streams exchanging heat in the same interval of the superstructure at different periods of operations.

Eqs. (1) and (2) are the objective functions used in determining the TAC for the first-step initial candidate multi-period network that is later tested for flexibility using various kinds of approaches in some of the existing methods [5, 9–14]. It can be seen that the utility cost calculation component of these equations will result in allotting equal contributions, in terms of utility usage durations, for each of the periods of operations present in the first-step candidate multi-period network. This implies that the candidate multi-period network that is designed at the initial step, and later tested for flexibility, may be limited based on the fact that it is designed with the assumption that these initial candidate critical points have equal-period durations. Since TAC is being solved for at the first step, the objective functions in Eqs. (1) and (2) will aim to simultaneously minimise both utility consumption and investment costs. The investment cost is influenced by the size of heat-exchanger areas and the number of units. Allowing this limitation at the first step of the flexible network synthesis process means that the flexibility analysis step needs to be sophisticated so as to compensate for this limitation. This is because some candidate networks that lie in the uncertain process parameter range that are tested in the flexibility step may be disqualified from being included in the flexible network feasible space due to the fact that Eqs. (1) and (2) were used as the objective functions for generating the initial candidate multi-period network. Furthermore, even at the flexibility testing stage, the feasible solution space may further be limited or constrained based on the fact that equal-period duration scenario was assumed. The authors of this chapter are of the view that adequately incorporating period durations at the stage of generating the candidate multi-period network is vital so as to reduce the degree of complexity of the flexibility tests that would be carried out subsequently. Moreover, it can be said that a HEN is flexible not only when it is able to feasibly transfer heat for scenarios where each of the potential possible operating points that lie within the range of disturbance/fluctuation has equal lengths of periods, but also when their lengths of periods are significantly different from each other, as well as being uncertain. This implies that the total annual cost of a flexible HEN is not fixed but depends on which operating points (including period durations) within the uncertain range of variability are active.

2. Problem statement

Given a set of hot-process streams and a set of cold-process streams, which have to be cooled and heated, respectively. Given also are the supply and target temperatures and the flow rates of these streams. Hot and cold utilities are also available. The task is to synthesise a flexible heat-exchanger network which is optimally operable (i.e. featuring a minimum TAC network) for any unforeseen process-operating parameter point lying within an uncertain operating range.

3. Methodology

The methodology adopted entails the use of a multi-period version of the simplified stage-wise superstructure (SWS) of Yee and Grossmann [19], as presented in Refs. [5, 8]. The SWS is shown in **Figure 1**. In this superstructure, each hot-process stream and each cold-process stream has the option of splitting within each interval and each period of operation where it exists so as to exchange heat with streams of the opposite kind in intervals 2 and 3. The hot- and cold-process streams are then taken to their final temperatures in intervals 1 and 4 through heat exchange with utilities of the opposite kind in each period of operation where the process streams exist. The details of the superstructure can be found in Refs. [5, 7].

3.1. Model equations

The detailed multi-period HENS models used in this chapter are shown in the appendix. For detailed explanations of each of these equations, the reader is referred to the multi-period SWS HENS model of Verheyen and Zhang [5] and Isafiade et al. [7]. The maximum area approach as introduced by Verheyen and Zhang [5] for HENS is also used in this chapter for flexible HENS as shown in Eq. (3)

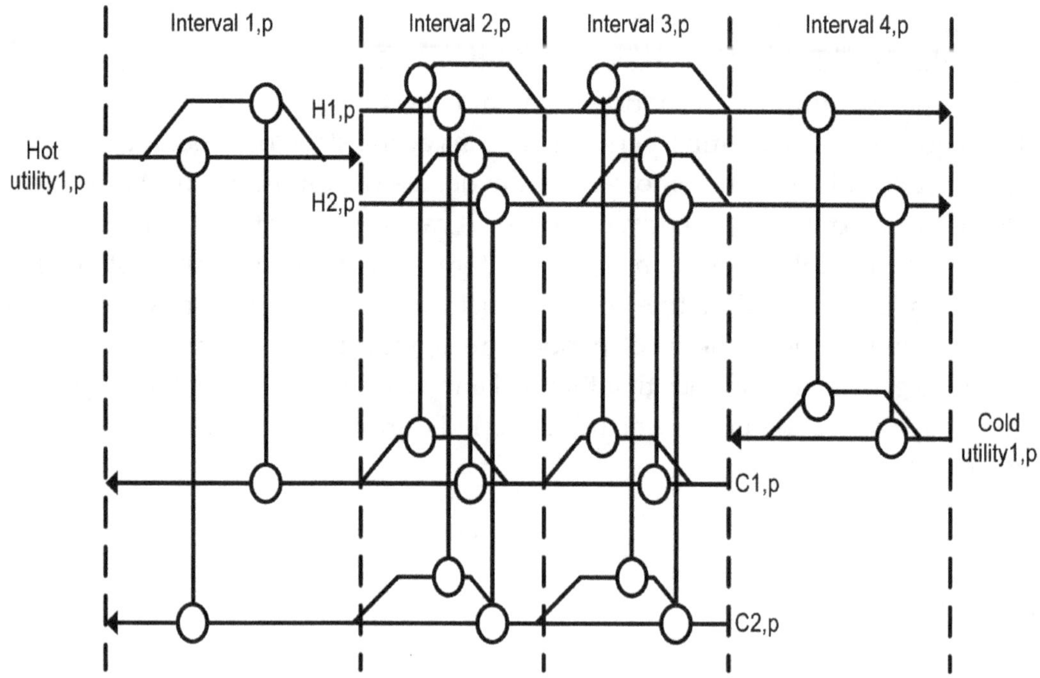

Figure 1. Multi-period version of SWS model.

$$A_{i,j,k} \geq \frac{q_{i,j,k,p}}{(LMTD_{i,j,k,p})(U_{i,j})} \tag{3}$$

The maximum area, $A_{i,j,k}$, is then included in the objective function shown in Eq. (4). However, it should be known that Eq. (4), which is the objective function used in this study, was introduced by Isafiade and Fraser [6] for multi-period networks having specified process parameter points

$$\min\left\{\left[\frac{DOP_p}{\sum_{p=1}^{NOP} DOP_p}\sum_{i\in H} CUC \cdot q_{i,j,k,p} + \frac{DOP_p}{\sum_{p=1}^{NOP} DOP_p}\sum_{i\in H} CUH \cdot q_{i,j,k,p}\right] + AF\left[\sum_{i\in H}\sum_{j\in C}\sum_{k\in K} CF_{ij} \cdot z_{i,j,k} + \sum_{i\in H}\sum_{j\in C}\sum_{k\in K} AC_{i,j,k} \cdot A_{i,j,k}^{AE_{i,j}}\right]\right\} \forall i \in H, j\in C, k\in K, p\in P \tag{4}$$

It is worth mentioning that the terms in the first square bracket of Eq. (4) are the annual operating cost terms. The presentation of these terms by Isafiade and Fraser [6] adequately allocates the contribution of each hot/cold utility to the annual operating cost of the flexible network based on each operating periods of duration. This is unlike the objective functions shown in Eqs. (1) and (2), which are used in most existing methods, and which make an implicit assumption that each operating parameter point within the uncertain range would operate at an equal/average period duration. Whereas this may not always be true because any of the parameter points may dominate at any point in time, hence the network needs to be flexible enough to handle unforeseen period durations.

For the example solved in this chapter, the solver DICOPT, which uses CPLEX for the MILP and CONOPT for the NLP sub-problems, has been used. The solver environment used is GAMS [21]. The machine used operates on Microsoft® Windows 7 Enterprise™ 64 bit, Intel® Core™ i5-3210M processor running at 2.50 GHz with 4 GB of installed memory.

3.2. Solution approach

Based on the foregoing explanations, in this chapter, the following procedure is adopted in generating the candidate initial multi-period network and the subsequent flexibility tests:

1. Identify a large set of critical/extreme candidate-operating parameter points which lie within the full disturbance range. The candidate points are then translated into a multi-period problem having a large number of periods with specified period durations. The resulting set of periods of operations should represent critical points that include the nominal conditions, maximum heat-load-required conditions, minimum heat-load-required conditions and a combination of each of these scenarios.

2. Solve the identified critical set of periods in Step 1 as a multi-period mixed integer non-linear programming (MINLP) problem with equal-period durations for each of the periods of operations.

3. Use the selected matches for the equal-period duration scenario in Step 2, as well as the areas of these matches to initialise the multi-period problem created in Step 1. This implies that the areas of the matches in the multi-period model of this third step should be fixed to the areas obtained for the equal-period scenario in Step 2. Solve the resulting model as an MINLP a number of times. For each time that the model is solved, each period of operation is made to dominate the total period durations, for example, 99.1% of the time, while the remaining time length is shared equally among the other periods of operations. This is necessary so as to examine the flexibility of the network obtained in Step 2 to feasibly transfer heat in a cost-efficient manner irrespective of the period durations of the participating critical operating points. However, if the network obtained in Step 2, which now has fixed areas, is infeasible for any of the critical periods being tested, the fixed areas would be adjusted until all heat loads in all periods are satisfied in terms of heat exchange. The adjustment should be done using the exchanger sizes obtained in Step 2 for the period concerned. The adjustment should continue until a compromise minimum TAC is obtained. The purpose of this step is to identify the critical exchanger areas (i.e. the maximum) that would be able to transfer heat in the final flexible network not only based on operating parameters but also based on unforeseen fluctuations in period durations of any of the potential operating points. It should be known that in order to identify the contribution of each dominant period of operation to the total utility usage, as well as the structure of the final flexible network, the objective function for multi-period HENS as used by Isafiade and Fraser [6] needs to be employed.

4. This step, which is the flexibility analysis step, entails generating a large number of operating points, as large as 100 periods of operations for small- to medium-scale problems, which are then appended to the multi-period network having fixed areas obtained

in Step 3. The model at this stage is solved with further minor adjustments to exchanger areas, if needed, so as to accommodate as many operating points as possible.

4. Example

One example, which was adapted from Chen and Hung [10–12], has been used to illustrate how the newly presented methodology works. The example was first presented by Floudas and Grossmann [3] and has also been solved by other authors [14, 15]. It has two hot streams, two cold streams, one hot utility (steam) and one cold utility (cooling water). **Table 1** shows the stream parameters and other costing details.

In the example, for hot stream 1, the heat capacity flow rate fluctuates around a nominal value of 1.4 kW/K by a magnitude of ±0.4, while its supply temperature fluctuates around a nominal value of 583 K by a magnitude of ±10. For cold stream 2, the heat capacity flow rate fluctuates around a nominal value of 2.0 kW/K by a magnitude of ±0.4, while its supply temperature fluctuates around a nominal value of 388 K by a magnitude of ±5. The objective in this example is to develop a flexible HEN that can feasibly transfer heat for the specified disturbance range in a minimum TAC network. In solving this problem, Chen and Hung [10] identified four extreme operating points within the uncertain process parameter range which include those for nominal conditions, maximum area, maximum cooling load and maximum heating load. The second, third and fourth sets of operating points, that is, the maximum area, maximum cooling load and maximum heating load, respectively, were appended one after the other, to the candidate network generated using the nominal operating conditions.

In solving this problem using the new method developed in this chapter, the first step entails identifying 10 sets of operating points (termed periods) that would be used to generate the base candidate multi-period network. The parameters for the 10 periods are shown in **Table 2**. The parameters listed in **Table 2** are then solved in Step 2 as a multi-period problem having 10 periods of operations with equal-period durations and unequal-period durations using Eqs. (3)

Stream	Heat capacity flow rate FCp (kW/K)	Supply temperature $T^S(K)$	Target temperature $T^t(K)$	Cost ($/kWh)
Hot-process stream 1, $H1$	1.4 ± 0.4	583±10	323	-
Hot-process stream 2, $H2$	2.0	723	553	-
Cold-process stream 1, $C1$	3.0	313	393	-
Cold-process stream 2, $C2$	2.0 ± 0.4	388 ± 5	553	-
Hot utility, $HU1$	-	573	573	171.428×10^{-4}
Cold utility, $CU1$	-	303	323	60.576×10^{-4}

Operating hours = 8600 (hours/year), heat exchanger capital cost function = $4333A^{0.6}$ ($/year), capital annualisation factor (AF) = 0.2, A in m^2, overall heat transfer coefficient (U) for all matches = 0.08 kW/($m^2\cdot K^{-1}$), $\Delta T_{\min}$ = 10 K.

Table 1. Stream, cost and capital equipment data for the example.

Period	Stream	$T^S(K)$	$T^t(K)$	FCp (kW/K)	Enthalpy (kW)	Period	Stream	$T^S(K)$	$T^t(K)$	FCp (kW/K)	Enthalpy (kW)
1	*H*1	583	323	1.4	364	6	*H*1	593	323	1.8	486
	C2	388	553	2	330		C2	388	553	2.0	330
2	*H*1	593	323	1.8	486	7	*H*1	573	323	1.0	250
	C2	383	553	2.4	408		C2	388	553	2.0	330
3	*H*1	573	323	1.0	250	8	*H*1	583	323	1.4	364
	C2	393	553	1.6	256		C2	383	553	2.4	408
4	*H*1	593	323	1.8	486	9	*H*1	583	323	1.4	364
	C2	393	553	1.6	256		C2	393	553	1.6	256
5	*H*1	573	323	1.0	250	10	*H*1	593	323	1.0	270
	C2	383	553	2.4	408		C2	383	553	2.0	340

Table 2. Periods of operations used to generate candidate network.

Dominant period	Equal	1	2	3	4	5	6	7	8	9	10
Period duration	10%	99.1	99.1	99.1	99.1	99.1	99.1	99.1	99.1	99.1	99.1
TAC ($)	38,133	38,340	34,458	38,715	46,573	41,852	41,219	31,025	32,425	38,964	30,689

Table 3. Total annual cost for each dominant period.

and (4) as well as Eq. (A1) in the appendix. For the unequal-period duration scenario, each of the selected periods of operations is made to dominate the total length of periods for all the 10 periods by a significant amount. The purpose of this is to ensure that the final flexible network is able to cater for all possible scenarios including the worst-case scenarios in terms of heat-exchanger area and utility requirement, irrespective of the duration of period for each of the parameter points lying within the full disturbance range. The resulting solution for each scenario is shown in **Table 3**. The average solution generation time for each of the solutions in **Table 3** is 5 S of CPU time. The TAC shown in the first column of **Table 3** is for a case where all periods have equal duration, that is, each period contributes 10% of the total period duration. The second column is for a case where the parameter points of period 1 dominates the total period durations by 99.1%. The same scenario applies for the rest of the columns, that is, each period concerned dominates the total period duration by 99.1%.

Step 3 requires that the selected matches, as well as their areas, for the equal-period duration solution network (TAC = 38,133 $ in **Table 3**) and the network obtained for the most dominant period (i.e. period 4, TAC = 46,573 $ in **Table 3**) be identified from **Table 3**. **Table 4** shows the selected matches, as well as their areas, for the two cases. It should be known that the solution network for the equal-period scenario has more units compared with that of a case where the dominant period (i.e. period 4) is considered. This implies that the solution for the dominant period will result in a simpler network with fewer units, but with higher operating cost. In this chapter, the solution of the equal-period scenario is used in subsequent steps so as to get a TAC

Equal-period scenario		Period 4 dominating	
Match	Area (m^2)	Match	Area (m^2)
HU1,C1,1	4.71	HU1,C2,1	43.04
H1,C1,5	32.74	H1,C1,5	14.08
H1,C2,4	6.04	H1,CU1,6	57.34
H1,CU1,6	57.34	H2,C1,5	4.77
H2,C1,5	4.77	H2,C2,3	17.26
H2,C2,2	27.36		
Total area	132.96		136.49

Table 4. Selected matches for equal-period duration and most dominant period.

Dominant period	Equal	1	2	3	4	5	6	7	8	9	10
Period duration	10%	99.1	99.1	99.1	99.1	99.1	99.1	99.1	99.1	99.1	99.1
TAC ($/yr)	38,134	39,396	35,064	37,127	47,235	41,479	43,419	31,247	33,669	40.943	31,763

Table 5. TAC for equal period and one dominating period at a time in the flexible network tested for 10 periods.

that competes with those presented in the literature. However, the resulting network of the equal-period scenario is still tested for flexibility to feasibly transfer heat in scenarios of unequal-period durations.

Step 3 further requires that the matches selected (including their areas) in the equal-period case be used to initialise the multi-period MINLP model of the 10-period problem data shown in **Table 2**, by fixing the areas of the matches to those of the equal-period scenario shown in **Table 4**. Note that the model is still solved as an MINLP at this stage, despite the fact that the matches and areas are fixed, because in getting compromise solutions for a flexible network, not all matches in **Table 4** (for the equal-period case) may be selected, in fact matches which do not exist on the table may even be added to the network. Solving the 10-period candidate multi-period network of **Table 2**, using the fixed areas of the matches shown in **Table 4** for the equal-period case, gives a range of TACs for each period dominating one at a time as shown in **Table 5**.

In Step 4, the flexibility of the network obtained in Step 3 was then further tested for more randomly generated parameter points lying within the full disturbance range by solving a large multi-period model with equal-period durations. The model at this stage is initialised using the matches, as well as their areas, shown for the equal-period case in **Table 4**. The network was not feasible for a case having 100 periods of equal durations, so the areas of the network were adjusted to obtain new set of areas as shown in **Table 6**.

After the adjustments, the total network area increased from 132.96 m^2 in **Table 4** to 135.4 m^2 in **Table 6**. The resulting network was then feasible to transfer heat in a cost-efficient manner

Match	Area (m^2)
HU1,C1,1	4.8
H1,C1,5	32.8
H1,C2,4	6.1
H1,CU1,6	59.7
H2,C1,5	4.8
H2,C2,2	27.4
Total area	135.4

Table 6. Heat-exchanger areas for final flexible network for the example.

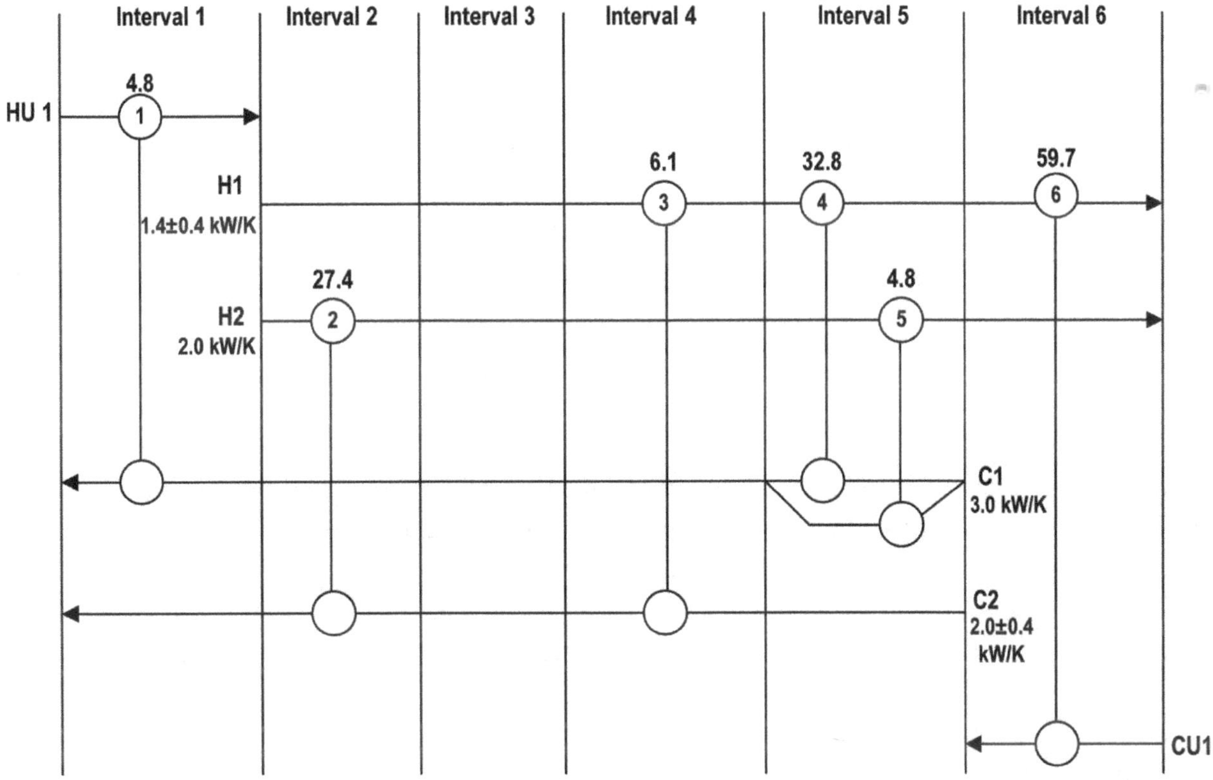

Figure 2. Final flexible structure of this study.

for all of the 100 possible periods of operations. The TAC of the 100-period scenario, which was obtained in 20.94 S of CPU time, is 38,992 $. The flexible network, which is shown in **Figure 2**, is deemed flexible at this stage; hence, it is selected as the final flexible network. In this figure, the areas of each of the heat exchangers are indicated on the units. It should be known that only the final flexible network obtained in Step 4 is shown because the network structure remains unchanged in each of the steps. **Table 7** shows a comparison of the final solution obtained in this example with those of other papers. It is worth mentioning that the solutions of other works, which are shown in **Table 7**, are presented for a case where the periods have equal duration. However, in

Costs	Floudas and Grossmann [3]	Chen and Hung [10]	Li et al. [14]	This work
Annual operating cost ($)	10,499	11,772	11,866	8554
Annual capital cost ($)	39,380	30,104	28,626	30,438
Total annual cost ($)	49,879	41,876	40,492	38,992

Table 7. Comparison of solutions for the example.

Dominant period	Equal	1	2	3	4	5	6	7	8	9	10
Period duration	10%	99.1	99.1	99.1	99.1	99.1	99.1	99.1	99.1	99.1	99.1
TAC ($/yr)	38,429	39,698	35,365	37,428	47,537	41,759	43,720	31,549	33,926	41,245	32,065

Table 8. Total annual cost for each dominant period in final flexible network for the example.

reality, the period durations may not be equal; hence, there is a need to also test or demonstrate the flexibility of the final network for unequal-period durations as done in this chapter. The final network obtained in this study has been tested for a 10-period scenario where period durations may be unequal and the TAC that should be anticipated for cases where each of the periods in the set of 10 periods dominates by 99.1% of the time is shown in **Table 8**. It is expected that the TAC that would be obtained when any of the 100 possible periods of operations dominates significantly by 99.1% of the time, or less, will not be too different from what is shown in **Table 8**.

Figures 3 and **4** show the final flexible networks obtained by other authors as found in Refs. [12, 14]. What is common to these two structures is that they both have six units, out of

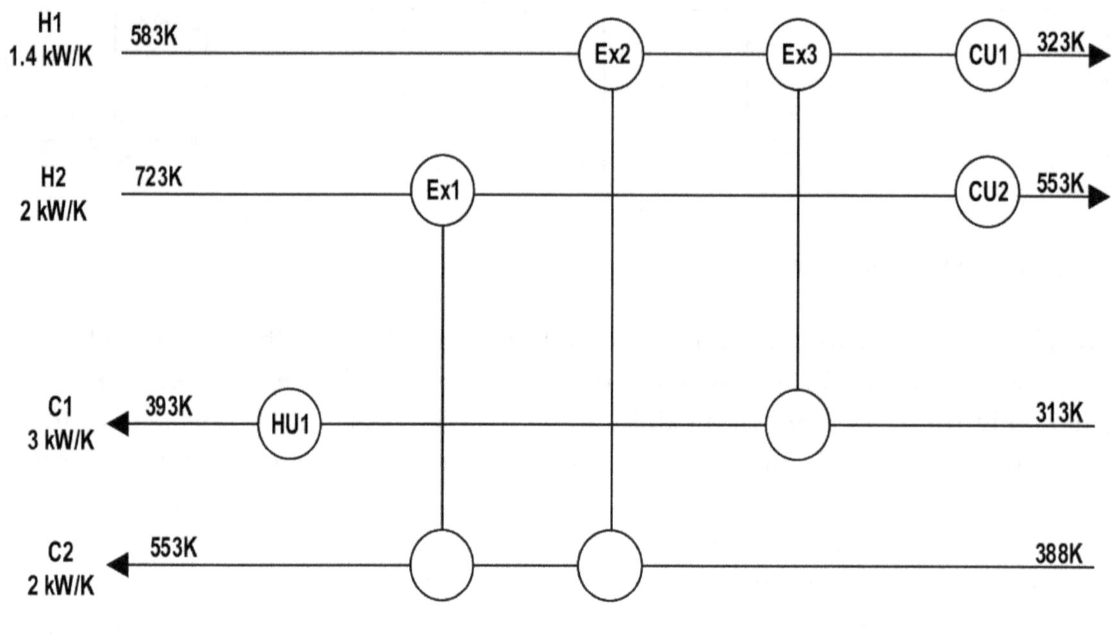

Figure 3. Final flexible HEN structure of Li et al. [14].

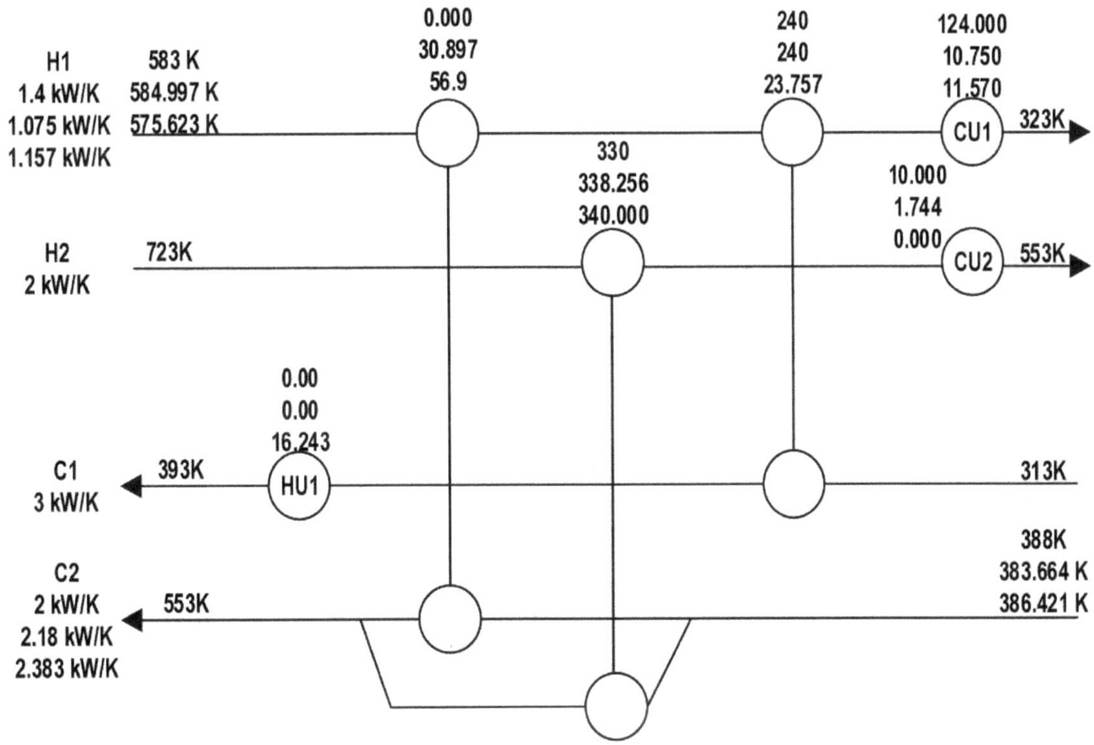

Figure 4. Final flexible HEN structure of Chen and Hung [12].

which two are coolers. This is unlike the structure obtained in this chapter which has only one cooler.

5. Conclusions

This chapter has presented a new simplified synthesis method for designing small- to medium-sized flexible heat-exchanger networks using a mixed integer non-linear programming multi-period synthesis approach. The new method improves on currently available methods in the literature based on the fact that the final flexible network is selected considering the possibility of one or more sets of process parameter points dominating the total period durations more than others. This is essential so as to effectively plan for utility management. Key limitations of the new method are its tediousness of application, especially in large-scale problems, due to the fact that the impact on the overall network TAC of the possibility of each set of selected critical points dominating the total period duration needs to be investigated in a sequential manner. Also, during the flexibility tests, heat-exchange areas need to be manually adjusted, and there is no specific criterion to consider for the adjustments.

Appendix

The set of equations shown in Eq. (A1) was used in the multi-period model of this study. The details of each of these equations can be found in Verheyen and Zhang [5] and Isafiade et al. [7].

$$\left\{\begin{array}{l}
\left.\begin{cases}(T^s_{i,p}-T^t_{i,p})F_{i,p} = \sum\limits_{k\in K}\sum\limits_{j\in C} q_{i,j,k,p} i\in H p\in P \\ (T^t_{j,p}-T^s_{j,p})F_{j,p} = \sum\limits_{k\in K}\sum\limits_{i\in H} q_{i,j,k,p} j\in C p\in P\end{cases}\right\}\text{overal energy balances} \\
\left.\begin{cases}(t_{i,k,p}-t_{i,k+1,p})F_{i,p} = \sum\limits_{j\in C} q_{i,j,k,p} k\in K i\in H p\in P \\ (t_{j,k,p}-t_{j,k+1,p})F_{j,p} = \sum\limits_{i\in H} q_{i,j,k,p} k\in K j\in C p\in P\end{cases}\right\}\text{stage energy balances} \\
\left.\begin{cases}t_{i,k=2,p} = T^s_{i,p} i\in H p\in P \\ t_{j,K-1,p} = T^s_{j,p} j\in C p\in P\end{cases}\right\}\text{assignment of superstructure inlet temperatures} \\
\left.\begin{cases}t_{i,k,p}\geq t_{i,k+1,p} i\in H p\in P \\ t_{j,k,p}\geq t_{j,k+1,p} j\in C p\in P\end{cases}\right\}\text{temperature feasibility} \\
\left\{q_{i,j,k,p}-\Omega_p z_{i,j,k} \leq 0\right\}\text{logical constraint for heat load} \\
\left.\begin{cases}dt_{i,j,k,p} \leq t_{i,k,p}-tc_{i,j,k,p} + \phi(1-z_{i,j,k}) k \in K i\in H j \in C p \in P \\ dt_{i,j,k+1,p} \leq th_{i,j,k+1,p}-t_{j,k+1,p} + \phi\ (1-z_{i,j,k})\ k \in K i \in H j \in C p \in P\end{cases}\right\}\text{exchanger approach temperatures} \\
\left\{dt_{i,j,k,p} \geq \varepsilon\ k \in K i \in H j \in C p \in P\right\}\text{bound for approach temperature} \\
\left\{LMTD_{i,j,k,p} = \frac{2}{3}\left((dt_{i,j,k,p})(dt_{i,j,k+1,p})\right)^{1/2} + \frac{1}{3}\left(\frac{(dt_{i,j,k,p}) + (dt_{i,j,k+1,p})}{2}\right)\right\}\text{logarithmic mean temperature difference} \\
z_{i,j,k}\in\{0,1\} \\
t_{i,k,p}, t_{j,k,p}, q_{i,j,k,p}, dt_{i,j,k,p}\geq 0
\end{array}\right\} \tag{A1}$$

Nomenclature

CU	Cold utility
HENS	Heat-exchanger network synthesis
HU	Hot utility
MILP	Mixed integer linear programme
NLP	Non-linear programme
MINLP	Mixed integer non-linear programme
SWS	Stage-wise superstructure

Indices

- i Hot-process streams and hot utilities
- j Cold-process streams and cold utilities
- p Index representing period of operation (p = 1,…NOP)
- k Stage boundaries

Sets

H Hot-process streams and hot utilities
C Cold-process streams and cold utilities
P Period of operation
K Stage boundaries or temperature locations (where $K-1$ is the set defining the stages)

Parameters

$T^s_{i,p}$	Supply temperature of hot streams (process and utility streams) for period p, K
$T^t_{i,p}$	Target temperature of hot streams (process and utility streams) for period p, K
$T^s_{j,p}$	Supply temperature of cold streams (process and utility streams) for period p, K
$T^t_{j,p}$	Target temperature of cold streams (process and utility streams) for period p, K
$F_{i,p}$	Heat capacity flow rate of hot stream i in period p, W/K
$F_{i,p}$	Heat capacity flow rate of cold stream j in period p, W/K
$U_{i,j}$	Overall heat transfer coefficient, W/(m^2·K)
AF	Annualisation factor
$CF_{i,j}$	Fixed cost for heat exchangers, \$/yr
$AC_{i,j}$	Area cost coefficient
AE	Area cost index
DOP_p	Duration of each period p
NOP	Number of periods
CUH_i	Cost per unit of hot utility i, \$/(W·yr)
CUC_j	Cost per unit of cold utility j, \$/(W·yr)
Ω_p	Upper bound for heat exchanged between match i and j in period p, W
ϕ	Upper bound for driving force in match i,j in period p, K

Variables

$A_{i,j,k}$	Area of match between hot stream i and cold stream j in interval k, m^2
$y_{i,j,k}$	Binary variable representing a match between hot stream i and cold stream j in stage k
$q_{i,j,k,p}$	Heat load of a match between hot stream i and cold stream j in stage k and in period p, W/K
$t_{i,k,p}$	Temperature of hot stream i at stage boundary k in period p, K
$t_{j,k,p}$	Temperature of cold stream j at stage boundary k in period p, K
$th_{i,j,k,p}$	Exit temperature of hot stream i from match i,j,k in period p, K
$tc_{i,j,k,p}$	Exit temperature of cold stream j from match i,j,k in period p, K
$fh_{i,j,p,k}$	Heat capacity flow rate split hot stream i in match i,j,k in period p, W/K
$fc_{i,j,p,k}$	Heat capacity flow rate split cold stream j in match i,j,k in period p, W/K
$dt_{i,j,k,p}$	Driving force in match i,j at stage boundary k and period p, K

TAC Total annual cost of the network, including annualised capital cost and the cost of hot and cold utilities, $/y

Acknowledgements

This study is supported by the National Research Foundation of South Africa (Grant numbers: 85536 and 87744) and the Research Office of the University of Cape Town, South Africa, and Sasol. The funds are gratefully acknowledged.

Author details

Adeniyi Jide Isafiade[1]* and Alireza Bahadori[2]

*Address all correspondence to: aj.isafiade@uct.ac.za

1 Department of Chemical Engineering, University of Cape Town, South Africa

2 School of Environment, Science and Engineering, Southern Cross University, Lismore, NSW, Australia

References

[1] Floudas CA, Grossmann IE. Synthesis of flexible heat exchanger networks for multi-period operations. *Comp. Chem. Engr.* 1986;**10**(2):153–168.

[2] Floudas CA, Grossmann IE. Automatic generation of multi-period heat exchanger network configuration. *Comp. Chem. Engr.* 1987;**11**(2):123–142.

[3] Floudas CA, Grossmann IE. Synthesis of flexible heat exchanger networks with uncertain flowrates and temperatures. *Comp. Chem. Engr.* 1987;**11**(4):319–336.

[4] Aaltola J. Simultaneous synthesis of flexible heat exchanger networks [PhD thesis]. Helsinki University of Technology, Department of Mechanical Engineering; 2002.

[5] Verheyen W, Zhang N. Design of flexible heat exchanger network for multi-period operation. *Chem. Eng. Sci.* 2006;**61**:7760–7753.

[6] Isafiade AJ, Fraser DM. Interval based MINLP superstructure synthesis of heat exchanger networks for multi-period operations. *Chem. Eng. Res. Des.* 2010;**88**(10):1329–1341.

[7] Isafiade A, Bogataj M, Fraser D, Kravanja Z. Optimal synthesis of heat exchange networks for multi-period operations involving single and multiple utilities. *Chem. Eng. Sci.* 2015:175–188.

[8] Ahmad MI, Zhang N, Jobson M, Chen L. Multi-period design of heat exchanger networks. *Chem. Eng. Res. Des.* 2012;**90**(11):1883–1895.

[9] Jiang D, Chang C-T. A new approach to generate flexible multi-period heat exchanger network designs with timesharing mechanisms. *Ind. Eng. Chem. Res.* 2013;**52**:3794–3804.

[10] Chen C-L, Hung P-S. Simultaneous synthesis of flexible heat exchange networks with uncertain source-stream temperatures and flow rates. *Ind. Eng. Chem. Res. Des.* 2004;**43**:5916–5928.

[11] Chen C-L, Kuo C-H, Hung P-S. A novel strategy for synthesis of flexible-heat exchange networks. *J. Chin. Inst. Chem. Engrs.* 2005;**36**(5):421–432.

[12] Chen C-L, Hung P-S. Synthesis of flexible heat-exchange networks and mass exchange networks, *Comp. Chem. Engr.* 2007;**31**:1619–1632.

[13] Escobar M, Trierweiler JO, Grossmann IE. Simultaneous synthesis of heat exchanger networks with operability considerations: flexibility and controllability. *Comp. Chem. Eng.* 2013;**55**:158–180.

[14] Li J, Du J, Zhao Z, Yao P. Structure and area optimisation of flexible heat exchanger networks. *Ind. Eng. Chem. Res.* 2014;**53**:11779–11793.

[15] J. Li J, Du J, Zhao Z, Yao P. Efficient method for flexibility analysis of large-scale non convex heat exchanger networks. *Ind. Eng. Chem. Res*, 2015;**54:**10757–10767.

[16] Isafiade AJ, Short M. Simultaneous synthesis of flexible heat exchanger networks for unequal multi-period operations. Process Safety and Environmental Protection, 2016. Vol 103, Part B, Pages 377–390. doi:10.1016/j.psep.2016.04.021

[17] Papoulias SA, Grossmann IE. A structural optimization approach to process synthesis-II. Heat recovery networks. *Comp. Chem. Eng.* 1983;**7**:707–721.

[18] Floudas CA, Ciric AR, Grossmann IE. Automatic synthesis of optimum heat exchanger network generation. *AICHE J.* 1986; **32**(2):276–290.

[19] Yee T, Grossman IE. Simultaneous optimisation models for heat integration-II. Heat exchanger network synthesis, *Comp. Chem. Eng.* 1990;**14**(10):1165–1184.

[20] Grossmann IE, Floudas CA. Active constraint strategy for flexibility analysis in chemical processes, *Comp. Chem. Eng.* 1987;**11**(6):675–693.

[21] Rosenthal RE. *GAMS – A User's Guide*, 2012. Washington, DC, USA: GAMS Development Corporation.

3

Unsteady Mixed Convection from Two Isothermal Semicircular Cylinders in Tandem Arrangement

Erick Salcedo, César Treviño, Juan C. Cajas and

Lorenzo Martínez-Suástegui

Abstract

In this chapter, two-dimensional mixed convection heat transfer in a laminar cross-flow from two heated isothermal semicircular cylinders in tandem arrangement with their curved surfaces facing the oncoming flow and confined in a channel is studied numerically. The governing equations are solved using the control-volume method on a nonuniform orthogonal Cartesian grid. Using the immersed-boundary method for fixed Reynolds number of $Re_D = u_D D/\upsilon = 200$, Prandtl number of $Pr = 7$, blockage ratio of $BR = D/H = 0.2$ and nondimensional pitch ratio of $\sigma = L/D = 3$, the influence of buoyancy and the confinement effect are studied for Richardson numbers in the range $-1 \leq Ri \leq 1$. Here, u_D is the average longitudinal velocity based on the diameter of the semicylinder. The variation of the mean and instantaneous nondimensional velocity, vorticity and temperature distributions with Richardson number is presented along with the nondimensional oscillation frequencies (Strouhal numbers) and phase-space portraits of flow oscillation from each semicylinder. In addition, local and averaged Nusselt numbers over the surface of the semicylinders are also obtained. The results presented herein demonstrate how the buoyancy and wall confinement affect the wake structure, vortex dynamics and heat transfer characteristics.

Keywords: bluff bodies, tandem arrangement, blockage ratio, interference effects, wall effects

1. Introduction

The flow and heat transfer past bluff bodies of various cross-sectional geometries is important because of advances in heat exchanger technology, cooling of electronic components and chips of different shapes and sizes. Although the majority of these studies have focused on studying the cross-flow past bluff bodies such as cylinders of circular [1–6], elliptic [7–10], rectangular [11–15] and square cross-sections [16–20], there are fewer studies on the semicircular cylinder geometry [21–24]. Gode *et al.* [25] studied numerically the momentum and heat transfer characteristics of a two-dimensional (2D), incompressible and steady flow over a semicircular cylinder and pointed out that the wake ceases to be steady somewhere in the range of $120 \leq Re \leq 130$. Boisaubert and Texier [26] performed solid tracer visualizations to assess the effect of a splitter plate on the near-wake development of a semicircular cylinder for Reynolds numbers of $Re = 200$ and 400 and three splitter plate configurations. Their results show that for $Re = 400$, the splitter plate causes an increase in near-wake length, a decrease in near-wake maximum width, a secondary vortex formation and a decrease of the maximum velocity in the recirculating zone, while for $Re = 200$, the near-wake keeps its symmetry and vortex shedding is inhibited. Nalluri *et al.* [27] solved numerically the coupled momentum and energy equations for buoyancy-assisted mixed convection from an isothermal hemisphere in Bingham plastic fluids and reported results for streamline and isotherm contours, local and mean Nusselt number as a function of the Reynolds, Prandtl, Richardson and Bingham numbers. Bhinder *et al.* [28] studied numerically the wake dynamics and forced convective heat transfer past an unconfined semicircular cylinder at incidence using air as the working fluid for Reynolds numbers in the range of $80 \leq Re \leq 180$ and angles of incidence in the range of $0^o \leq \alpha \leq 180^o$. Based on the flow pattern and the angle of incidence, they identified three flow distinct zones and proposed a correlation for the Strouhal and averaged Nusselt number as a function of Re and α. Chandra and Chhabra [29] performed a numerical study to assess the flow and thermal characteristics from a heated semicircular cylinder immersed in power-law fluids under laminar free and mixed convection for the case of buoyancy-assisted flow. Their results show that as the value of the Richardson and Reynolds numbers increase, the drag coefficient shows a monotonic increase and that the average Nusselt number increases with an increase in the value of the Reynolds, Prandtl and Richardson numbers.

The foregoing survey of literature reveals that although the great majority of research for the flow and heat transfer past a heated hemisphere in cross-flow has been made for an unbounded domain, there are relatively few studies that deal with the investigation of the blockage constraints present in the confined hemisphere problem. Kumar *et al.* [30] performed a numerical analysis to investigate the forced convection of power-law fluids (power-law index varying from 0.2 to 1.8) around a confined heated semicircular cylinder for Reynolds numbers between 1 and 40 and Prandtl number of 50. They assessed the effects of blockage ratios ranging from 0.16 to 0.50 and found that for a fixed value of Re, the length of the recirculation zone decreased with an increase in the value of n and that the drag coefficients and the averaged Nusselt number increased with increasing blockage ratio for any value of n.

From the foregoing discussion, it is clear that no prior results are available on the flow and heat transfer characteristics past a confined tandem hemisphere array under buoyancy-assisted and buoyancy-opposing conditions. This study aims to fill this void in the existing literature. In this work, we numerically investigate the transient fluid flow and thermal characteristics in the mixed convection regime around two isothermal semicylinders of the same diameter in tandem arrangement confined inside a vertical channel of finite length using fixed Reynolds and Prandtl numbers, fixed blockage ratio and gap width and several values of the buoyancy parameter (Richardson number).

2. Formulation of the problem

2.1. Governing equations and boundary conditions

Consider a 2D steady, Newtonian, incompressible Poiseuille flow fluid with a mean mainstream velocity u_0 at the entrance of a vertical duct over infinitely long semicylinders of diameter D placed symmetrically between two parallel plane walls as shown schematically in **Figure 1**. A system of Cartesian coordinates (x, y) is used with its origin located at the centre point of the upstream hemisphere. The length and height of the computational domain are defined in terms of the axial and lateral dimensions ($L_{tot} = 30D$ and H, respectively). The pitch-to-diameter ratio is $\sigma = L/D = 3$ and the blockage ratio $BR = D/H = 0.2$, where L is the longitudinal spacing between semicylinders. The upstream hemisphere is placed at a distance of $5.5D$ from the inlet to its centre and at a distance $S_1 = 24.5D$ from its centre to the outflow boundary. These values were chosen as they were estimated to be sufficiently large to allow the wake behind the downstream semicylinder to develop properly and to exit the domain without producing observable reflections. The forced flow enters the channel at ambient temperature T_0, and the semicylinders have a wall temperature of T_w. Flow rectifiers are placed at the channel exit producing a parallel flow at $x = S_1$. The thermophysical properties of the fluid are assumed to be constant except for the variation of density in the buoyancy term of the axial momentum equation (Boussinesq approximation) and the effect of viscous dissipation is neglected. Using the vorticity ($\Omega = \partial V/\partial X - \partial U/\partial Y$) and stream function formulation ($U = \partial\psi/\partial Y$, $V = -\partial\psi/\partial X$), the flow is described by the nondimensional equations

$$\frac{\partial^2\psi}{\partial X^2} + \frac{\partial^2\psi}{\partial Y^2} = -\Omega, \tag{1}$$

$$\frac{\partial\Omega}{\partial\tau} + \frac{\partial\psi}{\partial Y}\frac{\partial\Omega}{\partial X} - \frac{\partial\psi}{\partial X}\frac{\partial\Omega}{\partial Y} = \frac{1}{Re}\left(\frac{\partial^2\Omega}{\partial X^2} + \frac{\partial^2\Omega}{\partial Y^2}\right) + Ri\frac{\partial\theta}{\partial Y}, \tag{2}$$

$$\frac{\partial\theta}{\partial\tau} + \frac{\partial\psi}{\partial Y}\frac{\partial\theta}{\partial X} - \frac{\partial\psi}{\partial X}\frac{\partial\theta}{\partial Y} = \frac{1}{RePr}\left(\frac{\partial^2\theta}{\partial X^2} + \frac{\partial^2\theta}{\partial Y^2}\right). \tag{3}$$

where $\overline{V} = (U, V)$ is the dimensionless velocity vector and θ is the dimensionless temperature. In Eqs. (1)–(3), U and V are the X and Y components of $\overline{V}$, respectively. All velocity components are scaled with the oncoming mean bulk velocity u_0; the longitudinal and transverse coordinates are scaled with the semicylinder diameter D; the time is scaled with the residence time D/u_0, $\tau = tu_0/D$; the temperature is normalized as $\theta = (T - T_0)/(T_w - T_0)$. In the above equations, the nondimensional parameters are the Reynolds number, $Re = u_0 D/\nu$, the Prandtl number $\Pr = \nu/\alpha$ and the Richardson number,

$Ri = g\beta(T_w - T_0)D/u_0^2$, respectively (frequently, instead of using the Richardson number, the Grashof number is employed, $Gr = RiRe^2 = g\beta(T_w - T_0)D^3/\nu^2$). Here, g is the acceleration due to gravity, α is the thermal diffusivity, β is the thermal expansion coefficient of the fluid and ν is the kinematic viscosity.

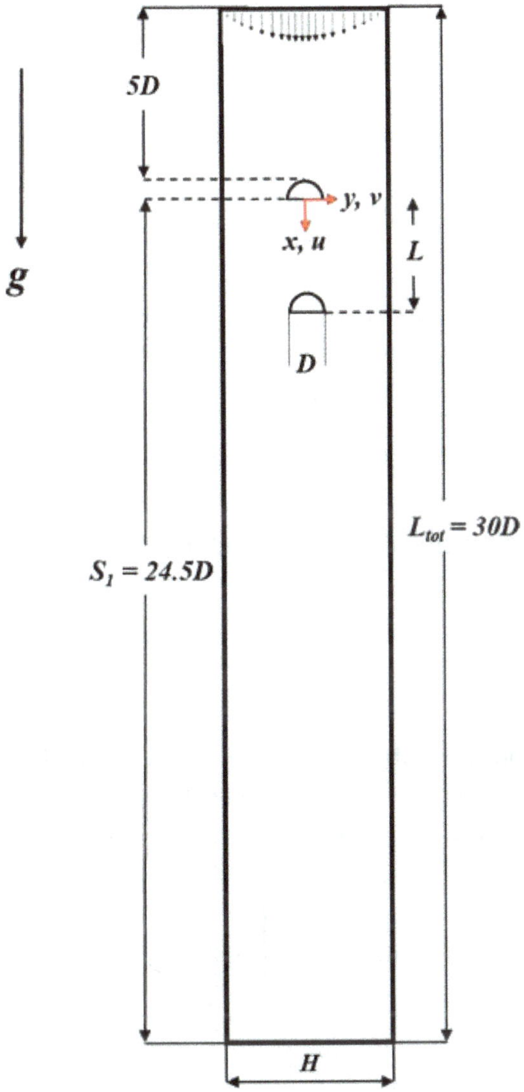

Figure 1. Schematic diagram of the computational domain and the configuration of the semicylinders inside the channel at $BR = 0.2$ and $\sigma = 3$.

Eqs. (1)–(3) have to be solved with the following boundary conditions:

The inflow boundary is specified by a developed velocity profile at the channel inlet

$$\psi - 1/2[1/BR + 3Y - 4(BR)^2Y^3] = \Omega - 12(BR)^2Y = \theta = 0, \tag{4}$$

For the channel walls, $\psi = 0, \ +1/BR$ at the left $(Y = -1/(2BR))$ and right walls $(Y = +1/(2BR))$, respectively. Vorticity at the walls is evaluated using Thom's first-order formula [31],

$$\Omega_w = 2(\psi_{w+1} - \psi_w)/\Delta n^2, \tag{5}$$

where Δn is the grid space normal to the wall. Adiabatic channel walls are assumed, $\partial\theta/\partial Y = 0$.

Homogeneous Neumann-type boundary conditions are adopted at the channel exit, provided that the outlet boundary is located sufficiently far downstream from the region of interest.

$$\partial\psi/\partial X|_{x=s_1} = \partial^2\psi/\partial X\partial Y|_{x=s_1} = \partial\theta/\partial X|_{x=s_1} = 0, \tag{6}$$

At the surface of the semicylinders,

$$\Omega - 2(\psi_{w+1} - \psi_w)/\Delta n^2 = \theta - 1 = 0. \tag{7}$$

No-normal and no-slip boundary conditions are enforced at the surface of each semicylinder. Due to the fact that the value of the stream function is an unknown constant along the surface of each hemisphere, its value is determined at each time step as part of the solution process [32].

With the temperature field known, the rate of heat flux q_j is obtained in nondimensional form with the local Nusselt number Nu_j, with $j = 1, 2$ for the upstream and downstream semicylinder, respectively. The local Nusselt numbers are evaluated from the following equation

$$Nu_j(S,\tau) = \frac{|q_j(S,t)|D}{(T_w - T_0)k} = \left|\frac{\partial\theta}{\partial n}\right|_S \tag{8}$$

where k is the thermal conductivity of the fluid and S is the surface of the immersed semicylinders. The surface-averaged (mean) Nusselt number is obtained by integrating the local Nusselt number along the surface of each semicylinder

$$\overline{Nu_j}(\tau) = \frac{1}{S}\int_S Nu_j(S,\tau)dS \tag{9}$$

2.2. Numerical solution

The governing equations are discretized using the power-law scheme described by Patankar [33] using a nonuniform staggered Cartesian grid with local grid refinements near the immersed semicylinders and near the channel walls. Eqs. (1)–(3) along with their corresponding boundary conditions are solved using a finite volume-based numerical method developed in Fortran 90 using parallel programming (OpenMP). Internal flow boundaries in the flow field are specified using the immersed boundary method [34]. For all computations, water is used as the cooling agent ($Pr = 7$). A stringent convergence criteria of the dependent variables of 1×10^{-7} is used, with an optimal time step of $\Delta\tau = 5\times10^{-4}$. A fully developed base flow is assigned as the initial value to each grid point in the domain, which physically means that both semicylinders are introduced into an isothermal fully developed cross-flow. For a given value of the Richardson number, computation is started immediately after the sudden imposition of a uniform wall nondimensional temperature from 0 to 1 on both semicylinders at time $\tau = 0$. Transient calculations are performed up to 500 nondimensional time units. In order to make comparisons with experimental results obtained on what are effectively unbounded domains, Chen *et al.* [35] defined a Reynolds number, $Re_D = u_D D/\nu$, where

$$u_D = \frac{1}{D}\int_{-D/2}^{D/2} u(y)dy. \tag{10}$$

In Eq. (10), u is the vertical component of the velocity field specified on the upstream boundary and u_D is the average longitudinal velocity based on the diameter of the semicylinder. The accuracy of the numerical algorithm was tested by comparing results of the mean Nusselt number against available analytical [2] and numerical results [35] for the standard case of a symmetrically confined isothermal circular cylinder in a plane channel. Details about the numerical solution, validation of the algorithm and the grid employed can be found elsewhere [36, 37].

3. Results and discussion

The numerical results presented in this work correspond in all cases to $Re_D = 200$, $Pr = 7$, $BR = 0.2$, and $\sigma = L/D = 3$. In this section, results are presented for the mean and instantaneous flow and thermal characteristics under varying thermal buoyancy. For clarity, only a portion of the computational domain is shown. The images display (from left to right) the nondimensional longitudinal and transverse velocity components with superimposed streamlines, the nondimensional vorticity field and the temperature field with superimposed velocity profiles. The color scales below each image map the velocity, vorticity and temperature contours, with red/yellow coloration representing positive vorticity or counterclockwise fluid rotation and the green regions reflecting a lack of rotational motion.

3.1. Response characteristics for assisting flow ($Ri = -1$)

In this section, the response characteristics for assisting flow are presented. **Figure 2** shows the resulting nondimensional mean flow and thermal profiles at $Ri = -1$ ($Gr = -4\times10^4$), illustrating how the relatively narrow wake of the upstream semicylinder reattaches at the forebody of the downstream semicylinder. Here, the near wake of the latter is clearly shorter and narrower and an increase in the longitudinal velocity component is observed at the central part of the channel ($Y = 0$) toward the downstream direction. The third strip illustrates how for the cooling process, the flow pattern is slightly asymmetric and the peak vorticity values are particularly large.

Figure 3 shows typical instantaneous flow and thermal patterns for $Ri = -1$, illustrating how small amplitude flow oscillation takes place within the gap, while Kármán vortices of relatively small size are shed from the rear face of the downstream semicylinder. The third strip shows how the interaction between the shear layers generated at the surface of both semicylinders and the channel walls increases toward the downstream direction and reaches a peak at a location of $X \approx 7$.

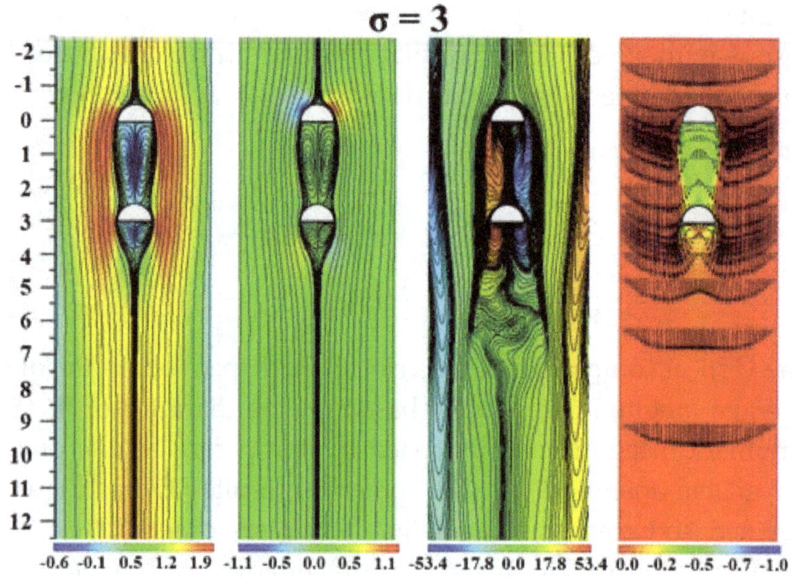

Figure 2. Nondimensional mean flow values at $Re_D = 200$, $BR = 0.2$, $\sigma = 3$ and $Ri = -1$ ($Gr = -4{\times}10^4$). From left to right: U and V velocity, Ω vorticity and θ temperature fields, respectively.

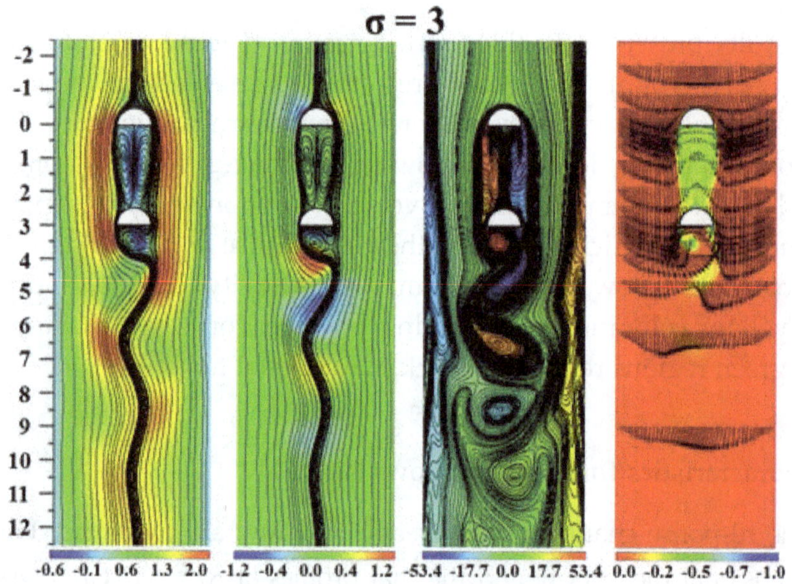

Figure 3. Nondimensional near wake patterns of instantaneous velocity, vorticity and temperature contours at $Re_D = 200$, $BR = 0.2$, $\sigma = 3$, and $Ri = -1$ ($Gr = -4{\times}10^4$). From left to right: U and V velocity, Ω vorticity and θ temperature fields, respectively.

Figure 4 shows the time variations of the nondimensional longitudinal and transverse velocity components at the symmetry plane and selected positions inside the channel. Clearly, the velocity fluctuations depict a harmonic behavior after a short induction time of $\tau \sim 100$. The inset of the top and bottom left images illustrates how the recirculation zone within the gap depicts small amplitude oscillations at a location of $(X, Y) = (1.5, 0)$, while the maximum amplitude of the velocity fluctuations is reached at a downstream position of $(X, Y) = (4.5, 0)$.

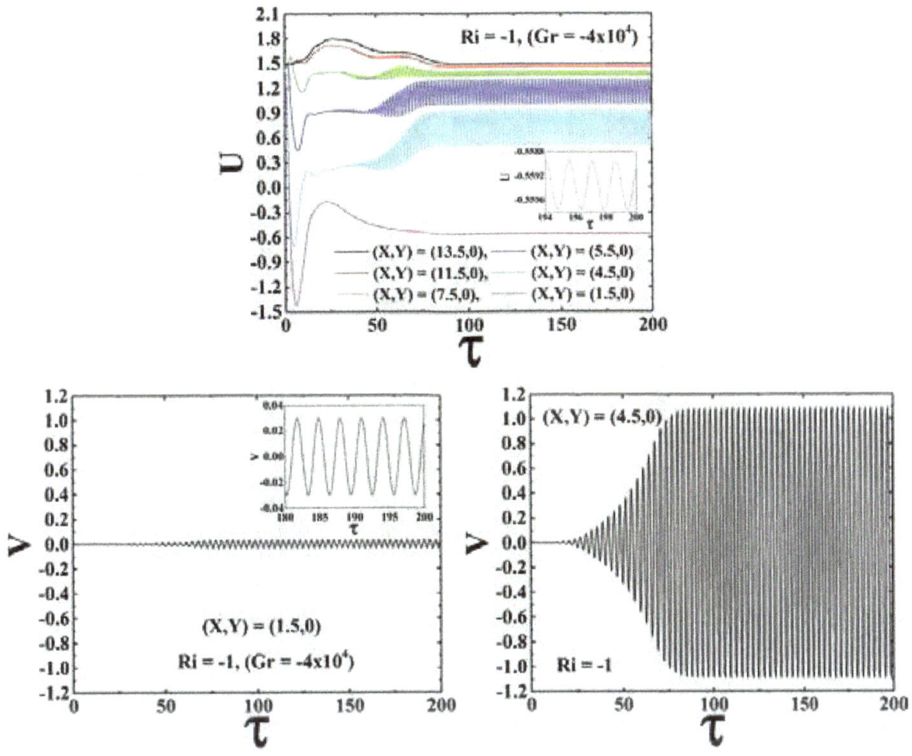

Figure 4. Time variations of the nondimensional longitudinal and transverse velocity components as a function of the nondimensional time at $Ri = -1$ ($Gr = -4{\times}10^4$). The extracted data is obtained at the symmetry plane and several X positions.

3.2. Response characteristics for isothermal flow ($Ri = 0$)

Figure 5 shows the nondimensional mean flow values for an isothermal flow ($Ri = 0$). In the absence of buoyancy, the mean flow solution is symmetric. Although the recirculation zone of the upstream semicylinder still occupies the total space within the gap, its width is now larger than the semicylinder diameter. In addition, the length of the near wake of the downstream semicylinder extends to $X{\approx}4.5$ and a slight decrease in vorticity strength takes place.

Figure 6 shows typical instantaneous patterns of velocity and vorticity illustrating how vortex shedding takes place at the rear of the downstream semicylinder. The third strip illustrates how in the absence of buoyancy, the interaction between the shear layers generated by the upstream semicylinder and the confining walls reduces.

Figure 7 shows the time variations of the nondimensional longitudinal and transverse velocity components at the symmetry plane and selected positions inside the channel. This image shows how after an induction time of $\tau{\sim}120$, flow oscillation within the gap and downstream of the lower semicylinder depict a nice harmonic behavior. Here, the amplitude of the oscillations reaches a peak at a location of $(X, Y) = (5.5, 0)$ and decreases toward the downstream direction.

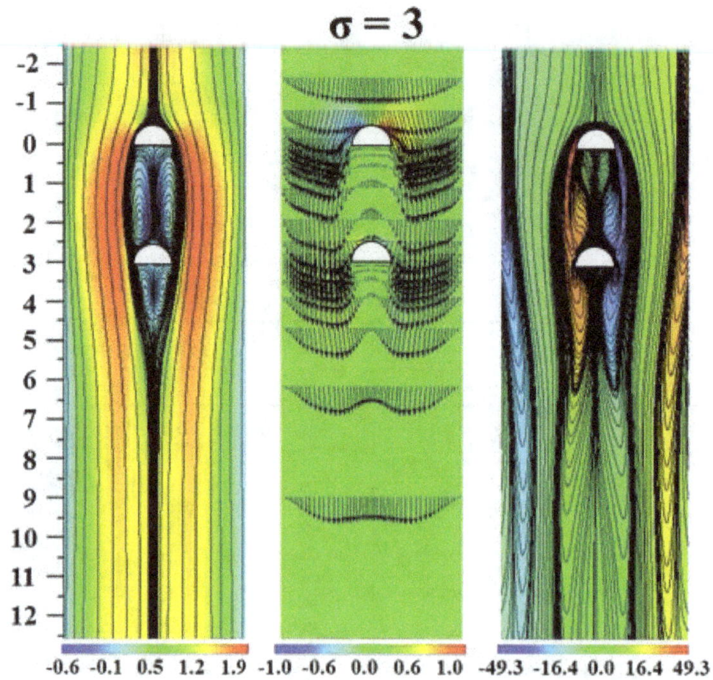

Figure 5. Nondimensional mean flow values for the unheated semicylinders at $Re_D = 200$, $BR = 0.2$, $\sigma = 3$, and $Ri = 0$. From left to right: U and V velocity and Ω vorticity fields, respectively.

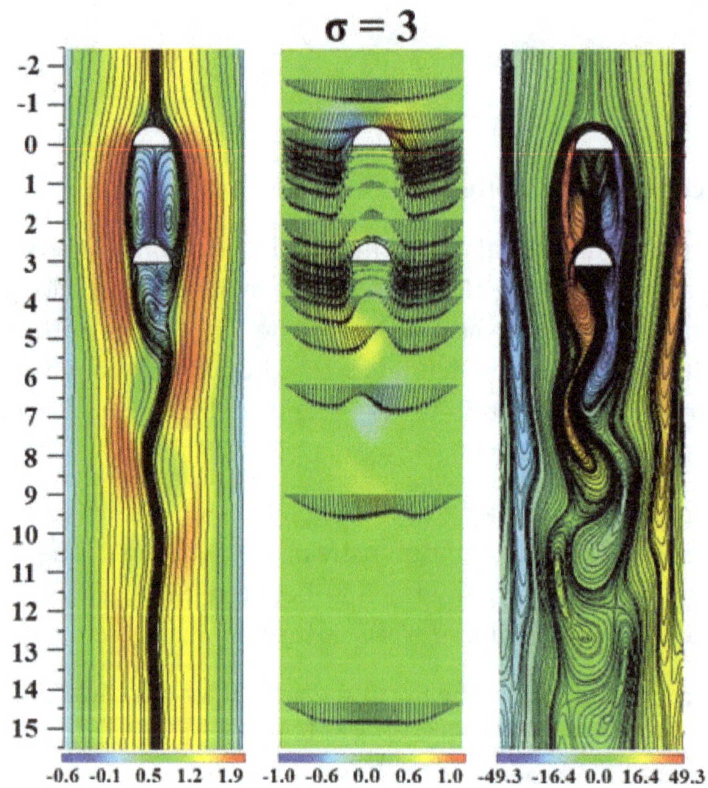

Figure 6. Nondimensional near-wake patterns of instantaneous velocity and vorticity contours for the unheated semicylinders at $Re_D = 200$, $BR = 0.2$, $\sigma = 3$, and $Ri = 0$. From left to right: U and V velocity and Ω vorticity fields, respectively.

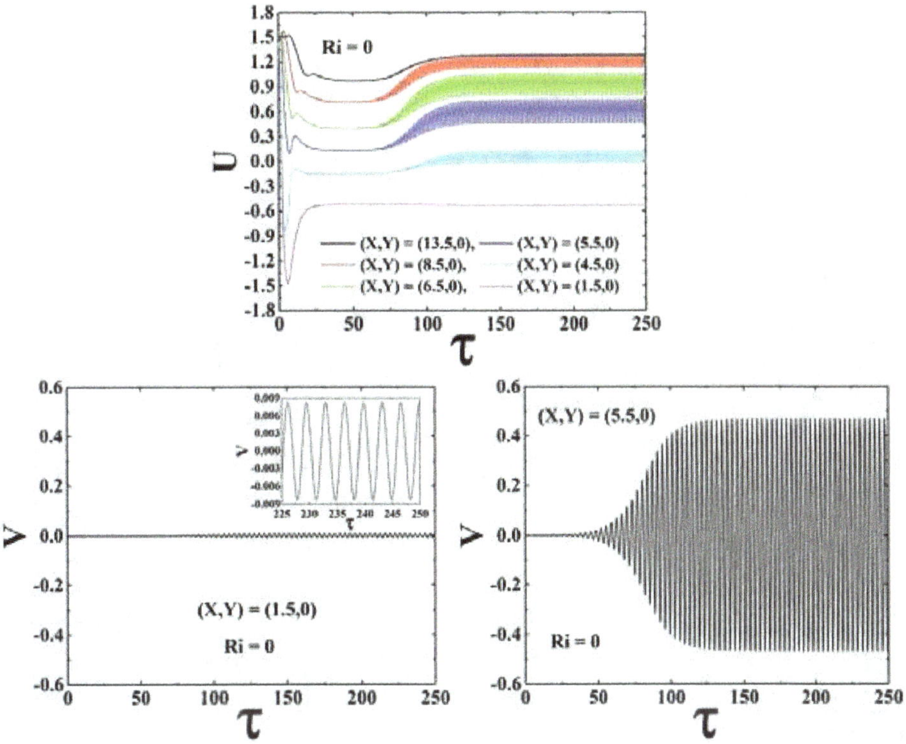

Figure 7. Time variations of the nondimensional longitudinal and transverse velocity components as a function of the nondimensional time at $Re_D = 200$, $BR = 0.2$, $\sigma = 3$, and $Ri = 0$. The extracted data is obtained at the symmetry plane and several X positions.

3.3. Response characteristics for opposing flow ($Ri = 1$)

In this section, the response characteristics for opposing flow are presented. **Figure 8** shows the nondimensional mean flow values at $Ri = 1$ ($Gr = 4 \times 10^4$). Clearly, because of the presence of flow reversal, the width of the symmetric recirculation zone present within the gap and at the rear of the downstream semicylinder increases. As a result, the blockage effect is enhanced and the longitudinal velocity component reaches peak values close to the semicylinders. Note how due to secondary flow, both recirculation zones behind each semicylinder have approximately the same size. Also, because of the presence of relatively strong upward flow within the gap, a bridge that reconnects the thermal layers of both semicylinders increases buoyancy strength and vorticity strength reduces.

Figure 9 shows a typical instantaneous flow and thermal pattern at $Ri = 1$ ($Gr = 4 \times 10^4$), illustrating how the shedding process changes in the presence of flow reversal. Here, the recirculation zone within the gap impinges the forebody of the downstream semicylinder and pairs periodically with the vortices shed by the downstream semicylinder. Note how because of the presence of relatively high upward flow, the downstream semicylinder sheds typical Kármán vortices of relatively large size. The third strip illustrates how vorticity contours become more complex toward the downstream direction. Here, A highlights how wall vorticity merges with downstream vortices with the same sign. The fourth strip shows how the total surface of the downstream semicylinder is completely surrounded by upward flow that produces a thermal plume at the upper stagnation point of the lower hemisphere. As such, heat transfer decreases because of the presence of relatively high temperature fluid within the gap.

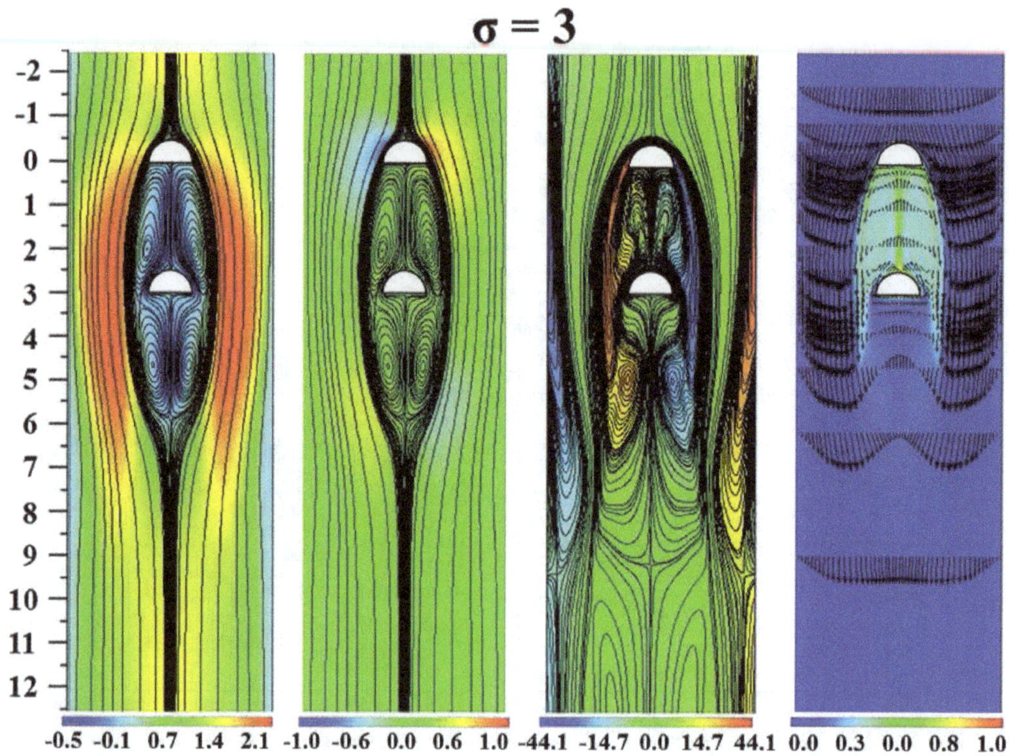

Figure 8. Nondimensional mean flow vales at $Re_D = 200$, $BR = 0.2$, $\sigma = 3$, and $Ri = 1$ ($Gr = 4\times10^4$). From left to right: U and Vvelocity, Ω vorticity and θ temperature fields, respectively.

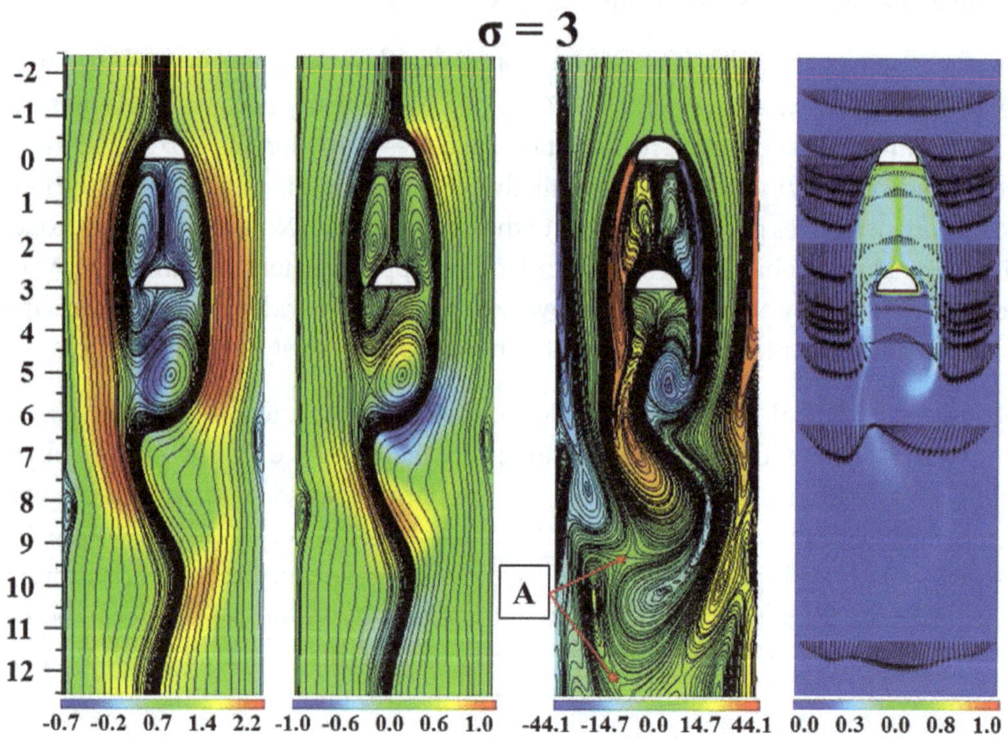

Figure 9. Nondimensional near-wake patterns of instantaneous velocity, vorticity and temperature contours at $Re_D = 200$, $BR = 0.2$, $\sigma = 3$, and $Ri = 1$ ($Gr = 4\times10^4$). From left to right: U and Vvelocity, Ω vorticity and θ temperature fields, respectively.

Figure 10 shows the time variations of the nondimensional longitudinal and transverse velocity components at the symmetry plane and selected longitudinal positions inside the channel. Clearly, time-periodic flow oscillation sets in after an induction time. The inset of the lower left image shows how the recirculation zone within the gap depicts periodic flow oscillation of relatively small amplitude.

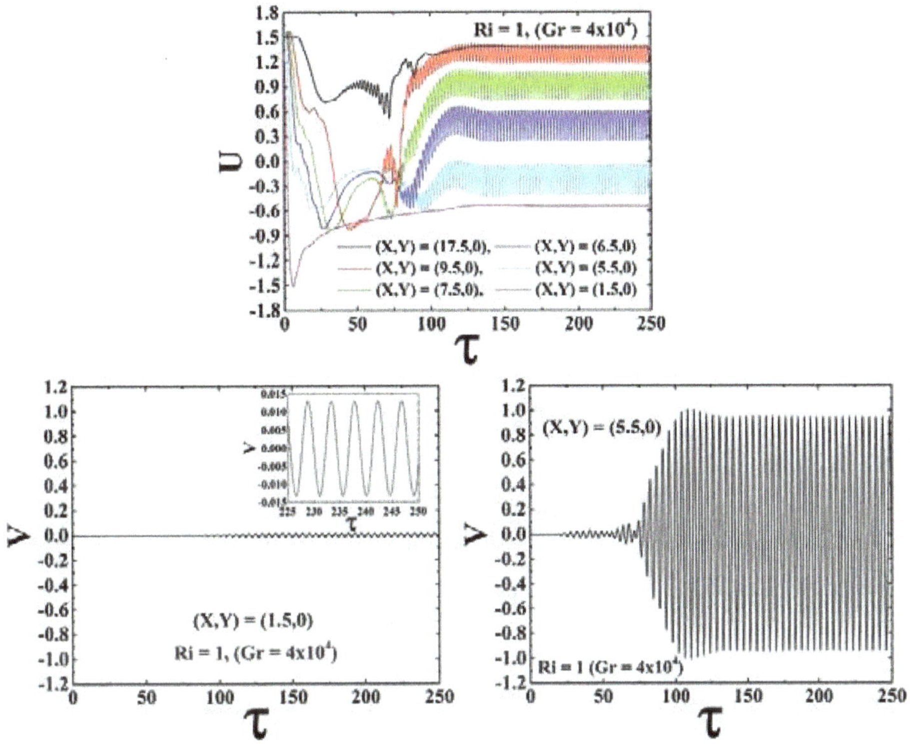

Figure 10. Time variations of the nondimensional longitudinal and transverse velocity components as a function of the nondimensional time at $Re_D = 200$, $BR = 0.2$, $\sigma = 3$, and $Ri = 1$ ($Gr = 4 \times 10^4$). The extracted data is obtained at the symmetry plane and several X positions.

3.4. Strouhal number and phase space plots

The left images in **Figure 11** show (from top to bottom) the normalized spectrum of the transverse velocity component as a function of the nondimensional frequency (Strouhal number), $St = fD/u_0$ for $Ri = -1$, 0 and 1, respectively.

These images show how for $Ri = -1$, 0 and 1 and for selected locations within the gap and downstream of the lower semicylinder, there is a sharp peak at $St = 0.32111$, 0.29448, and 0.22295, respectively, indicating that the wake vortex shedding of both semicylinders is time-periodic and is dominated by a single fundamental frequency. These images exemplify how for the three values of the buoyancy parameter studied, the recirculation zone of the upstream semicylinder locks on to the shedding frequency of the downstream one. In addition, these images show how the Strouhal number decreases for increasing values of the buoyancy parameter. The right images in **Figure 11** show the corresponding phase-space relation between the longitudinal and transverse velocity signals after the vortex shedding reaches an established periodicity. The inset of these figures describe the fluctuations at a location of $(X, Y) = (1.5, 0)$. For all cases, the single orbit with a double loop illustrates how the periodic alternate shedding of vortices takes place at the space within the gap and downstream of the lower semicylinder.

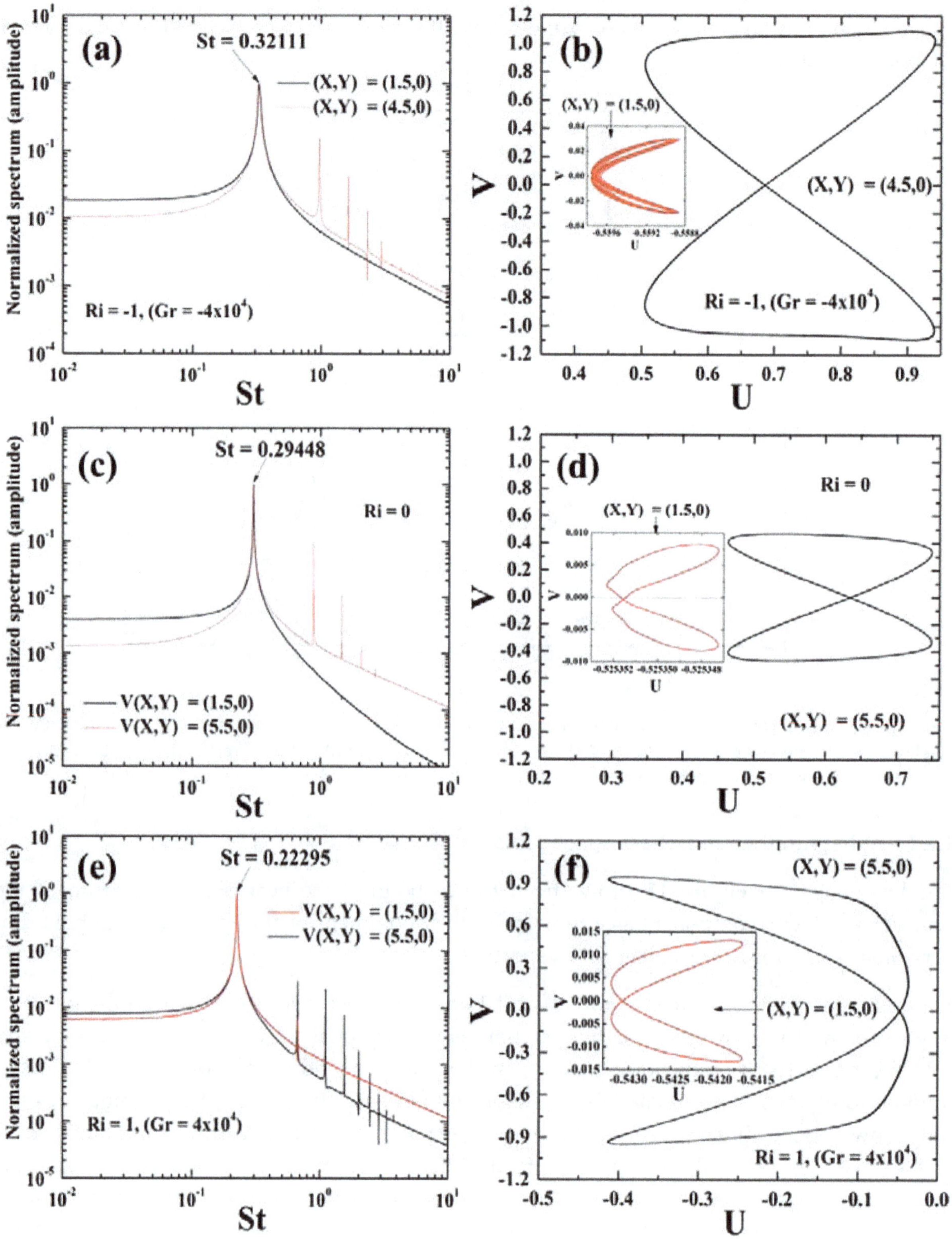

Figure 11. $Re_D = 200$, $BR = 0.2$, $\sigma = 3$, and $Ri = -1$, 0 and 1. Left images: Normalized spectrum of the longitudinal and transverse velocities. Right images: Phase-space plot of the longitudinal velocity signal as a function of the transverse velocity signal.

4. Heat transfer

In this section, the heat transfer characteristics of the semicylinder array are presented for buoyancy assisting and opposing flow.

4.1. Local Nusselt numbers

Figures 12a and **b** show representative distributions of the local Nusselt number defined in Eq. (8) over the curve length ***A-B-C-D*** (body contour of each semicylinder) for $Ri = -1$ and $Ri = 1$, respectively. In these figures, the broken and continuous lines correspond to the upstream and downstream semicylinder, respectively. For assisting/opposing buoyancy, when the warm/cold downward flow impinges the front stagnation point of the upstream semicylinder, the temperature gradient is maximum and the local Nusselt number reaches its peak value at point *C*. Beyond point *C*, as the warm/cold downward-flowing fluid travels through the front half of the semicylinder along the surface ***B-C-D***, it yields/picks up thermal energy and the local Nusselt number gradually decreases toward points ***B*** and ***D***. The cold/warm upward flow present between both semicylinders impinges the rear of the upstream one, a local maximum is reached at point ***A*** and a progressive increase in the local Nusselt number is observed over the curve length ***B-D***. Depending on whether buoyancy assists/opposes the flow and because of the presence of the recirculation zone within the gap that yields/picks up thermal energy from the wake of the upstream semicylinder, a local minimum of the local Nusselt number is reached at the front stagnation point of the downstream semicylinder. Thus, the local Nusselt number beyond point ***C*** gradually increases toward points ***B*** and ***D***. As the flow detaches from the tip of the downstream semicylinder (points ***B*** and ***D***), the local Nusselt number reaches a local/global maximum for assisting/opposing buoyancy, respectively.

4.2. Overall Nusselt number

Figure 13 shows the time variation of the surface-averaged Nusselt number of both semicylinders with Richardson number. In these figures, the broken and continuous lines correspond to the upstream and downstream semicylinder, respectively. **Figure 13** shows how the presence of the upstream semicylinder has a significant effect on the heat transfer characteristics of the downstream semicylinder and lower heat transfer rates are achieved by the latter. For clarity, in the inset of **Figure 13**, the value of the mean Nusselt number of both semicylinders is plotted in a limited range of the nondimensional time, from $\tau = 180$ to 200. It is worth to mention that the discernible periodic oscillations of the mean Nusselt number of the lower semicylinder are closely related to flow oscillation due to vortex shedding for both cases.

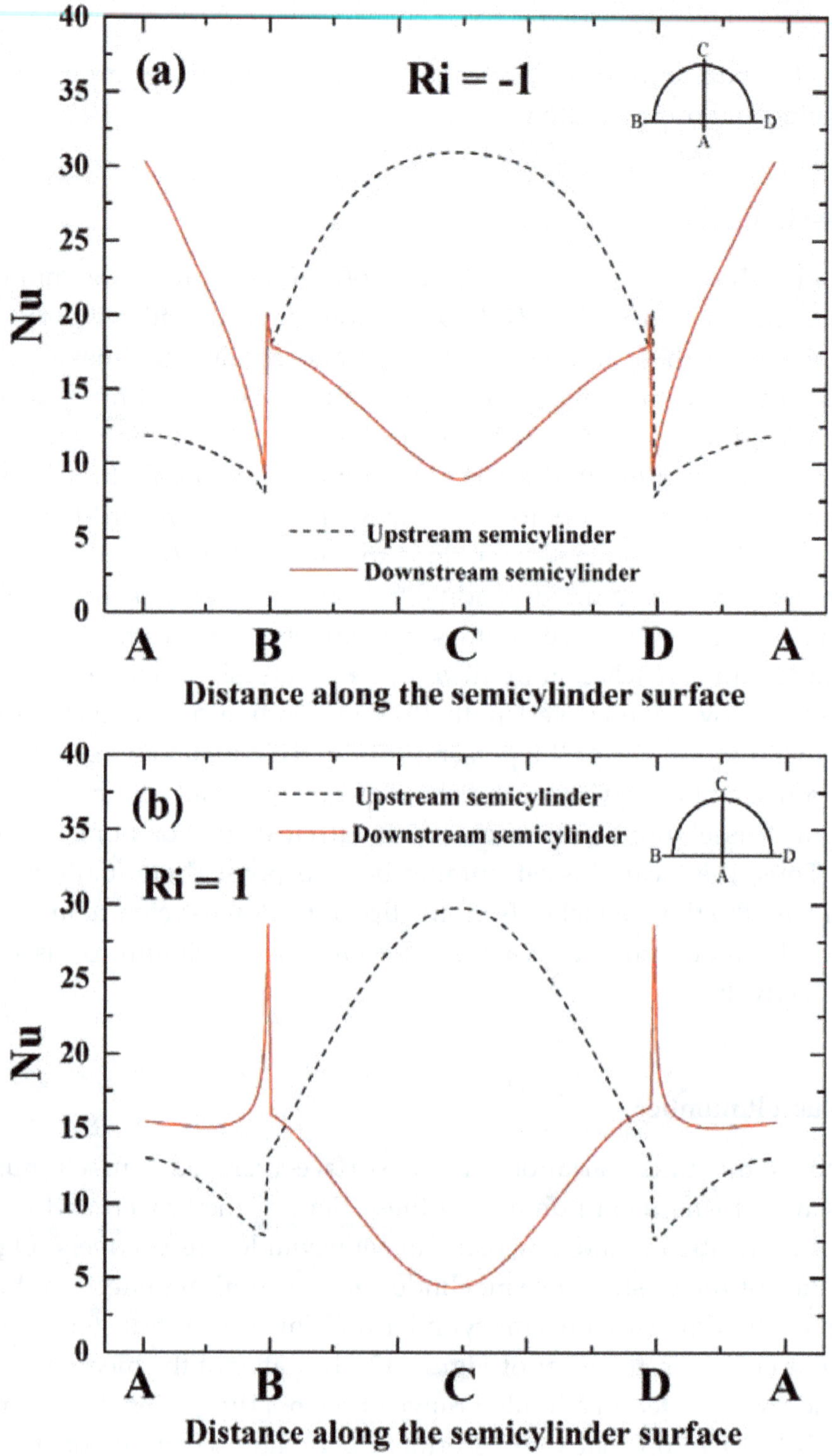

Figure 12. Distribution of the local Nusselt number on the surface of each semicylinder versus distance along each semicylinder surface for $Re_D = 200$, $BR = 0.2$, $\sigma = 3$, and $Ri = -1$ and 1, respectively.

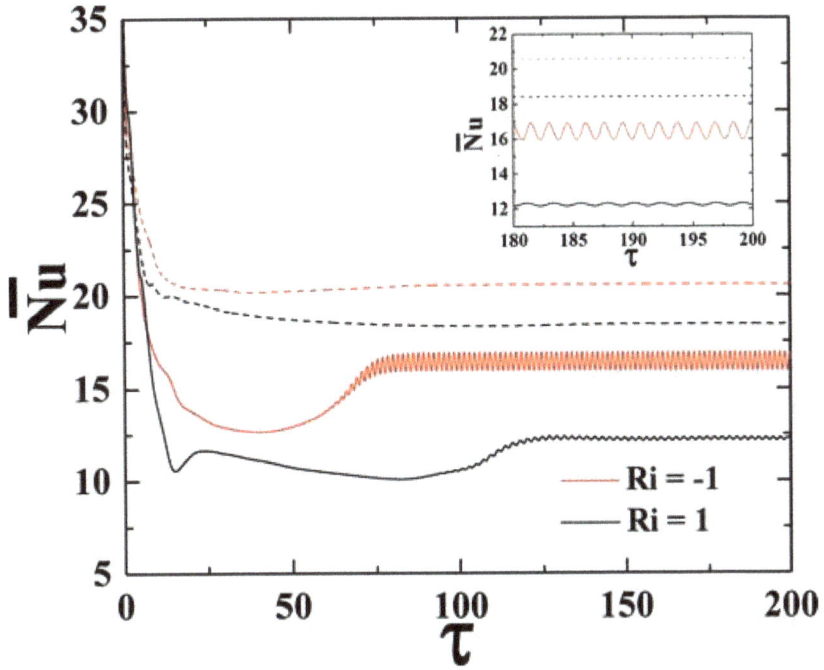

Figure 13. Time-evolution of the overall Nusselt numbers at $Re_D = 200$, $BR = 0.2$, $\sigma = 3$, and $Ri = -1$ and $Ri = 1$ for the upstream (broken lines) and downstream (continuous lines) semicylinders, respectively.

5. Conclusions

In this work, numerical simulations have been carried out to study the unsteady flow and heat transfer characteristics around two identical isothermal semicylinders arranged in tandem and confined in a channel. The blockage ratio, Prandtl number and pitch-to-diameter are fixed at $BR = 0.2$, $\Pr = 7$ and $\sigma = 3$, respectively. Numerical simulations are performed using the control-volume method on a nonuniform orthogonal Cartesian grid. The immersed-boundary method is employed to identify the semicylinders confined inside the channel. The influence of buoyancy has been assessed on the resulting mean and instantaneous flow, vortex shedding properties, nondimensional oscillation frequencies (Strouhal numbers), phase-space portraits of flow oscillation, thermal fields and local and overall nondimensional heat transfer rates (Nusselt numbers) from each semicylinder. Results show that in this parameter space, the flow patterns reach a time-periodic oscillatory state, the recirculation zone of the upper semicylinder completely fills the space within the gap and vortex shedding from the lower semicylinder occurs. For values of the Richardson number of for $Ri = -1$ and $Ri = 1$, steady-state and time periodic oscillations of the mean Nusselt number are observed for the upstream and downstream semicylinder, respectively.

Acknowledgements

This research was supported by the Consejo Nacional de Ciencia y Tecnología (CONACYT), Grant No. 167474.

Nomenclature

BR	blockage ratio, D/H
D	semicylinder diameter (characteristic length)
f	vortex shedding frequency (Hz)
g	gravity acceleration
Gr	Grashof number based on semicylinder diameter, $Gr = g\beta(T_w - T_0)D^3/\nu^2$
h	local heat transfer coefficient
H	width of computational domain
k	thermal conductivity of fluid
L	pitch (centre-to-centre distance between two semicylinders)
L_{tot}	length of computational domain
n	normal direction
Nu	local Nusselt number (see Eq. (8))
$\overline{Nu}$	Average Nusselt number (see Eq. (9))
Pe	Peclet number, $Pe = u_0 D/\alpha$
Pr	Prandtl number, $Pr = \nu/\alpha$
Re	Reynolds number based on u_0, $Re = u_0 D/\nu$
Re_D	Reynolds number based on u_D, $Re = u_D D/\nu$
Ri	Richardson number based on semicylinder diameter, $Ri = Gr/Re^2$
S	surface area of immersed semicylinders
S_1	length from the origin to the channel outlet
St	Strouhal number based on semicylinder diameter, $St = fD/u_0$
t	time
T	temperature
T_0	fluid temperature at the channel inlet
T_w	temperature at the surface of the semicylinders
u, v	longitudinal and transverse velocity components, respectively
u_0	mean fluid velocity at the channel inlet
u_D	average longitudinal velocity over the semicylinders (see Eq. (10))
U	nondimensional longitudinal velocity component, $U = u/u_0$
V	nondimensional transverse velocity component, $V = v/u_0$
x, y	Cartesian rectangular coordinates
X	nondimensional longitudinal coordinate, $X = x/D$
Y	nondimensional transverse coordinate, $Y = y/D$

Greek symbols

α	thermal diffusivity of fluid
β	thermal volumetric expansion coefficient
μ	dynamic viscosity
ν	kinematic viscosity
ψ	nondimensional stream function
Ω	nondimensional vorticity
σ	nondimensional pitch-to-diameter ratio, $\sigma = L/D$

θ nondimensional temperature, $\theta = (T - T_0)/(T_w - T_0)$
τ nondimensional time

Subscripts

0 ambient or reference
1, 2 refers to the upstream and downstream semicylinder, respectively
w at the surface of the semicylinders

Author details

Erick Salcedo[1], César Treviño[2], Juan C. Cajas[3] and Lorenzo Martínez-Suástegui[4*]

*Address all correspondence to: lamartinezs@ipn.mx

1 Departamento de Termofluidos, Facultad de Ingeniería, UNAM México, Ciudad de México, Mexico

2 UMDI, Facultad de Ciencias, Universidad Nacional Autónoma de México, Sisal, Yucatán, Mexico

3 Barcelona Supercomputing Center (BSC), Barcelona, Spain

4 ESIME Azcapotzalco, Instituto Politécnico Nacional, Ciudad de México, Mexico

References

[1] Hu H, Koochesfahani MM. Thermal effects on the wake of a heated circular cylinder operating in mixed convection regime. Journal of Fluid Mechancis. 2011;**685**:235–270. DOI: http://dx.doi.org/10.1017/jfm.2011.313.

[2] Khan WA, Culham JR, Yovanovich MM. Fluid flow and heat transfer from a cylinder between parallel plates. Journal of Thermophysics and Heat Transfer. 2004;**18**(3):395–403. DOI: http://dx.doi.org/10.2514/1.6186.

[3] Perng SW, Wu HW. Buoyancy-aided/opposed convection heat transfer for unsteady turbulent flow across a square cylinder in a vertical channel. International Journal of Heat and Mass Transfer. 2007;**50**(19–20):3701–3717. DOI: http://dx.doi.org/10.1016/j.ijheatmasstransfer.2007.02.026.

[4] Sarkar S, Dalal A, Biswas G. Unsteady wake dynamics and heat transfer in forced and mixed convection past a circular cylinder in cross flow for high Prandtl numbers. International Journal of Heat and Mass Transfer. 2011;**54**(15–16):3536–3551. DOI: http://dx.doi.org/10.1016/j.ijheatmasstransfer.2011.03.032.

[5] Sharma N, Dhiman AK, Kumar S. Mixed convection flow and heat transfer across a square cylinder under the influence of aiding buoyancy at low Reynolds numbers. International Journal of Heat and Mass Transfer. 2012;**55**:2601–2614. DOI: http://dx.doi.org/10.1016/j.ijheatmasstransfer.2011.12.034.

[6] Lima E Silva ALF, Silveira-Neto A, Damasceno JJR. Numerical simulation of two-dimensional flows over a circular cylinder using the immersed boundary method. Journal of Computational Physics. 2003;**189**:351–370. DOI: http://dx.doi.org/10.1016/S0021-9991(03)00214-6.

[7] Kondjoyan A, Daudin JD. Effects of free stream turbulence intensity on heat and mass transfer at the surface of a circular and elliptic cylinder, axis ratio 4. International Journal of Heat and Mass Transfer. 1995;**38**:1735–1749. DOI: 10.1016/0017-9310(94)00338-V.

[8] Liao CC, Lin CA. Influences of a confined elliptic cylinder at different aspect ratios and inclinations on the laminar natural and mixed convection flows. International Journal of Heat and Mass Transfer. 2012;**55**:6638–6650. DOI: http://dx.doi.org/10.1016/j.ijheatmasstransfer.2012.06.073.

[9] Paul I, Prakash KA, Vegadesan S. Forced convective heat transfer from unconfined isothermal and isoflux elliptic cylinders. Numerical Heat Transfer, Part A: Applications. 2013;**64**:648–675. DOI: http://dx.doi.org/10.1080/10407782.2013.790261.

[10] Richter A, Nikrityuk PA. New correlations for heat and fluid flow past ellipsoidal and cubic particles at different angles of attack. Powder Technology. 2013;**249**:463–474. DOI: http://dx.doi.org/10.1016/j.powtec.2013.08.044.

[11] Bruno L, Salvetti MV, Ricciardelli F. Benchmark on the aerodynamics of a rectangular 5:1 cylinder: an overview after the first four years of activity. Journal of Wind Engineering and Industrial Aerodynamics. 2014;**126**:87–106. DOI: http://dx.doi.org/10.1016/j.jweia.2014.01.005.

[12] Igarashi T. Fluid flow and heat transfer around rectangular cylinders (the case of a width/height ratio of a section of 0.33–1.5). International Journal of Heat and Mass Transfer. 1987;**30**:893–901. DOI: doi:10.1016/0017-9310(87)90008-1.

[13] Mannini C, Soda A, Schewe G. Unsteady RANS modelling of flow past a rectangular cylinder: investigation of Reynolds number effects. Computers and Fluids. 2010;**39**(9):1609–1624. DOI: http://dx.doi.org/10.1016/j.compfluid.2010.05.014.

[14] Okajima A, Yi D, Sakuda A, Nakano T. Numerical study of blockage effects on aerodynamic characteristics of an oscillating rectangular cylinder. Journal of Wind Engineering and Industrial Aerodynamics. 1997;**67–68**:91–102. DOI: 10.1016/S0167-6105(97)00065-2.

[15] Patruno L, Ricci M, de Miranda S, Ubertini F. Numerical simulation of a 5:1 rectangular cylinder at non-null angles of attack. Journal of Wind Engineering and Industrial Aerodynamics. 2016;**151**:146–157. DOI: http://dx.doi.org/10.1016/j.jweia.2016.01.008.

[16] Dhiman AK, Chhabra RP, Sharma A., Eswaran V. Effects of Reynolds and Prandtl numbers on heat transfer across a square cylinder in the steady flow regime. Numerical Heat Transfer, Part A. 2006;**49**:717–731. DOI: doi/abs/10.1080/10407780500283325.

[17] Igarashi T. Heat transfer from a square prism to an air stream. International Journal of Heat and Mass Transfer. 1985;**28**(1):175–181. DOI: 10.1016/0017-9310(85)90019-5.

[18] Sohankar A, Norberg C, Davidson L. Low Reynolds number flow around a square cylinder at incidence: study of blockage, onset of vortex shedding and outlet boundary condition. International Journal for Numerical Methods in Fluids. 1998;**26**:39–56. DOI: 10.1002/(SICI)1097-0363(19980115)26:1<39::AID-FLD623>3.0.CO;2-P.

[19] Tong JCK, Sparrow EM, Minkowycz WJ, Abraham JP. A new archive of heat transfer coefficients from square and chamfered cylinders at angles of attach in crossflow. International Journal of Thermal Sciences. 2016;**105**:218–223. DOI: http://dx.doi.org/10.1016/j.ijthermalsci.2016.03.008.

[20] Yoo SY, Goldstein RJ, Chung MK. Effects of angle of attack on mass transfer from a square cylinder and its base plate. International Journal of Heat and Mass Transfer. 1993;**36**:371–381. DOI: 10.1016/0017-9310(93)80013-K.

[21] Chandra A, Chhabra RP. Flow over and forced convection heat transfer in Newtonian fluids from a semi-circular cylinder. International Journal of Heat and Mass Transfer. 2011;**54**:225–241. DOI: http://dx.doi.org/10.1016/j.ijheatmasstransfer.2010.09.048.

[22] Chandra A, Chhabra RP. Laminar free convection from a horizontal semi-circular cylinder to power-law fluids. International Journal of Heat and Mass Transfer. 2012;**55**:2934–2944. DOI: http://dx.doi.org/10.1016/j.ijheatmasstransfer.2012.02.034.

[23] Sasmal C, Shyam R, Chhabra RP. Laminar flow of power-law fluids past a hemisphere: momentum and forced convection heat transfer characteristics. International Journal of Heat and Mass Transfer. 2013;**63**:51–64. DOI: http://dx.doi.org/10.1016/j.ijheatmasstransfer.2013.03.059.

[24] Tiwari AK, Chhabra RP. Laminar natural convection in power-law liquids from a heated semi-circular cylinder with its flat side oriented downward. International Journal of Heat and Mass Transfer. 2013;**58**:553–567. DOI: http://dx.doi.org/10.1016/j.ijheatmasstransfer.2012.11.051.

[25] Gode A, Sahu AK, Chhabra RP. Two-dimensional steady flow over a semi-circular cylinder: drag coefficient and Nusselt number. International Journal of Advances in Engineering Sciences and Applied Mathematics. 2011;**3**(1–4):44–59. DOI: 10.1007/s12572-011-0039-6.

[26] Boisaubert N, Texier A. Effect of a splitter plate on the near-wake development of a semi-circular cylinder. Experimental Thermal and Fluid Science. 1998;**16**:100–111. DOI: http://dx.doi.org/10.1016/S0894-1777(97)10009-7.

[27] Nalluri SV, Patel SA, Chhabra RP. Mixed convection from a hemisphere in Bingham

plastic fluids. International Journal of Heat and Mass Transfer. 2015;**84**:304–318. DOI: http://dx.doi.org/10.1016/j.ijheatmasstransfer.2014.12.059.

[28] Bhinder APS, Sarkar S, Dalal A. Flow over and forced convection heat transfer around a semi-circular cylinder at incidence. International Journal of Heat and Mass Transfer. 2012;**55**:5171–5184. DOI: http://dx.doi.org/10.1016/j.ijheatmasstransfer.2012.05.018.

[29] Chandra A, Chhabra RP. Mixed convection from a heated semi-circular cylinder to power-law fluids in the steady flow regime. International Journal of Heat and Mass Transfer. 2012;**55**:214–234. DOI: http://dx.doi.org/10.1016/j.ijheatmasstransfer.2011.09.004.

[30] Kumar A, Dhiman A, Baranyi L. CFD analysis of power-law fluid flow and heat transfer around a confined semi-circular cylinder. International Journal of Heat and Mass Transfer. 2015;**82**:159–169. DOI: http://dx.doi.org/10.1016/j.ijheatmasstransfer.2014.11.046.

[31] Thom A. The flow past circular cylinders at low speeds. Proceedings of the Royal Society A. 1933;**41**:651–669. DOI: 10.1098/rspa.1933.0146.

[32] Mittal R, Iaccarino G. Immersed boundary methods. Annual Review of Fluid Mechanics. 2005;**37**:239–261. DOI: 10.1146/annurev.fluid.37.061903.175743.

[33] Patankar S. Numerical Heat Transfer and Fluid Flow. 1st ed. United States of America: Hemisphere Publishing Corporation; 1980. 214 p. DOI: ISBN 9780891165224.

[34] Mittal R, Dong H, Bozhurttas M, Najjar F, Vargas A, von Leobbecke A. A versatile sharp interface immersed boundary method for incompressible flows with complex boundaries. Journal of Computational Physics. 2008;**227**(10):4825–4852. DOI: 10.1016/j.jcp.2008.01.028.

[35] Chen JH, Pritchard WG, Tavener SJ. Bifurcation of flow past a cylinder between parallel plates. Journal of Fluid Mechanics. 1995;**284**:23–41. DOI: http://dx.doi.org/10.1017/S0022112095000255.

[36] Martínez-Suástegui L, Treviño C. Transient laminar opposing mixed convection in a differentially and asymmetrically heated vertical channel of finite length. International Journal of Heat and Mass Transfer. 2008;**51**(25–26):5991–6005. DOI: 10.1016/j.ijheatmasstransfer.2008.04.055.

[37] Salcedo E, Cajas JC, Treviño C, Martínez-Suástegui L. Unsteady mixed convection heat transfer from two confined isothermal circular cylinders in tandem: buoyancy and tube spacing effects. International Journal of Heat and Fluid Flow. 2016;**60**:12–30. DOI: http://dx.doi.org/10.1016/j.ijheatfluidflow.2016.04.001.

4

Transient Effectiveness Methods for the Dynamic Characterization of Heat Exchangers

Tianyi Gao, Bahgat Sammakia and James Geer

Abstract

This chapter introduces transient effectiveness methods for dynamic characterization of heat exchangers. The chapter provides a detailed description and review of the transient effectiveness methodology. In this chapter, all the transient effectiveness–related knowledge/works are summarized. The goal of this chapter is to provide a thorough understanding of the transient effectiveness for the reader and to provide guidance for utilizing this methodology in related heat exchanger transient characterization studies. Basically, there are three important applications for transient effectiveness methodology: (1) characterization of heat exchanger dynamic behaviors; (2) characterization of the transient response of closed-coupled cooling/heating systems with multiple heat exchanger units; and (3) development of compact transient heat exchanger models. This innovative modeling method can be used to assist in the development of physics-based predictive, capabilities, performance metrics, and design guidelines, which are important for the design and operation of highly reliable and energy efficient mechanical systems using heat exchangers.

Keywords: transient effectiveness, inlet temperature variation, fluid mass flow rate variation, heat exchanger dynamic performance, compact transient modeling, system level characterization

1. Introduction

Transient effectiveness methodology is a new analytical method which is developed for studying the dynamic performance of a heat exchanger. The concept was originally introduced by Cima and London in 1958 and used as a signature in representing the heat exchanger transient performance. The concept was then used for developing generalized transient effectiveness for plotting the transient response of a counter-flow heat exchanger [1]. In some of the recent studies [2, 3], the transient effectiveness concept is used for developing a new methodology for dynamic characterization of cross-flow heat exchangers. In this chapter, a complete summary and review of the transient effectiveness method is provided, including the methodology development, transient effectiveness characterization, modeling validation, as well as the three major application and usefulness of the transient effectiveness. The heat exchanger configuration considered in most of the studies as well as in this chapter is an unmixed-unmixed cross flow one. It needs to be mentioned here that the majority of the work and results are summarized and published in different scientific journals by the same group of authors. This work provides a complete connection of all the existing research efforts and major results related to the transient effectiveness methodology. The readers can obtain a clear idea of this methodology and utilize it in the corresponding research and studies directly.

2. Transient effectiveness

2.1. Governing equations and numerical solution

Effectiveness which is defined as the ratio of actual heat transferred rate over the maximum heat transfer rate is introduced for characterizing heat exchanger steady-state performance. Cima and London [1] extended this concept to a time-dependent one in [1]. In their study, a generalized transient effectiveness was developed based on Eqs. (1a) and (1b), and then used as a means for representing the transient analog results for a counter-flow heat exchanger instead of using outlet temperatures.

$$\varepsilon_h(t) = \frac{c_{p_h}[T_{h,in}(t) - T_{h,out}(t)]}{c_{p_{\min}}[T_{h,in}(t) - T_{c,in}(t)]} \tag{1a}$$

$$\varepsilon_c(t) = \frac{c_{p_c}[T_{c,out}(t) - T_{c,in}(t)]}{c_{p_{\min}}[T_{h,in}(t) - T_{c,in}(t)]} \tag{1b}$$

The transient effectiveness concept and its governing equations were introduced and directly used for characterizing dynamic performance of a cross-flow heat exchanger in references [2, 3]. In these studies, the transient effectiveness governing equations are solved numerically by coupling them with thermal dynamic heat exchanger equations which are shown in Eqs. (2)–(4). These three sets of governing equations are widely used in most of the existing literature [4–9] for solving similar problems. A full numerical solution for these equations and a comprehensive heat exchanger transient behavior characterization using numerical modeling are conducted in [10–13]. Most of the variation scenarios were covered in these studies, including single fluid temperature variations, fluid mass flow rate variations, as well as multiple variation combinations.

$$\frac{\partial T_{wall}}{\partial t} = r_c^{\beta} \cdot T_c + R \cdot r_h^{\beta} \cdot T_h - (r_c^{\beta} + R \cdot r_h^{\beta}) \cdot T_{wall} \tag{2}$$

$$V_c \frac{\partial T_c}{\partial t} = r_c^{\beta} \cdot T_{wall} - r_c^{\beta} \cdot T_c - r_c \frac{\partial T_c}{\partial X} \tag{3}$$

$$\frac{V_h}{R}\frac{\partial T_h}{\partial t} = r_h^{\beta} \cdot T_{wall} - r_h^{\beta} \cdot T_h - r_h \frac{\partial T_h}{\partial Y} \tag{4}$$

2.2. Transient effectiveness method verification

The methodology and the numerical solution are verified by comparison with several published results in [1, 14, 15]. First, several published analytical solutions and analog solutions for the transient effectiveness of a 1D contour-flow heat exchanger are used [14]. The equivalent method was used and the same transient effectiveness equations were integrated into the numerical code and then compared to the results presented in a form as generalized transient effectiveness. **Figure 1(a)** shows a comparison of the numerical solutions and the analytical data points [14, 15]. This case represents a response of a heat exchanger under a fluid

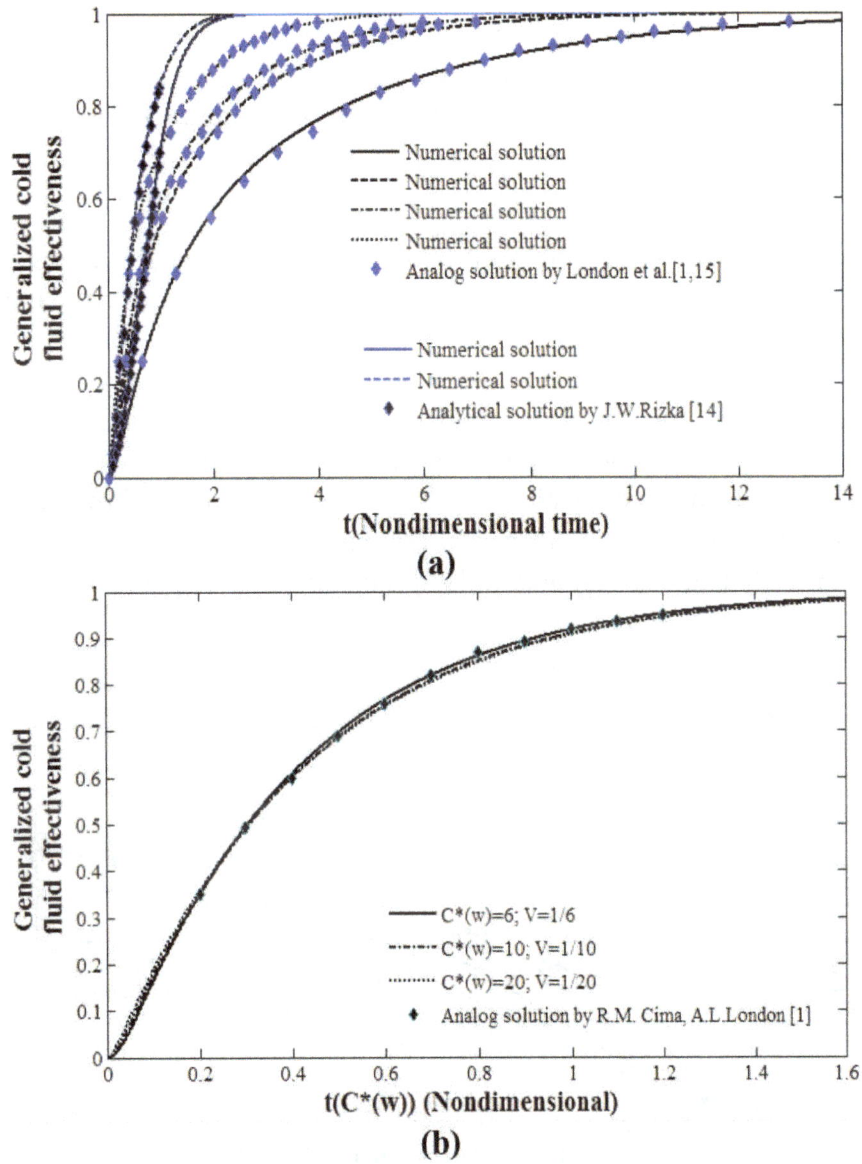

Figure 1. (a) Comparison of the numerical solutions with the analytical results in [14] and analog results in [1, 15]; (b) Comparison of the numerical solutions with the analog results in [1] for *NTU* 1–1.5 and *NTU* 1.5–1.

inlet temperature step change. **Figure 1(b)** shows comparison between the analog solution [1] and numerical solution under a step change in the fluid mass flow rate. The mass flow rate step increase results in the *NTU* variation from 1 to 1.5, and the step decrease results in the *NTU* variation from 1.5 to 1, which is also mentioned in the figure caption. It can be seen from both figures that the results are in good agreement, and different scenarios including fluid mass flow rate change and inlet temperature change are validated.

2.3. Parametric study

A detailed study of the characterizing transient effectiveness under different variation conditions including both inlet temperature change and mass flow rate change is presented in [2, 3]. It is found in these studies that the transient effectiveness can be used as a measure of the heat exchanger dynamic performance from one steady state to the new equilibrium state under certain inputs. In addition, the impact of modeling physical parameters, including *NTU*, E, R, and V, can be represented on the effectiveness curves. *NTU* results are chosen as an example to discuss in this section. More detailed parametric results are summarized in references [2, 3].

2.3.1. Inlet temperature variation

The inlet temperature variation does not influence the final steady-state values of the effectiveness curve lines. This means that the effectiveness curve always returns to the initial value after a certain transient variation. **Figure 2** shows the transient effectiveness of two fluids plotted versus nondimensional time for a wide range of *NTU* values for the step change. It can be seen from the figure that *NTU* governs both transient and steady-state variation of effectiveness. The larger the *NTU* value, the longer time is taken to reach the final steady state. When comparing the hot fluid transient effectiveness curve and the cold fluid transient effectiveness curve, a time lag is seen on the hot fluid curves. This time lag indicates that the corresponding fluid takes some time to begin to respond at the outlet, after the variation is applied at the inlet.

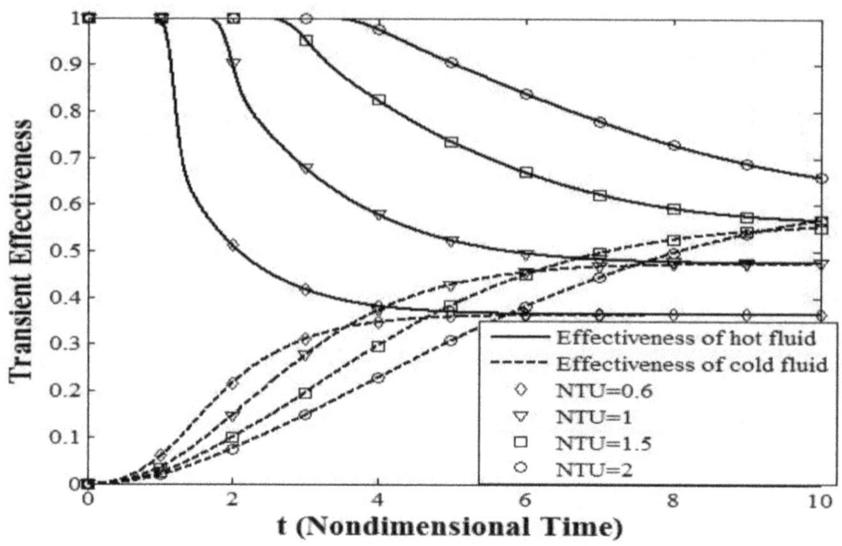

Figure 2. Effect of *NTU* on the transient effectiveness results with $E = 1$, $V_h = V_c = 1$, $R = 1$; step change to the hot fluid inlet temperature.

The larger the *NTU* value, the longer the time lag. A larger *NTU* value can be simply understood as a larger heat exchanger physical size. Therefore, the time lag is longer for a larger *NTU* value.

2.3.2. Fluid mass flow rate variation

The characteristics of the effectiveness under fluid mass flow rate change are discussed in this section. **Figure 3** shows that the steady-state conditions of the effectiveness curve changes due to a change in fluid mass flow rate. The difference is clearly seen from the transient effectiveness between a cold fluid mass flow rate change and a hot fluid mass flow rate change.

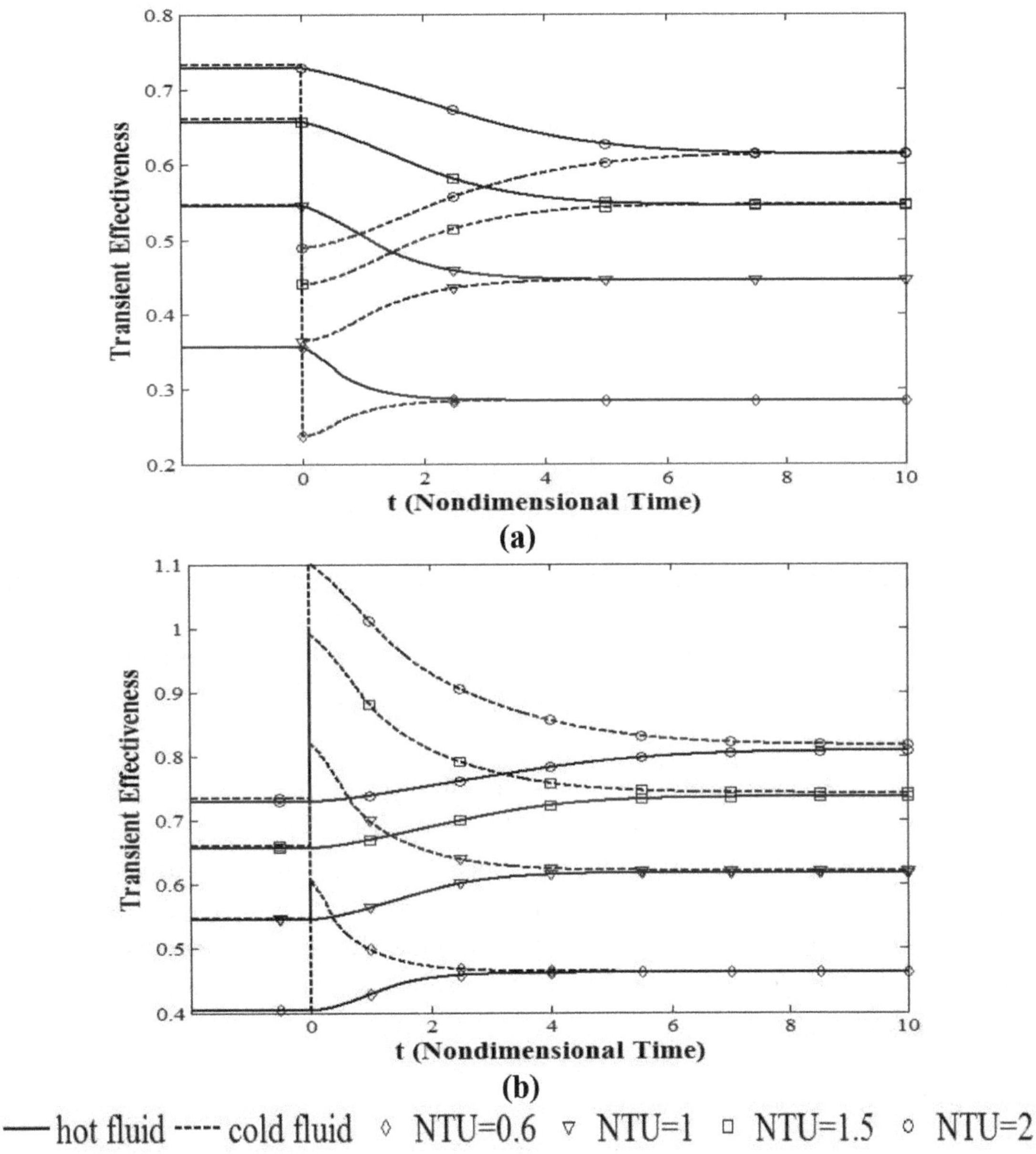

Figure 3. Effect of *NTU* on the transient effectiveness results with $E = 0.5$, $R = 1$, $V = 1$; (a) step change to hot fluid mass flow rate; (b) step change to cold fluid mass flow rate.

A step change is seen in all the cold fluid curves before the curves move smoothly and reach a steady state. In Eq. (1b), $T_{c,out}(t)$, and C_c/C_{min} (C_{min} is considered as a hot fluid capacity rate, which gives $C_c/C_{min} = E$) govern the cold fluid effectiveness variations. The previously mentioned step change at the very beginning is due to the step change of $C_c(E)$. In terms of a mass flow rate ramp variation scenario, the variation in transient effectiveness curves at the beginning is dominated by E and $T_{c,out}(t)$. The transient effectiveness also illustrated the combination impact of the fluid mass flow rate variation and the physical parameters. As an example, **Figure 3** shows the transient effectiveness results versus different NTU values. The impact of the NTU value on the steady-state and transient performance of the transient effectiveness can be clearly seen in the curves of the figure. As an example, the larger the NTU value the longer time the heat exchanger takes to reach steady state. The difference of the variation of the cold fluid and the hot fluid is clearly distinguished in the same figure.

It can be seen that the transient effectiveness curves are able to represent the transient response of heat exchangers under different variation conditions by comparing the curves shown in **Figure 3(a)** and **(b)**. This means that a transient effectiveness curve represents more physical information than an outlet temperature curve, since the curves are distinguished clearly when different boundary conditions are applied. In addition, the transient effectiveness curves also reflect the influences of the physical parameters on the transient and steady-state responses of the heat exchangers.

2.4. Experimental verification

Experimental measurements on a liquid to air cross-flow heat exchanger cores are presented in reference [16], in which the liquid mass flow rate or inlet temperature varied in time following controlled functional forms (step jump and ramp). The specific design enables the control of transient variations in the inlet temperature and mass flow rate on both the air and water flow streams supplied to the heat exchanger device. More details regarding the entire experimental setup and tests can be found in reference [16]. The experimental data were used to characterize and validate the transient effectiveness methodology and the transient numerical solution in reference [17], and the more comprehensive understanding of the characteristics of the transient effectiveness is obtained.

For modeling a specific heat exchanger, the modeling physical parameters (E, NTU, R, V_h, V_c) need to be extracted and calculated using the heat exchanger hardware data and one set of steady-state experimental data for modeling a specific heat exchanger and specific dynamic physical scenarios. The procedure can be interpreted as integrating the hardware data into the nondimensional mathematical model (Eqs. (2)–(4)) to model a specific heat exchanger device. One of the methodologies can be referenced to calculate the physical parameter and is presented in reference [18].

2.4.1. Inlet temperature variation

Several functional forms are designed to vary the water inlet temperature and water flow rate. Ramp functions for water inlet temperature change and step functional forms of water flow rate change are selected to present here. The detailed information of each experimental case designed is shown in **Table 1**. The physical parameters used in the numerical solution for each case are also summarized in **Table 1**.

Water inlet temperature change (°C)	Water flow rate (GPM)	Air flow rate (m^3/s)	Air inlet temperature (°C)	E	NTU	R	V_{water}	V_{air}
Case 1: Ramp change 22.22–50.28	2	0.2787	22.85	1.62	0.2613	4.78	1.456	0.01
Case 2: Ramp change 23.06–50.26	2	0.3716	23.06	1.2153	0.2186	4.344	1.456	0.01
Water flow rate change (GPM)	**Air flow rate (m^3/s)**	**Water inlet temperature (°C)**	**Air inlet temperature (°C)**	E	NTU	R	V_{water}	V_{air}
Case 3: Step function 2–5	0.2870	50.40	20.95	1.62	0.264	4.78	1.456	0.01
Case 4: Step function 2–5	0.5574	50.39	21.74	0.79	0.211	3.77	1.456	0.01

Table 1. Test cases.

Two important characteristics of the transient effectiveness are discussed in this section. In case 1, the water inlet temperature is lower than the air inlet temperature at the beginning, and then becomes the hot fluid after the variation. This is the scenario that the cold fluid becomes the hot fluid due to the temperature change. When plotting the transient effectiveness curves, a mathematical singularity point is seen. In Eqs. (1a) and (1b), the term $T_{h,in} - T_{c,in}$ will vary from a positive value to a negative value. This performance is characterized by both simulation modeling and experimental testing. By comparing with the regular fluid temperature curve, the transient effectiveness curves can capture this special scenario. At the same time, they contain all the steady-state and transient characteristics. In case 2, since the water and air are at the same temperature, no singularity point is generated. By comparing the results of cases 1 and 2, the difference between the boundary conditions applied and the transient response is clearly reflected on the transient effectiveness curves. The initial conditions of case 2 can be considered as an idle condition. When plotting the transient effectiveness curve for this special case, there will be a sensitive region after the variation is applied at the time between 10 and 15 s, which is shown in **Figure 5**. Since the two fluid inlet temperatures are same, the numerator and denominator in Eqs. (1a) and (1b) equal 0. Then, even a very miner error in either temperature data may result in a major difference in the effectiveness value. In both **Figures 4** and **5**, the numerical results show a faster response than the experimental results. This is because axial dispersion and longitudinal conduction are neglected in the numerical modeling. When using the transient effectiveness curve plotting the fluid inlet temperature variation cases, the variation form and some of the corresponding characteristics in the fluid inlet temperature can be represented at the same time.

2.4.2. *Mass flow rate variation*

In cases 3 and 4, variations are applied to the water fluid mass flow rate. **Figures 6** and **7** show the transient effectiveness results of these the two cases, respectively. In terms of the steady-state results, the increase in the water (C_{max} fluid) leads to an increase in the effectiveness value, which is shown in **Figure 6**. In case 4, a special scenario is considered in which the

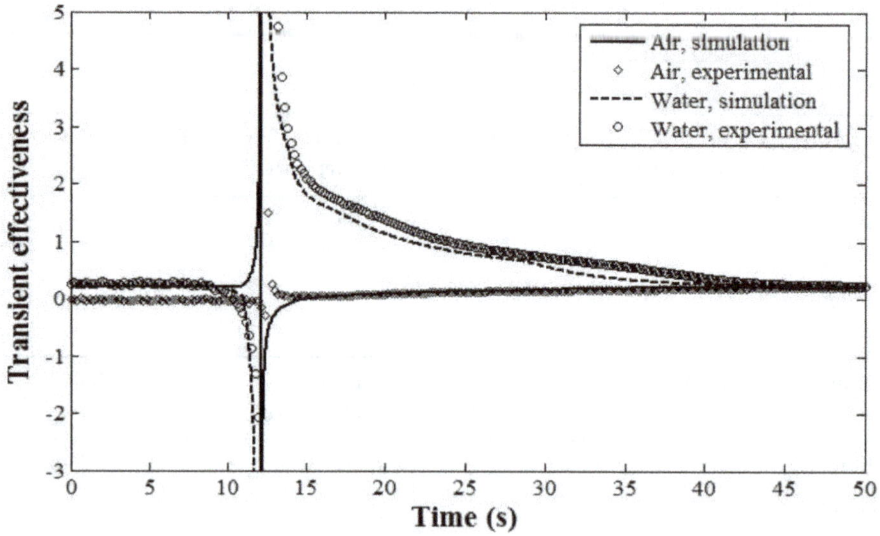

Figure 4. Transient effectiveness results of case 1.

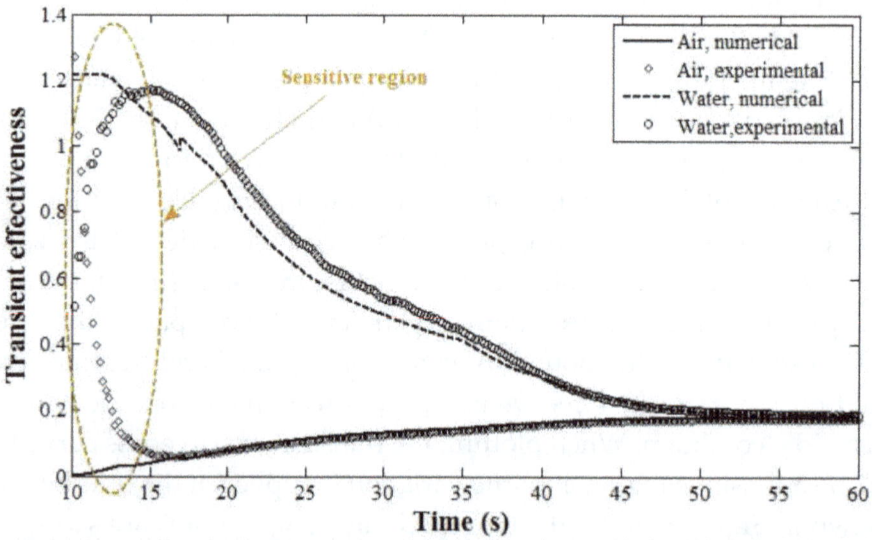

Figure 5. Transient effectiveness of case 2.

minimum capacity fluid (water) becomes the maximum capacity fluid, due to the change in the fluid mass flow rate. Then the air becomes the minimum capacity fluid due to the change. This scenario may be seen in an actual heat exchanger industrial application, especially in certain failure scenarios. It can be seen in **Figure 7** that step changes are seen on both of the curves, before the curves move smoothly and reach the final steady state. By comparing the results shown in **Figures 6** and 7, it can be found that the transient effectiveness results are clearly different when different fluid mass flow rate variation scenarios are applied. Again, this transient effectiveness methodology can present the heat exchanger dynamic performance in a more comprehensive manner in fluid mass flow rate variation scenarios. It contains the information of the two fluids dynamic responses, and the corresponding variations applied as well (both the fluid inlet temperature and the fluid mass flow rate).

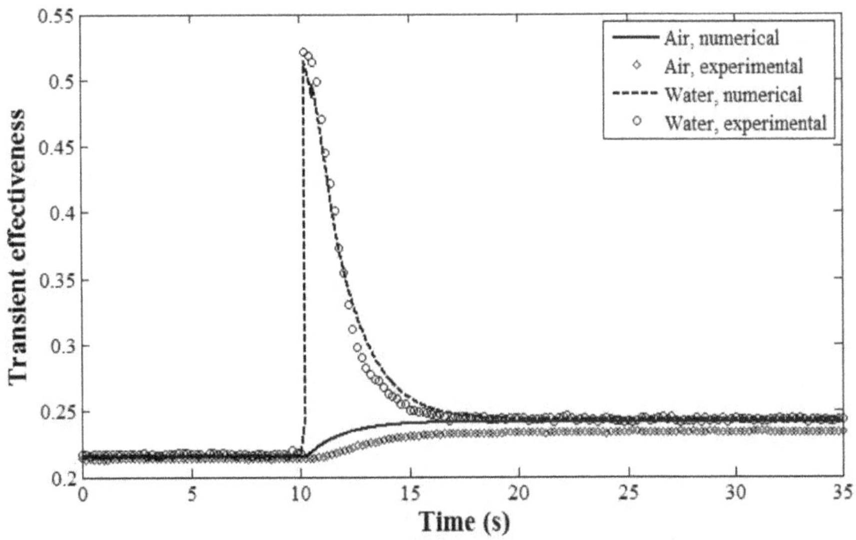

Figure 6. Transient effectiveness results of case 3.

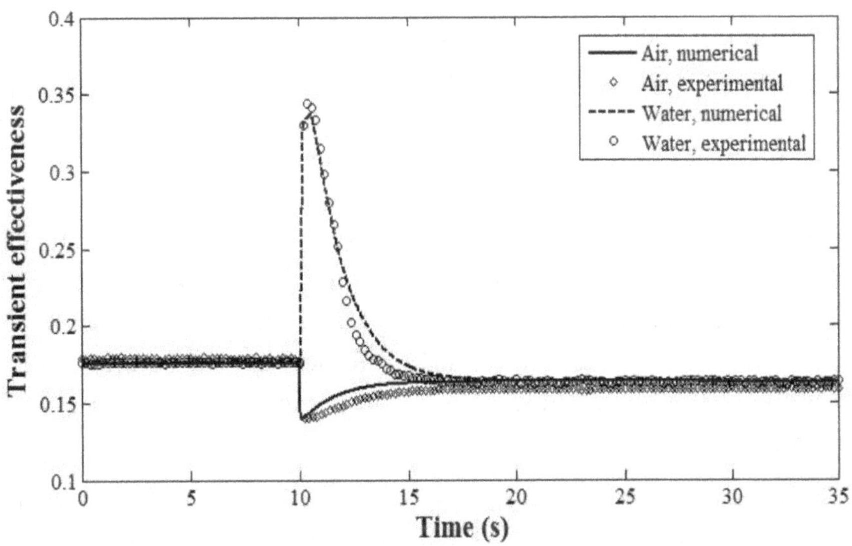

Figure 7. Transient effectiveness results of case 4.

2.4.3. *Summary*

This section provides several important characteristics of the transient effectiveness for dynamic characterization of a heat exchanger transient performance. Several experimental test cases are selected and analyzed. Two cases of fluid inlet temperature change and two cases of fluid mass flow rate change cases provide a more complete understanding of the transient effectiveness method in characterizing the dynamic performance of the heat exchanger. The transient effectiveness methodology can be used as an alternative for representing the dynamic performance of the heat exchanger. It is a more effective way than using the fluid temperature results, and it contains more information, including the variation condition applied to the heat exchanger, initial conditions and some special circumstances such as the cold fluid becoming as the hot fluid, C_{min} fluid becoming C_{max} fluid, and so on.

3. Characterization of a liquid cooling system using transient effectiveness

This section illustrates an example of investigating a liquid cooling system which has several heat exchanger units using the transient effectiveness method and its corresponding characteristics. Several experimental tests are conducted on a data center liquid cooling test facility and the results are reported in reference [19]. The transient effectiveness method is used to analyze the performance of heat exchangers and the dynamic performance of the entire test facility. The transient effectiveness method provides an analyzing method for investigating and characterizing the transient performance of heat exchangers which are working in the cooling and heating systems with multiple coupled heat transfer loops, in which multiple heat exchanger units are used.

3.1. Description of the test facility and experimental test scenarios

Figure 9 shows the liquid cooling chiller-less data center test facility designed by IBM. Basically the entire system contains a liquid cooling server rack, a liquid to liquid heat exchanger, and a dry cooler. The rack was fully populated with liquid cooled volume servers. Each server dissipates approximately 350 W. Then the total maximum rack power can reach 15 kW. There is also a side car heat exchanger contained within the rack on the side for cooling the rack circulated air. The air is recirculated inside the rack driven by server fans. The CPU and DIMM are cooled using cold plate and cold rails, which are directly attached to them. The heat captured by the rack circulated air and the liquid are then transferred to the atmosphere through the sidecar heat exchanger, the liquid-to-liquid heat exchanger, and the dry cooler. More details in terms of the description of each component and the entire test facility design are reported in references [19, 20].

Temperature sensors (T_1–T_{10}) are located at various locations, including the inlet and outlet of each component, including the cold plates, servers, sidecar heat exchanger, buffer unit, and dry cooler, as shown in **Figure 8**. The detailed description of the sensor locations and functions, as well as the data collection and data processing, is summarized in reference [19]. Three transient test cases were designed and conducted and the detailed experimental designs for the three cases are shown in **Table 2**.

3.2. Transient effectiveness

3.2.1. Transient effectiveness calculation

The experimental data for the sidecar heat exchanger, the buffer unit, and the dry cooler for all the three cases are used to generate the transient effectiveness curves. The corresponding results are shown in **Figure 9(a)–(c)** for the buffer unit, the side card heat exchanger, and the dry cooler, respectively. In the current study, C_{min} fluid of each heat exchanger unit is used in the current study.

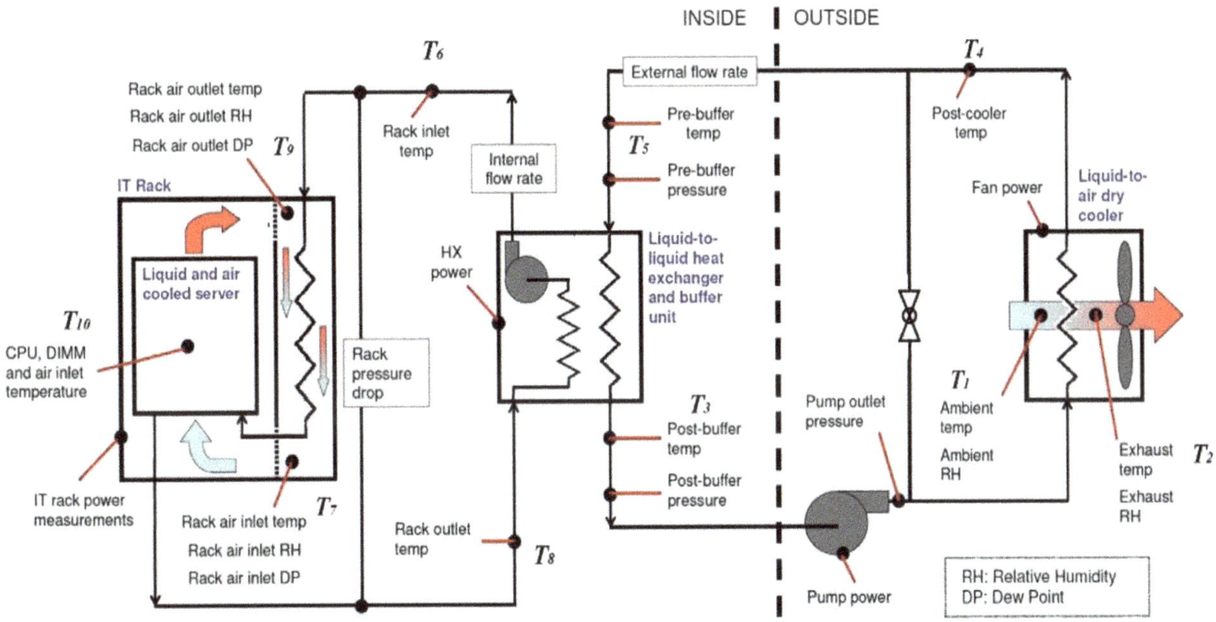

Figure 8. Schematic representation of the test facility and experimental setup.

Test cases	Server power	Internal flow rate (GPM)/water	External flow rate (GPM)/propylene glycol	Dry cooler blower fan speed set point (RPM)	Ambient air temperature (°C)
Case 1: Server power increase	Idle-Full	4	6.01	150	20.1–20.6
Case 2: Flow rate increase	Full	4–7.7	6.43	150	20.6–19.2
Case 3: Server power decrease	Full-Idle	7.7	6.43	150	19.2–18.2

Table 2. Transient test cases.

For the sidecar heat exchanger:

$$\varepsilon(\tau) = \frac{c_{air}(\tau) \cdot [T_{Rack\ outlet\ air}(\tau) - T_{Rack\ inlet\ air}(\tau)]}{c_{\min}(\tau) \cdot [T_{Rack\ outlet\ air}(\tau) - T_{Prerack}(\tau)]} \tag{5}$$

For the buffer unit:

$$\varepsilon(\tau) = \frac{c_{internal}(\tau) \cdot [T_{Postrack}(\tau) - T_{Prerack}(\tau)]}{c_{\min}(\tau) \cdot [T_{Postrack}(\tau) - T_{Prebuffer}(\tau)]} \tag{6}$$

For the dry cooler:

$$\varepsilon(\tau) = \frac{c_{air}(\tau) \cdot [T_{Ambient\ \ air}(\tau) - T_{Exhaust\ \ air}(\tau)]}{c_{\min}(\tau) \cdot [T_{Ambient\ \ air}(\tau) - T_{Postbuffer}(\tau)]} \tag{7}$$

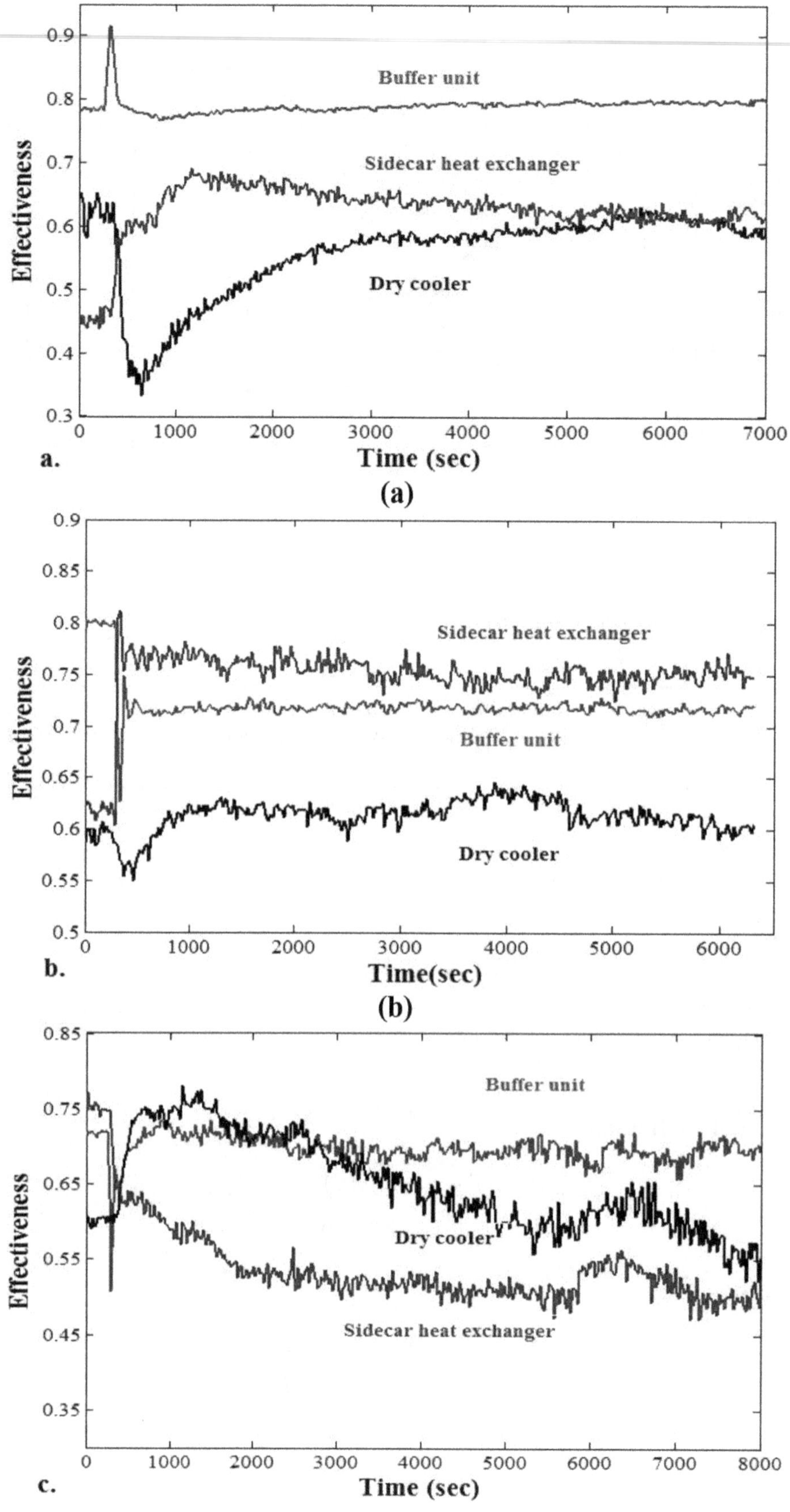

Figure 9. Transient effectiveness curves, (a) case 1; (b) case 3; and (c) case 2.

3.2.2. Transient response analysis using the transient effectiveness

The transient effectiveness for each component is plotted in **Figure 9(a)–(c)**, with the results of the three test cases. **Figure 9(a)** shows the transient effectiveness for test case 1. It can be seen by comparing the transient effectiveness curves that the transient response time of the buffer unit is very short compared to the other two heat exchangers. Since the two fluids of the buffer unit are constant, the final effectiveness is kept as the same value of 0.8. The dry cooler shows a relatively long response time, since its corresponding transient effectiveness curve takes longer time to approach a steady state. Since there is no fluid mass flow rate changes, the final steady states are the same as the initial one of 0.6. In terms of the side car heat exchanger, it can be seen that a new steady state is reached. This illustrates that the air mass flow has varied in this test case. **Figure 9(b)** shows the transient effectiveness results for case 2. In this test case, the internal fluid mass flow rate is varied. Therefore, the final steady-state values of the sidecar heat exchanger and the buffer unit are changed. In terms of the buffer unit, the effectiveness curves show a rapid response and rapidly approaches a new steady-state condition. The sidecar heat exchanger shows a similar fast response performance. The dry cooler takes much longer to reach the same steady-state condition (since the dry cooler has no fluid mass flow rate variation), compared to the other two heat exchangers. **Figure 9(c)** shows the transient effectiveness results for test case 3. Similar to test case 1, the dry cooler and the sidecar heat exchanger take longer time before they settle down and approach a steady state. The buffer unit variation time is much smaller, as shown in the curves. It is also illustrated in the transient effectiveness curves that the air flow within the server rack is varied in this test case, since the sidecar heat exchanger reaches a different final steady-state value. More analysis regarding the cause of the variation in rack air flow is presented in reference [19]. By plotting the transient effectiveness curves, the dynamic performance of each heat exchanger component and the time taken to approach a new steady state can be seen clearly. In addition, based on the characteristics of the transient effectiveness curves, more dynamic performance related to the variation applied to the heat exchanger is illustrated.

For a closed coupled system, especially when multiple heat exchanger units are used, the transient effectiveness can be used to characterize the thermal capacitance effects of each unit. **Figure 9(a)–(c)** shows that the buffer unit effectiveness reaches steady state much faster than the other two heat exchanger units. The dry cooler takes the longest time, which is seen in all three cases. This illustrates that the thermal capacitance of the buffer unit is much less than that of the dry cooler. Actually, the dry cooler is a much larger unit located outside of the building and the buffer unit is a small plate heat exchanger. The time taken for the sidecar heat exchanger to reach steady state in cases 1 and 3 is long. However, the sidecar heat exchanger takes a much shorter time in case 2. Here are some explanations: in cases 1 and 3, which involve variations in server power, the server thermal mass is involved. The impact of the thermal mass extends the time taken for the rack recirculated air and the entire rack side air dynamic to reach steady state. Then the time taken for the side card heat exchanger to reach steady state is longer in cases 1 and 3. The server thermal mass is not involved in case 2. Therefore, only the thermal capacitance of the side card heat exchanger is dominate in the transient response. Based on this analysis, it can be seen that the thermal capacitance of the sidecar heat exchanger and the buffer unit are much smaller compared with the one of the dry cooler heat exchanger. The temperature results are collected at different locations, capturing a detailed response sequence. However, since the heat exchanger units are connected to each other using the internal loop and external loop, it is very difficult to characterize the response time of certain heat exchanger by using any temperature result. The temperature results vary during the entire test run. The transient effectiveness method provides a way to observe individual component performance, even though it is in a closed coupled system, by fliting the influence of the neighbored components. The buffer unit transient effectiveness curves have reached steady-state conditions, while the temperatures are still varying. This illustrates that the buffer unit itself has reached a steady-state thermal-exchange condition during a transient event. This can be understood as a self "steady-state" condition in a transient environment. In this condition, even though the corresponding fluid temperatures vary with time, the heat exchanger has approached a steady-state condition.

3.3. Summary

This section illustrates that the transient effectiveness can be used for characterizing the dynamic response of a closed coupled heat transfer loop, which has multiple heat exchanger units installed. It also represents the thermal capacitance impact of each component during different transient events. In addition, some detailed physical insights, which cannot be directly captured from temperature results, can be indicated by the transient effectiveness results.

4. CFD compact heat exchanger modeling

This section discusses another important application of the transient effectiveness concept and model, which can be used in developing computational fluid dynamics (CFD) compact transient heat exchanger modeling methodologies. There are two methods which are proposed in references [21, 22]. The methods can be used to model different types of heat exchangers, including a counter-flow heat exchanger and a cross-flow heat exchanger. In addition, the compact models developed can be used to model different variation scenarios, including fluid inlet temperature variation, fluid mass flow rate variation, and multiple combination variation scenarios.

4.1. Compact modeling methodology I

4.1.1. Modeling methodology development

It has been shown in previous studies that the transient effectiveness is able to characterize the dynamic response of heat exchangers. When studying heat exchanger dynamic response, the transient input can be either an inlet temperature variation or a mass flow rate variation. This case may become more complicated when considering multiple variation combination scenarios. Then the outlet temperature transient performance will be a complicated form, as shown in references [12, 23]. The transient effectiveness is correspondingly more complicated, due to the fact that the transient effectiveness is reflecting the variation in both the fluid inlet temperature and the outlet temperature. When comparing with the steady-state ε-NTU method, the first methodology was developed by extending the concept to a time-dependent effectiveness.

4.1.2. CFD compact heat exchanger model

The ε-*NTU* heat exchanger modeling methodology has been widely used in heat exchanger steady-state studies. This method and theoretical equations have been incorporated into most of the commercial CFD codes with a heat exchanger modeling option. For heat exchanger steady-state modeling, effectiveness performance data of the heat exchangers are used to obtain the corresponding term εC_{min}, and then the ε-*NTU* equation is used to represent the corresponding heat exchanger unit under certain flow rate operating conditions. The transient effectiveness concept discussed in the previous section is extended here to develop a compact transient heat exchanger model. The modeling methodology uses the transient effectiveness in the standard ε-*NTU* heat exchanger equation to extend the ε-*NTU* model to a compact transient heat exchanger form, as shown in Eq. (8). This transient effectiveness is denoted as $\varepsilon_T{}'$ in the current study. By applying transient effectiveness to the equation, the transient compact model can be developed.

$$Q' = \varepsilon' C'_{\min}(T'_{h,in} - T'_{c,in}) \tag{8}$$

A CFD compact transient heat exchanger model is developed based on this transient methodology using the commercial code FloTherm [24]. The basic methodology correlates a negative linear source function as in Eq. (9) to Eq. (8) to represent the heat exchanger model. In FloTherm, the *value* and the *coefficient* can be set as transient variables for this linear heat source function. Therefore, Eq. (9) can be correlated to the transient compact heat exchanger model shown in Eq. (8). The detailed description of the FloTherm linear source model and the correlation method can be found in reference [24].

$$Q = Coefficient \times (T_{c,in} - T_{h,out}) = Coefficient \times (T_{water,in} - T_{air,out}) \tag{9}$$

4.1.3. Verification with thermal dynamic model

Thermal dynamic modeling results and the experimental test results are used as the input for calculating the transient effectiveness, and the effectiveness is then integrated into the CFD model. Then outlet temperatures predicted by the CFD compact model are compared with the thermal dynamic modeling results and experimental results. The detailed validation study is presented in reference [21]. Here a multiple variation combination case is presented as an example. It can be seen in **Figure 10** that the CFD compact modeling results are in good agreement with the thermal dynamic modeling results.

4.1.4. Verification with experimental data

In this section, this CFD compact model is verified using experimental data. The experimental tests discussed in the previous section are used. The original data are summarized in reference [19]. The transient test results, including the fluid mass flow rate and temperature variations, are incorporated into Eqs. (1a) and (1b) to calculate the transient effectiveness for the heat exchanger unit under different scenarios. Then the transient effectiveness ($\varepsilon_T{}'$) is used in the CFD model. The dry cooler results are chosen as an example to discuss in this section. The detailed formulas for the transient effectiveness calculations are shown in Eq. (10).

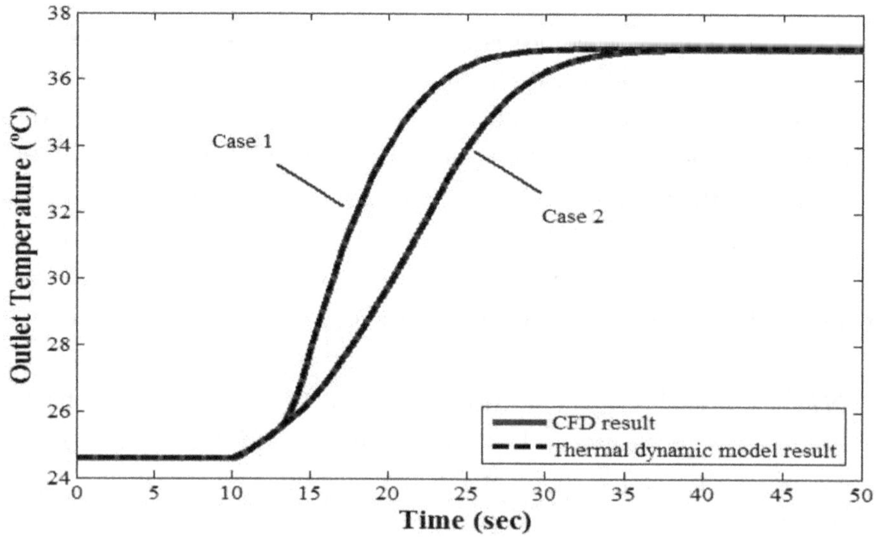

Figure 10. Hot fluid outlet temperature results, case 1: cot fluid inlet temperature step change and cold fluid mass flow rate ramp change; case 2: hot fluid inlet temperature step change and cold fluid mass flow rate step change.

For an air to liquid cross-flow heat exchanger—dry cooler:

$$\varepsilon_T{}' = \frac{c_{external}(\tau) \cdot [T_{Ambient\ air}(\tau) - T_{Exhaust\ air}(\tau)]}{c_{\min}(\tau) \cdot [T_{Ambient\ air}(\tau) - T_{Postbuffer}(\tau)]} \tag{10}$$

The test data discussed in Section 3 is used here for calculating the transient effectiveness of the dry cooler, and the three cases shown in **Table 2** are plotted in **Figure 11**. The comparison results are shown in **Figure 12**, and the two sets of results are in good agreement.

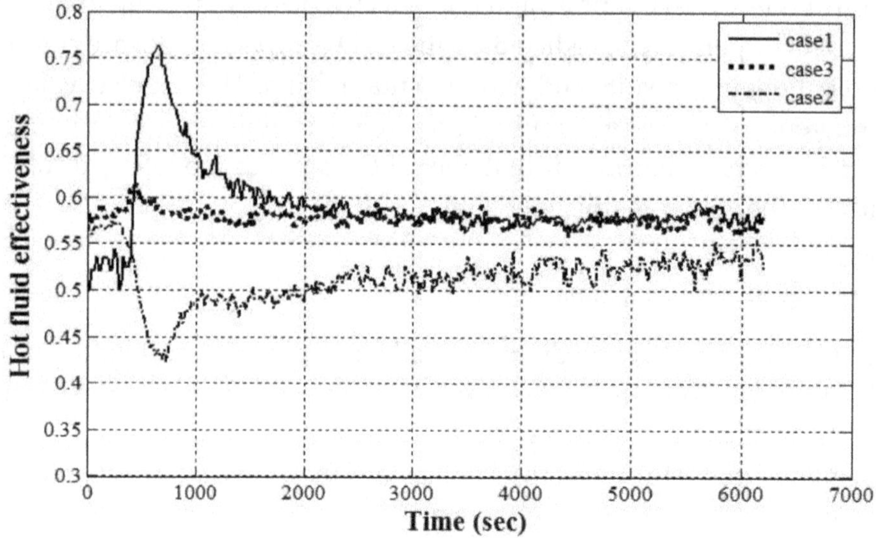

Figure 11. Hot fluid transient effectiveness in the three test cases.

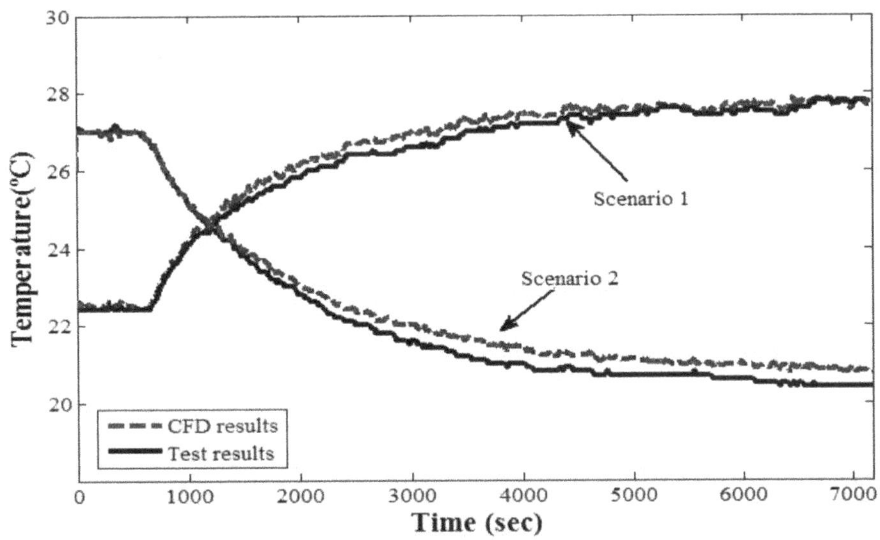

Figure 12. Comparison of the hot fluid outlet temperatures results (secondary fluid) in case 1 and case 2.

4.2. Compact modeling methodology II

The limitation of modeling methodology I is that the transient effectiveness, which used as the input for the CFD compact model, is generated based on the existing solutions from either the thermal dynamic model or experimental tests. This means that a transient effectiveness curve only represents a specific case and can only be used for modeling one certain transient case. Then the CFD model can be only used for modeling the cases with the same boundary condition, due to the limitation of the transient effectiveness used in the code. In addition, as discussed in the previous section, the transient effectiveness variation can be very complex for certain scenarios. This limitation results in the fact that this compact model may not be applied to a system level modeling work. Therefore, a derivative transient effectiveness method is developed.

4.2.1. Modeling methodology development

Eq. (11) is generated by adding the three governing partial differential equations and used as a simplified correlation in representing heat exchanger transient performance. By considering a single energy balance equation to represent the cross-flow heat exchanger using fluid inlet and out flow, Eq. (11) can be expressed in Eq. (12). The energy balance equation, together with the ε-*NTU* methodology, is shown in Eq. (13). Eq. (14) is then generated by substituting Eq. (13) into Eq. (12). Here, a new term $(pVc_p)_{\text{heat exchanger}}$ is introduced, which is a lumped thermal capacitance of the heat exchanger, including the capacitance of the heat exchanger metal and the two fluids. By lumping the thermal capacitances together, Eq. (14) is then expressed as Eq. (15). The term $T_{\text{heat exchanger}}$ represents the heat exchanger temperature, which can be understood as an averaged value of the two fluids and the heat exchanger metal. The negative sign represents a negative heat source. Eq. (15) is the governing equation of the second methodology. Eq. (16) is the expression of the fluid outlet temperature, and the hot fluid is used as an example.

$$MC\frac{\partial T_{wall}}{\partial t}+C_c^o\frac{\partial T_c}{\partial t}+C_h^o\frac{\partial T_h}{\partial t}+(m'c)_h\frac{\partial T_h}{\partial (y/L_h)}+(m'c)_c\frac{\partial T_c}{\partial (x/L_c)}=0 \tag{11}$$

$$MC\frac{\partial T_{wall}}{\partial t}+C_c^o\frac{\partial T_c}{\partial t}+C_h^o\frac{\partial T_h}{\partial t}+(m'c)_h\cdot(T_{h,out}-T_{h,in})+(m'c)_c\cdot(T_{c,out}-T_{c,in})=0 \tag{12}$$

$$\dot{Q}=(m'c)_c\cdot(T_{c,out}-T_{c,in})=(m'c)_h\cdot(T_{h,out}-T_{h,in})=\varepsilon C_{\min}\cdot(T_{h,in}-T_{c,in}) \tag{13}$$

$$MC\frac{\partial T_{wall}}{\partial t}+C_c^o\frac{\partial T_c}{\partial t}+C_h^o\frac{\partial T_h}{\partial t}+(m'c)_h\cdot(T_{h,out}-T_{h,in})+\varepsilon C_{\min}\cdot(T_{h,in}-T_{c,in})=0 \tag{14}$$

$$-\dot{Q}=-\varepsilon C_{\min}\cdot(T_{h,in}-T_{c,in})=\dot{m}c_p(T_{h,out}-T_{h,in})+(pVc_p)_{heat\ exchanger}\times\frac{\partial T_{heat\ exchanger}}{\partial t} \tag{15}$$

$$T_{h,\ out}=\frac{-\varepsilon C_{\min}\cdot(T_{h,in}-T_{c,in})-(pVc_p)_{heat\ exchanger}\times\frac{\partial T_{heat\ exchanger}}{\partial t}}{(\dot{m}c_p)_h}+T_{h,\ in} \tag{16}$$

4.2.2. CFD compact heat exchanger model

In reference [22], a CFD compact model is realized in the commercial CFD code FloTherm using methodology II. The detailed procedure regarding the model development is presented in reference [22]. The heat exchanger is modeled using a linear heat source module, as shown in Eq. (9), and server solid blocks, which are used to represent the thermal capacitance. Two heat source modules are used to represent the supply fluid inlet temperature and mass flow rate. The user is able to manipulate the parameters in the linear heat source module and material setting in the solid rods module to correlate it to the governing equation (Eq. (15)) of methodology II.

4.2.3. Model verification

It was mentioned that the lumped capacitance term is dominated by the capacitance of the heat exchanger coil and the two fluids, as well as their corresponding weight. For modeling verification purposes, a method was used to adjust the estimated thermal properties initially considered in the model, instead of deriving the actual lumped capacitance value. A method for lumping the three capacitance terms is a comprehensive study, which requires developing a complex physical correlation. In addition, it may vary from case to case. Basically, when using a lumped capacitance value, it should have the same impact on the heat exchanger transient response. Therefore, the curve adjustment method was used. The detailed procedure for adjusting the curve is presented in reference [22].

4.2.3.1. Inlet temperature variation scenario

The inlet temperature variation case is considered in this section, and the fluid mass flow rate is set to a constant value. For the cross-flow heat exchanger model considered in this work, the hot fluid is modeled as the supply fluid and as the $C_{\min}$ fluid. Based on Eq. (17), the variation in the fluid inlet temperature either T_h and/or T_c will impact the left side of the equation. The effectiveness and $C_{\min}$ are constant values, since the fluid mass flow rate is set at a constant value.

$$-\varepsilon C_{\min} \cdot (T_{h,in} - T_{c,in}) = -\varepsilon C_{\min} \cdot (T_{a,in} - T_{w,in}) = \dot{m}c_p(T_{a,out} - T_{a,in}) + (pVc_p)_{heat\ exchanger} \times \frac{\partial T_{heat\ exchanger}}{\partial t} \tag{17}$$

The analytical and numerical solutions of the thermal dynamic model shown in Eqs. (2)–(4) are used to verify the compact model shown in Eq. (17). A hot fluid inlet temperature step change scenario is used as an example in this section. **Figure 13** shows several sets of results, including the CFD modeling results, which are illustrated by solid lines, the analytical results presented in reference [8], which are indicated by discrete round black points, and the numerical results, which are plotted in dashed lines. The detailed information of each case is shown in the figure legend. It can be seen that the three sets of solutions are in good agreement for the case NTU = 1.5. It needs to be noted that axial dispersion is dismissed in both the numerical results and the analytical solution. It has been concluded in references [8, 25–27] that the primary fluid responds immediately, with no time delay to the sudden variation applied at the inlet. It also has been concluded that the axial dispersion has a clear impact on the fluid dynamic performance, when the NTU value is larger than 2. It can be seen in **Figure 13** that the CFD model results are in good agreement with the numerical solution. By comparing the two NTU = 2 cases (with and without axial dispersion), it can be seen that both the steady-state and transient performances of the outlet temperatures are influenced by the axial dispersion. Even for the same modeling case (NTU = 2), since the numerical results are used to calculate the ε or the coefficient value used in the CFD model, the CFD modeling results are different. It is seen that the CFD curve responds rapidly at the early response for the two NTU = 2 cases. Similar performance has been presented in reference [28].

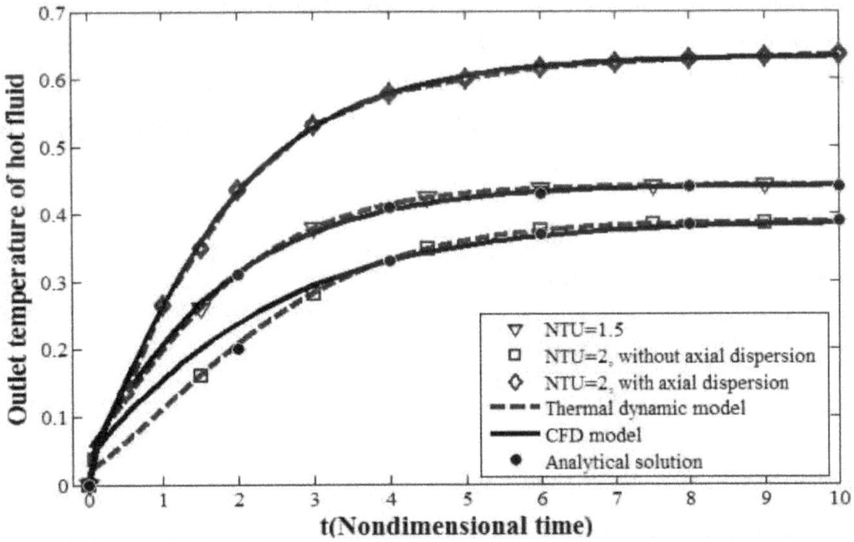

Figure 13. Comparison of the CFD modeling results with analytical and numerical solutions.

4.2.4. Fluid mass flow rate variation scenario

4.2.4.1. Mass flow rate variation-based transient effectiveness

It has been discussed in the previous section that modeling a case that involves fluid mass flow rate changes is more complicated than modeling a fluid inlet temperature variation, due to the changing in the heat transfer coefficient. Therefore, the impact of the fluid mass flow rate variation on the heat transfer coefficient should be considered. In this section, both the hot fluid and the cold fluid inlet temperatures are considered as constant. In Eq. (18), ε'_m is defined as a time-dependent variable and it represents the effectiveness changes due to fluid mass flow rate variations. The term $C_{\min}$ is also a variable in the cases that the $C_{\min}$ fluid mass flow rate changes.

$$-\varepsilon'_m C_{\min}\cdot(T_{h,in}-T_{c,in}) = -\varepsilon'_m C_{\min}\cdot(T_{a,in}-T_{w,in}) = \dot{m}c_p(T_{a,out}-T_{a,in}) + (\rho V c_p)_{heat\ exchanger} \times \frac{\partial T_{heat\ exchanger}}{\partial t} \tag{18}$$

It is important for modeling the heat exchanger transient response to correctly characterize the effectiveness due to variations in the fluid mass flow rate, and in the corresponding heat transfer coefficient. Based on the steady-state ε-*NTU* results, different steady-state mass flow rates and heat transfer coefficients govern the *NTU* values. Thus, the ε value changes due to the variation of *NTU*. This concept is extended to a "mass flow rate variation based" transient effectiveness. Due to the mass flow rate variation, the heat transfer coefficient changes are denoted by the NTU' value in Eq. (19). In addition, mass flow rate variations lead to changes in the heat capacity rate ratio (E'), as in Eq. (20). The detailed mathematical procedure is presented in reference [22]. Then the "mass flow rate based transient effectiveness (ε_m')" concept is defined by extending the theoretical steady-state correlation of ε and *NTU* to the transient case. The theoretical steady-state correlations are shown in Eqs. (21) and (23) for a cross-flow heat exchanger and for a counter-flow heat exchanger, respectively. By integrating the NTU' and E' equations (Eqs. (19) and (20)), the mass flow rate variation-based transient effectiveness can be expressed as Eqs. (22) and (24). They are designated as the ε-*NTU* transient theoretical correlations. The transient theoretical correlations are used to calculate the corresponding mass flow rate based transient effectiveness under the corresponding mass flow rate variations for the CFD heat exchanger models.

$$NTU' = \frac{(mc_p)_{\min}}{(m'c_p)_{\min}}\cdot r_c^{\beta}\cdot\frac{(r_h^{\beta}+R\cdot r_h^{\beta})}{r_c^{\beta}+R\cdot r_h^{\beta}}\cdot NTU \tag{19}$$

$$E' = \frac{(m'c_p)_h}{(m'c_p)_c} = \frac{r_h}{r_c}E \tag{20}$$

For a unmixed-unmixed cross-flow heat exchanger

$$\varepsilon = 1-\exp\left\{\frac{NTU^{0.22}}{E}\left[\exp\left(-E\cdot NTU^{0.78}\right)-1\right]\right\} \tag{21}$$

$$\varepsilon'_m = 1 - \exp\left\{\frac{NTU^{0.22}}{E}\left[\exp\left(-E \cdot NTU'^{0.78}\right)\right]\right\} \tag{22}$$

For a counter cross-flow heat exchanger

$$\varepsilon = \frac{1 - \exp[-NTU \cdot (1 - E)]}{1 - E' \cdot \exp[-NTU \cdot (1 - E)]} \tag{23}$$

$$\varepsilon'_m = \frac{1 - \exp[-NTU' \cdot (1 - E')]}{1 - E' \cdot \exp[-NTU' \cdot (1 - E')]} \tag{24}$$

Two methodologies have been developed based on the transient effectiveness methodology. The first transient effectiveness is the temperature-based transient effectiveness, or full transient effectiveness. The second transient effectiveness is denoted as the mass flow rate based transient effectiveness method, or partial transient effectiveness. The major difference between the two transient effectiveness models is that the partial transient effectiveness only considers the impact of the variations in the fluid mass flow rate and the corresponding heat transfer coefficient, and thermal capacitance effects are dismissed.

4.2.4.2. Verification with numerical solution of thermal dynamic model

An example is selected here to perform the CFD compact model verification in modeling fluid mass flow rate changes. A set of numerical solutions for the thermal dynamic models are used. Two variation cases are considered: they are a ramp increase in the cold fluid mass flow rate, and a ramp increase in the hot fluid mass flow rate. To show the difference between the two modeling methodologies, both the full transient effectiveness and partial transient effectiveness are presented together. This difference can be seen clearly in **Figure 14**, between the two effectiveness models which are calculated using Eqs. (1a) and (22) for the same variation case. It is found that the hot fluid mass flow rate variation leads a larger difference between the two final steady states, which is not seen for the cold fluid mass flow rate variation case. One possible reason is that the hot fluid is modeled as the C_{min} fluid. Therefore, Eq. (19) is used to calculate NTU'. Based on Eq. (19), r_h may result in a larger impact on the NTU' value than on $r_{c.}$ The temperature results are plotted in **Figure 15**, and the hot fluid outlet temperatures are used to compare with the previously verified numerical solutions. It can be seen that the compact modeling results are in good agreement with the numerical solutions.

4.2.4.3. Validation with transient experimental data

Several experimental data presented in reference [29] are used to validate the modeling methodology. It needs to be mentioned that the data shown in reference [29] is for a counter-flow heat exchanger. By considering the CFD model as a black box, the counter-flow heat exchanger is modeled using the same model as the cross-flow heat exchanger, with proper modification for the model dimensions. In terms of calculating the partial transient effectiveness, Eq. (24) is used.

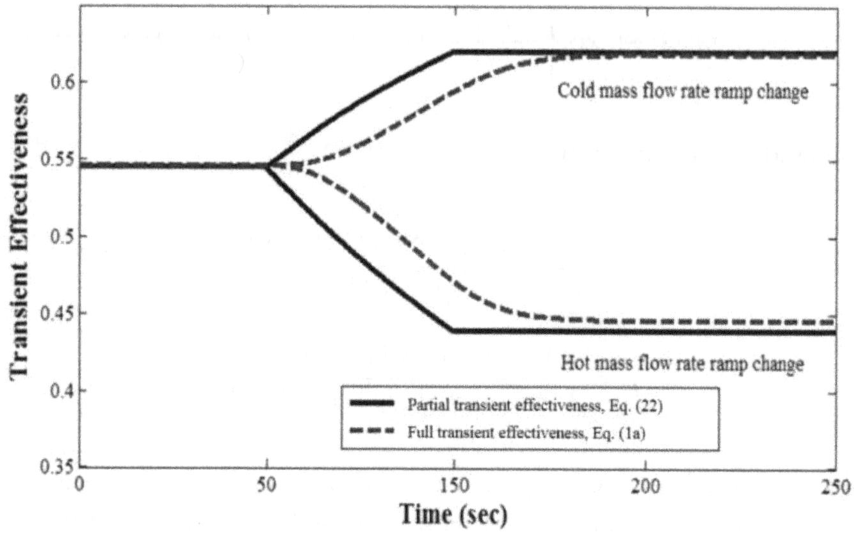

Figure 14. Transient effectiveness of the hot fluid under mass flow rate ramp change.

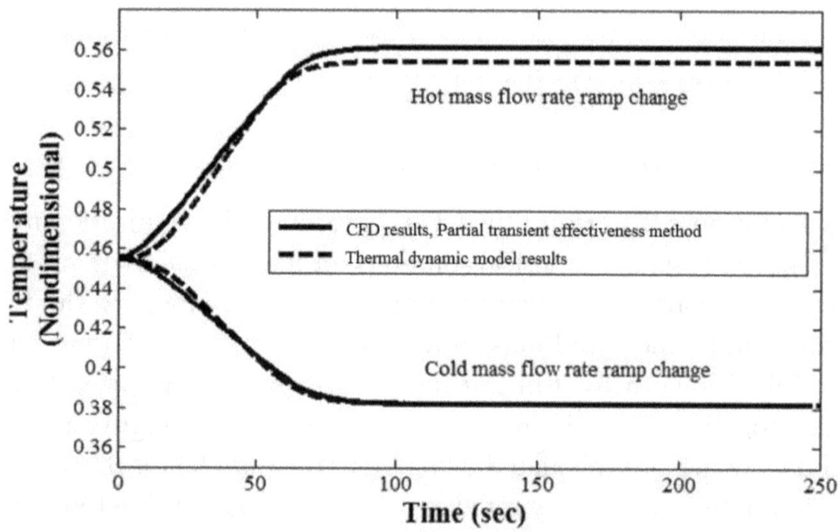

Figure 15. Outlet temperature of the hot fluid.

The mass flow rate variation magnitude was considered as r_h = 1.56/0.45 and applied to the hot fluid mass flow rate. The analytical solution presented in reference [29] is also plotted in the same figure. Therefore, **Figure 16** shows the experimental data, the CFD modeling results, and the analytical results. In addition, the effect of the lumped specific heat used in the current compact model is studied. Eq. (25) represents the nondimensional Peclet number. This number is used to represent the ratio of the thermal energy transported to the other fluid through convection to the energy conducted within the fluid. A small Pe_L value represents a stronger conduction effect. A large Pe_L value indicates that the impact of axial conductance is minimal. In the current CFD model, when using a relatively large heat exchanger specific heat, the axial dispersion effect can be reduced significantly. Therefore, the set point of the specific heat value has a major impact on the conductance. It can be seen in **Figure 16** that the solutions are in good agreement. When the axial dispersion impact is considered in the CFD model, the corresponding results are in good agreement with the experimental data. When the axial dispersion impact is neglected in the CFD model, the corresponding results are in good agreement with the analytical solution. The impact of axial dispersion can be seen clearly in delaying the transient response.

$$Pe_L = \frac{UL}{k/\rho C_p} = \frac{UL \cdot \dot{m} c_p}{k} \tag{25}$$

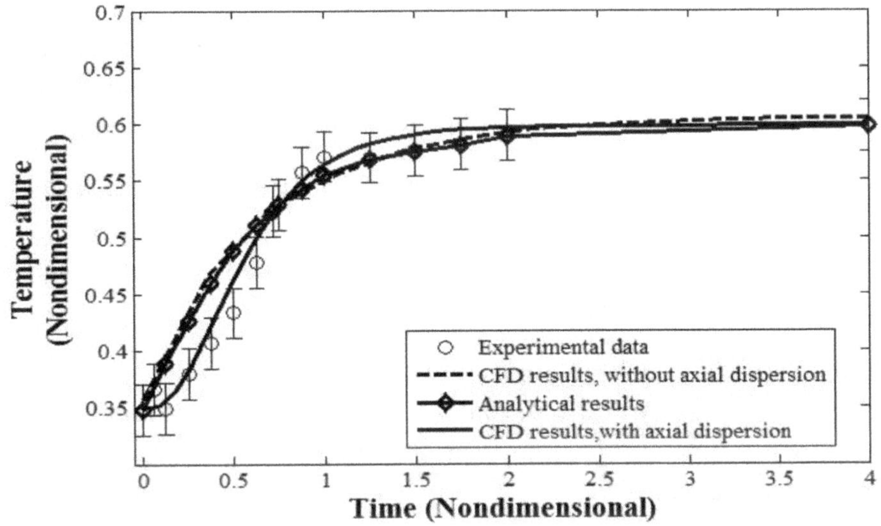

Figure 16. Fluid outlet temperature under mass flow rate step change, case 1.

4.3. Summary

In this section, the transient effectiveness concept is used to develop heat exchanger modeling methodologies. Detailed development procedures are provided. The first method is to extend the steady-state effectiveness concept to a transient concept, and the calculation of this transient effectiveness is based on the actual temperature results. This method can be used to integrate the numerical and analytical solutions and experimental data into the CFD model. The second modeling method is to extend the steady-state theoretical correlation ε-NTU to a transient correlation. This method is then used for developing CFD compact transient heat exchanger models for modeling the scenario that fluid mass flow rates change. This section provides a comprehensive summarization of the compact modeling methodology validation. Experimental data, analytical solutions, and numerical solutions are used to compare with the compact modeling results. The results show that the transient effectiveness-based CFD compact models are in good agreement with the experimental data and analytical solutions for different variation scenarios, including fluid inlet temperature changes, fluid mass flow rate changes, and combinations of multiple variations cases.

5. Conclusion

The aim of this chapter is to provide a comprehensive review of the transient effectiveness methodology for heat exchanger analysis. This chapter provides a thorough connection of all the transient effectiveness-related knowledge/work. Novel transient effectiveness methodologies for studying heat exchanger transient characterization are introduced, and a detailed analytical, numerical, and experimental study of these models is presented. Mathematical models, analytical and numerical analysis, experimental testing, and validating studies provide a better understanding of the transient effectiveness methodology. It is shown that the transient effectiveness methodology is very useful for thermal dynamic characterization of heat exchangers and the development of compact/CFD transient models. In addition, it is found that methodology is also useful for analyzing cooling system transient experimental results.

The transient effectiveness curves represent both the heat exchanger dynamic behavior and the corresponding boundary conditions on a single curve. It depicts the heat exchanger transient response in a more comprehensive manner, when compared with outlet temperature curves.

The transient effectiveness methodology is shown to be useful for characterizing the thermal capacitance effects of the entire system, as well as each component, during transient events. The transient effectiveness curves clearly capture the transient response and the impact of thermal capacitance on each heat exchanger unit.

Two CFD compact modeling methodologies are developed and validated, namely a full transient effectiveness methodology and a partial transient effectiveness methodology. These two compact models are accurate and fast, and can be integrated into large scale models, such as system/building level models.

Acknowledgements

The authors would like to state that the majority portion of this chapter was taken from previously published work by the same group of authors.

Nomenclature

Roman letter symbols

c_p	fluid specific heat, J/kg·K
C_{wall}	specific heat of the wall of HX, J/kg·K
$C_{\min}$	minimum capacity rate fluid
$C_{\max}$	maximum capacity rate fluid
E	heat capacity rate ratio, $(mc_p)_h/(mc_p)_c$
D	dimension
k	thermal conductivity, W/m·K
L	length of heat exchanger, m
$\dot{m}$	mass flow rate, kg/s
M	mass of the wall (core) of heat exchanger, kg
NTU	number of transfer units
NTU'	time dependent NTU due to mass flow rate variation
R	conductance ratio, $(hA)_h/(hA)_c$
r	mass flow rate variation ratio, $r = m'/m$
T	dimensionless temperature

V	capacitance ratio
C	flow-stream capacity rate
Q/q	heat transfer rate
X, Y	dimensionless length
t	dimensionless time
RPM	revolution per minute
GPM	gallon per minute
CFM	cubic feet per minute

Dimensionless groupings

Pe_L	Peclet number

Greek letter symbols

τ	time, s
ρ	density, kg/m^3
β	constant number, 0.8
ε	effectiveness
$\varepsilon_T{}'$	temperature dependent transient effectiveness/full transient effectiveness
$\varepsilon_m{}'$	mass flow rate dependent transient effectiveness/partial transient effectiveness

Subscripts

h	hot fluid
c	cold fluid
wall	coil and fin of heat exchanger
a	air
w	water
in	inlet
out/o	outlet
max	maximum
min	minimum

Author details

Tianyi Gao*, Bahgat Sammakia and James Geer

*Address all correspondence to: tgao1@binghamton.edu

SUNY Binghamton University, Binghamton, NY, USA

References

[1] Cima, R. M., London, A. L., "Transient Response of a Two-Fluid Counter-Flow Heat Exchanger-The Gas-Turbine Regenerator," Transactions of ASME, Vol. 80, pp. 1169–1179, 1958.

[2] Gao, T., Sammakia, B., Geer, J., Ortega, A., Schmidt, R., "Transient Effectiveness Characteristics of Cross Flow Heat Exchangers in Data Center Cooling Systems," In Thermal and Thermomechanical Phenomena in Electronic Systems (ITherm), May 27–30, 2014 14th IEEE Intersociety Conference on, IEEE, 2014.

[3] Gao, T., Sammakia, B., Geer, J., Ortega, A., Schmidt, R., "Dynamic Analysis of Cross Flow Heat Exchangers in Data Centers Using Transient Effectiveness Method," IEEE Transaction of Component Packaging and Manufacturing Technology, Vol. 4, no. 12, pp. 1925–1935, November 2014.

[4] Dusinberre, G. M., "Calculation of Transient Temperatures in Pipes and Heat Exchangers," Transactions of ASME, Vol. 76, pp. 421–426, 1954.

[5] Myers, G. E., Mitchell, J. W., Norman, R.F., "The Transient Response of Crossflow Heat Exchangers, Evaporators and Condensers," Transactions of ASME Journal of Heat Transfer, Vol. 89, pp. 75–80, 1967.

[6] Romie, F. E., "Transient Response of Gas-to-Gas Crossflow Heat Exchangers with Neither Gas Mixed," ASME Journal of Heat Transfer, Vol. 105, pp. 563–570, 1983.

[7] Gvozdenac, D. D., "Analytical Solution of the Transient Response of Gas-to-Gas Crossflow Heat Exchanger with Both Fluids Unmixed," ASME Journal of Heat Transfer, Vol. 108, pp. 722–727, 1986.

[8] Spiga, G., Spiga, M., "Two-Dimensional Transient Solutions for Crossflow Wheat Exchangers with neither Gas Mixed," Journal of Heat Transfer, Vol. 109, pp. 281–286, 1987.

[9] Chen, H. T., Chen, K. C., "Simple Method for Transient Response of Gas-to-Gas Crossflow Heat exchangers With Neither Gas Mixed," International Journal of Heat Mass Transfer, Vol. 34, no. 11, pp. 2891–2898, 1991.

[10] Gao, T., Sammakia, B., Geer, J., "A Review, Comparison, and Analysis of Cross Flow Heat Exchanger Transient Modeling in the Cases Fluid Mass Flow Rate and Supply Temperature Varies," Applied Thermal Engineering, Vol. 84, pp. 15–26, June 2015.

[11] Gao, T., Murray, B., Sammakia, B., "Analysis of Transient and Hysteresis Behaviors of Crossflow Heat Exchangers under Variable Fluid Mass Flow Rate for Data Center Cooling Application," Applied Thermal Engineering, Vol. 84, pp. 15–26, June 2015.

[12] Gao, T., Sammakia, B., Geer, J., "Dynamic response and control analysis of cross flow heat exchangers under variable temperature and flow rate conditions," International Journal of Heat and Mass Transfer, Vol. 81, pp. 542–553, February 2015.

[13] Gao, T., Sammakia, B., Murray, B., Ortega, A., Schmidt, R., "Cross Flow Heat Exchanger Modeling of Transient Temperature Input Conditions," IEEE Transactions of Component Packaging and Manufacturing Technology, Vol. 4, no. 11, pp. 1796 – 1807, September 2014.

[14] Rizika, J. W., "Thermal Lags in Flowing Incompressible Fluid Systems Containing Heat Capacitors," Transactions of ASME, Vol. 78, pp. 1407–1413, 1956.

[15] London, A. L., Biancardi, F. R., Mitchell, J. W., "The Transient Response of Gas-Turbine Plant Heat Exchangers-Regenerators, Intercoolers, Precoolers, and Ducting," Transaction of ASME, Journal of Engineering For Power, Vol. 81, Series A., pp. 443–448, 1959.

[16] DelValle, M., Ortega, A., "Experimental characterization of the transient response of air/water cross-flow heat exchangers for Data Centers cooling systems," Proceedings of the ASME 2015 International Technical Conference and Exhibition on Packaging and Integration of Electronic and Photonic Microsystems and ASME 2015 13th International Conference on Nanochannels, Microchannels, and Minichannels, Paper number InterPACKICNMM2015-48375, July 6–9, 2015, San Francisco, California

[17] Gao, T., Delvalle, M., Ortega, A., Sammakia, B., "Numerical and Experimental Characterization of Transient Effectiveness of a Water to Air Heat Exchanger in data center cooling systems," Proceedings of the ASME 2015 International Technical Conference on Packaging and Integration of Electronic and Photonic Microsystems (InterPACK 2015), Paper No. InterPACK2015-48371, San Francisco, CA, July 6–9, 2015.

[18] DelValle, M., Ortega, A., "Numerical and compact models to predict the transient behavior of cross-flow heat exchangers in data center applications," Thermal and Thermomechanical Phenomena in Electronic Systems (ITherm), pp. 698–705, 2014.

[19] Gao, T., David, M., Geer, J., Schmidt, R., Sammakia, B., "Experimental and Numerical Dynamic Investigation of an Energy Efficient Liquid Cooled Chiller-Less Data Center Test Facility," Energy & Buildings, Vol. 91C, pp. 83–96, March 2015

[20] David, P. M., Iyengar, K. M., Parida, P., Simons, E. R., Schultz, M., Gaynes, M., Schmidt, R., Chainer, T., "Impact of Operating Conditions on a Chiller-less Data Center Test Facility with Liquid Cooled Servers," In Thermal and Thermomechanical Phenomena in Electronic Systems (ITherm) 2012, San Diego, California, 2012.

[21] Gao, T., Sammakia, B., Geer, J., David, M., Schmidt, R., "Experimentally Verified Transient Models of Data Center Crossflow Heat Exchangers," IMECE, Paper number IMECE2014-36022, November 14–20, Montreal, Canada,2014.

[22] Gao, T., Geer, J., Sammakia, B., "Development and Verification of Compact Transient Heat Exchanger Models using Transient Effectiveness Methodologies." International Journal of Heat and Mass Transfer, Vol. 87, pp. 265–278, 2015.

[23] Gao, T., Geer, J., Sammakia, B., "Nonuniform Temperature Boundary Condition Effect on Data Center Cross Flow Heat Exchanger Dynamic Performance," International Journal of Heat and Mass Transfer, Vol. 79, pp. 1048–1058, December 2014.

[24] Gao, T., Sammakia B., Samadiani, E., Schmidt R., "Steady State and Transient Experimentally Validate Analysis of Hybrid Data Centers," Journal of Electronic Packaging, Vol. 137, no. 2, pp. 021007-021007-12, June 2015.

[25] Roetzel, W., Xuan, Y., Dynamic Behavior of Heat Exchangers, Computational Mechanics Publications, Southampton, UK: WIT Press, 1999.

[26] Shah, R. K., Sekulic, D. P., Hoboken, Fundamentals of heat exchanger design, NJ: John Wiley & Sons, 2003.

[27] Baclic, B. S., Heggs, P. J., "On the Search for New Solutions of the Single-Pass Crossflow Heat Exchanger Problem," International Journal of Heat and Mass Transfer, Vol. 28, no. 10, pp. 1965–1976, 1985.

[28] Mishra, M., Das, K. P., Sarangi, S., "Transient Behavior of Crossflow Heat Exchangers with Longitudinal Conduction and Axial Dispersion," Journal of Heat Transfer, Vol. 126, pp. 425–433, 2004.

[29] Abdelghani-Idrissi, M. A., Bagui, F., Estel, L., "Analytical and Experimental Response Time to Flow Rate Step Along a Counter Flow Double Pipe Heat Exchangers," International Journal of Heat and Mass Transfer, Vol. 44, no. 19, pp. 3721–3730, 2001.

5

Computational Modeling of Vehicle Radiators Using Porous Medium Approach

Barbaros Çetin, Kadir G. Güler and Mehmet Haluk Aksel

Abstract

A common tool for the determination of thermal characteristics of vehicle radiators is the experimental testing. However, experimental testing may not be feasible considering the cost and labor-time. Basic understanding of the past experimental data and analytical/computational modeling can significantly enhance the effectiveness of the design and development phase. One such computational modeling technique is the utilization of computational fluid dynamics (CFD) analysis to predict the thermal characteristics of a vehicle radiator. However, CFD models are also not suitable to be used as a design tool since considerable amount of computational power and time is required due to the multiple length scales involved in the problem, especially the small-scale geometric details associated with the fins. Although fins introduce a significant complexity for the problem, the repetitive and/or regular structure of the fins enables the porous medium based modeling. By porous modeling, a memory and time efficient computational model can be developed and implemented as an efficient design tool for radiators. In this work, a computational methodology is described to obtain the hydrodynamic and thermal characteristics of a vehicle radiator. Although the proposed methodology is discussed in the context of a vehicle radiator, the proposed methodology can be implemented to any compact heat exchanger with repetitive fin structures which is an important problem for many industrial applications.

Keywords: computational modeling, radiator, porous medium

1. Introduction

One of the main components of a cooling system of an engine is a radiator. Vehicle radiators are typically fin-and-tube-type compact heat exchangers (HXs) and composed of inlet manifolds, outlet manifolds, tubes and fins as shown in **Figure 1**. Simply, a radiator works with two fluids which are air and anti-freeze water mixture. Hot antifreeze water mixture flows through the tubes, whereas cooling air flows through the fins resulting in heat exchange between both streams.

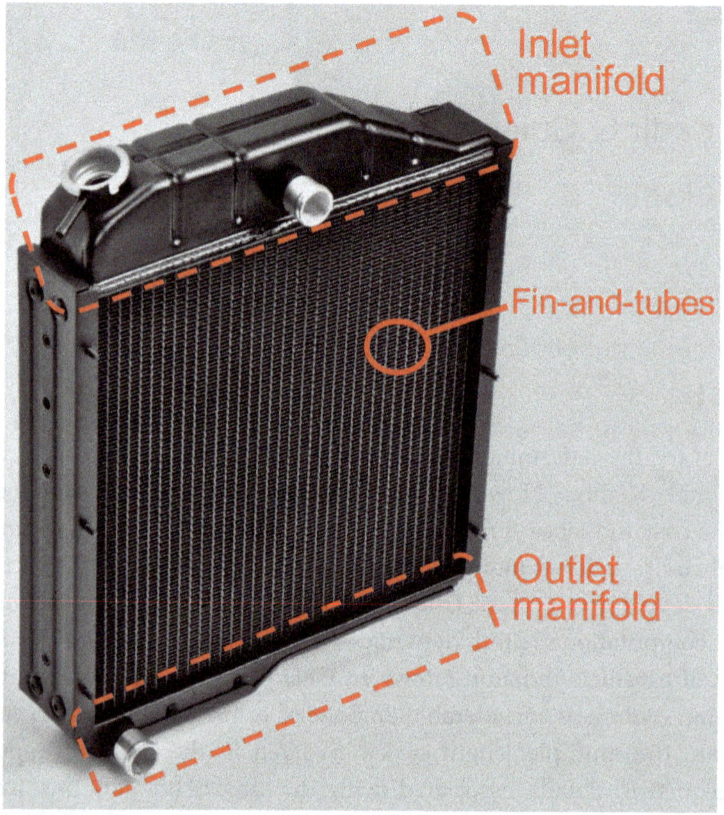

Figure 1. A typical 4-row tractor radiator.

Due to the strong competition in the automotive industry, radiators with better performance (higher cooling capacity, less hydrodynamic loss, less weight, etc.) have been desired. A common tool for the determination of thermal characteristics of vehicle radiators is the experimental testing. However, experimental testing may not be feasible considering the cost and labor-time. Basic understanding of the past experimental data and analytical/computational modeling can significantly enhance the effectiveness of the design and development phase. There are techniques available to analyze HXs such as log mean temperature difference (LMTD) and effectiveness-NTU (ε-NTU). However, these techniques require some parameters known *a priori* such as overall heat transfer coefficients and/or NTU relations for a given HX. There are no general expressions for overall heat transfer coefficients and/or ε-NTU relations valid for any HX. Therefore, these parameters need to be predicted either from analytical expressions [1], experimental data [2, 3] and/or computational models [3–6]. *A priori* knowledge of these parameters is required for the designer. Therefore, implementation of LMTD and/or ε-NTU is not feasible especially for vehicle radiators which may include custom-designed fin configurations. Alternatively, computational fluid dynamics (CFD) analysis can be applied to predict the thermal characteristics of a radiator. However, CFD analysis of a full-size HX is not feasible due to extremely high number of cells required to resolve the complex nature of the HXs; especially the fin structures. This point is more problematic when the number of fins is high in the case of heavy-duty vehicle radiators. Although fins introduce a significant complexity for the problem, the repetitive and/or regular structure of the fins enables the porous medium based modeling. From computational point of view, this approach offers some unique advantages. The complex fluid flow occurring through fins can be introduced into the model through porous parameters. Although the determination of these porous parameters requires a rigorous, detailed computational model with very fine mesh structure especially within the regions mainly responsible for the fluid friction and heat transfer, this modeling can be performed on a representative unit cell due to the repetitive nature of the fins. Once these effects are included through the porous parameters, the mesh structure simplifies dramatically and considering the whole geometry, the number of degree of freedom of the system drops down to a feasible number (in the order of 10 millions). Besides, the porous modeling does not require any boundary layer meshing since the friction and heat transfer parameters are already included through the porous parameters.

1.1. Porous modeling

Porous modeling is governed by three models. The simplest model is the Darcy's model which is suggested by Henry Darcy (1856) during his investigations on hydrology of the water supplies of Dijon [7]. Darcy's equation is expressed as:

$$\frac{\Delta p}{l} = -\frac{\mu}{\alpha} V \tag{1}$$

where, Δp is the pressure drop, l is the pipe length, V is the average velocity, μ is the dynamic viscosity and α is permeability of porous domain. Permeability depends on the fluid properties and the geometrical properties of the medium. The dependence of the pressure drop on velocity in the Darcy's equation is linear; therefore, Darcy's equation is applicable when the flow is laminar. As the velocity increases, the dependence of the pressure drop on velocity becomes non-linear due to drag caused by solid obstacles. At this point, there are two extended models proposed in the literature namely Forchheimer and Forchheimer-Brinkman model. For moderate Reynolds numbers, including nonlinear effects, pressure drop is defined as Forchheimer's equation [7]:

$$\frac{\Delta p}{l} = -\left(\frac{\mu}{\alpha} V + \frac{C_F}{\sqrt{\alpha}} \frac{1}{2} \rho V^2\right) \tag{2}$$

where C_F is the dimensionless form-drag constant and ρ is the density of the fluid. The first term denotes the viscous characteristics of porous flow and the second term (also called Forchheimer term) denotes the inertial characteristics. Lastly, Forchheimer-Brinkman model includes additional Laplacian term in addition to Forchheimer's equation. Forchheimer-Brinkman model is expressed as [7]:

$$\frac{\Delta p}{l} = -\left(\frac{\mu}{\alpha}V + \frac{C_F}{\sqrt{\alpha}}\frac{1}{2}\rho V^2 - \tilde{\mu}\nabla^2 V\right) \tag{3}$$

where $\tilde{\mu}$ is the effective viscosity. In general, added Laplacian term (also known as Brinkman term) resolves effects of the flow characteristics in a thin boundary layer at the near wall regions. Strictly speaking, the last term becomes important for large porosity (ratio of the fluid volume to the solid volume in a porous medium) values which means the effect is negligible for many practical applications where typically porosity value is relatively small. Eq. (3) without the quadratic term is known as extended Darcy (or Brinkman) model. Therefore, Forchheimer-Brinkman model is the most general model, but the inclusion of the Brinkman and Forchheimer term on the left-hand side can be questionable since the Brinkman term is appropriate for large porosity values, yet there exists uncertainty about the validity of the Forchheimer term at larger porosity values [7].

Velocity definition in porous modeling is specified by using two different descriptions: superficial formulation and physical velocity formulation. Superficial velocity formulation does not take the porosity into account during the evaluation of the continuity, momentum and energy equations. On the other hand, physical velocity formulation includes porosity during the calculation of transport equations [8]. The continuity and momentum transport equation for a porous domain using Forcheimer's model can be written as [2]:

$$\frac{\partial}{\partial t}(\gamma\rho) + \nabla.\left(\gamma\rho\vec{V}\right) = 0 \tag{4}$$

$$\frac{\partial}{\partial t}\left(\gamma\rho\vec{V}\right) + \nabla.\left(\gamma\rho\vec{V}\vec{V}\right) = -\gamma\nabla p + \nabla.(\gamma\bar{\tau}) + \gamma\vec{B}_f - \left(\frac{\gamma^2\mu}{\alpha}\vec{V} + \gamma^3\frac{C_2}{2}\rho\left|\vec{V}\right|\vec{V}\right) \tag{5}$$

where γ is the porosity, C_2 is the inertial coefficient for porous domain and $\overrightarrow{B_f}$ is the body force term.

Besides flow modeling, heat transfer modeling for porous flow is described by using two models which are (i) equilibrium model and (ii) nonequilibrium model. Equilibrium model (one-equation energy model) is used when the porous medium and fluid phase are in thermal equilibrium. However, in most cases, fluid phase and porous medium are not in thermal equilibrium. For such cases, nonequilibrium thermal model is more realistic. In the case of the radiator, this issue is important since the temperature difference between the solid (fins) and the fluid (air flowing through fins) is the driving mechanism for the heat transfer [4]. Therefore, the nonequilibrium model includes two energy equations (known as also two-equation energy model): one is for the fluid domain and the other is for the solid domain. The coupling of these two models is via the term which represents the heat transfer between the fluid and the solid domains. The conservation equations for the two energy model can be written as [2]:

$$\frac{\partial}{\partial t}\left(\gamma\rho_f E_f\right)+\nabla.\left(\vec{V}\left(\rho_f E_f+p\right)\right)=\nabla.\left[\gamma k_f \nabla T_f-\left(\sum_i h_i J_i\right)+\left(\bar{\bar{\tau}}\vec{V}\right)\right]+S_f^h+h_{fs}A_{fs}\left(T_s-T_f\right) \quad (6)$$

$$\frac{\partial}{\partial t}\left(\left(1-\gamma\right)\rho_s E_s\right)=\nabla.\left(\left(1-\gamma\right)k_s\nabla T_s\right)+S_s^h+h_{fs}A_{fs}\left(T_f-T_s\right) \quad (7)$$

where subscripts 's' and 'f' stand for solid and fluid, respectively. E is the total energy, T is the temperature, k is the thermal conductivity, S is the energy source term and $((\Sigma_i h_i J_i))$ stands for the effect of enthalpy transport due to the diffusion of species. The last term in both of the equations is the coupling term which models heat transfer between the fluid and solid domains. In this coupling term, h_{fs} denotes heat transfer coefficient for the fluid/solid interface and A_{fs} denotes the interfacial area density that is the ratio of the area of the fluid/solid interface and the volume of the porous zone.

Through Eqs. (3)–(7), there are many parameters which are material's property and fixed once the materials for the fluid and the solid are selected. On the other hand, there are some parameters (i.e. porous parameters) which are functions of material, geometry and the flow condition. These parameters are γ, α, C_2, h_{fs} and A_{fs}. Among these, γ and A_{fs} are purely geometric parameters and can be determined once the geometry of the porous structure is known. In the case of a radiator modeling, once the geometry of the fins is set, these two parameters can be determined beforehand. The other parameters are flow-dependent, meaning that they need to be determined for a specific flow condition. At this point, these parameters can be determined through some analytical expressions [9–11], experimental results (e.g. wind tunnel testing) [4, 12], empirical correlations [13] and/or computational models [14–21] typically valid for a representative unit cell. All these approaches were implemented in the literature for different studies for the analysis of micro/macro heat sinks and HXs.

1.2. Computational modeling of heat exchangers

Porous modeling can be implemented to any geometry which would resemble a porous structure. Moreover, if the porous structure has repetitive nature, the porous coefficients can be obtained through a detailed modeling of representative unit cell through analytical, experimental or computational means. Heat sinks are very good examples of this case and porous modeling approach has been implemented for the analysis of micro/macro heat sinks [9–12]. A two-equation energy model has been implemented to analyze a straight-finned heat sink [9] together with Darcy's model and implemented to perform an optimization for an internally finned tube [10] and to discuss the effect of aspect ratio and effective thermal conductivity on the thermal performance of a micro-heat sink [11] together with extended Darcy's model. The heat sinks contain a regular structure; therefore, there is a chance to derive analytical expressions to estimate the porous parameters [9–11].

Considering the HXs with complex fin structures, computational modeling is even more challenging; therefore, the computational models typically focus on specific subcomponents of HXs such as a representative unit cell for the fin structure [14–21], radiator fan [22] and inlet manifold [23, 24]. The thermal performance of a HX can be achieved by simply increasing the performance of the fin structure alone. A fin structure with higher heat transfer together with less pressure drop can significantly enhance the performance of the entire system. To investigate the thermal performance of a fin structure, experimental [14–18] and/or computational models [14, 17, 19–21] can be realized for different fin geometries. Moreover, improving the flow maldistribution at the inlet manifold may also increase the thermal performance. Computational modeling of the flow maldistribution may lead to performance enhancement for HXs [23, 24].

Analyzing subcomponents may lead to qualitative conclusion for the thermal performance of an HX, however to estimate the thermal performance quantitatively, a rigorous 3-D modeling of the entire HX is required. Since a rigorous modeling is not computationally feasible, a 2D model [4], hydraulic and thermal resistances-based models [12, 25] and 3D mesoscale models (considering macro control volumes) have been introduced in the literature to predict the thermal performance quantitatively [26–29]. A 2D model was developed to compare the equilibrium model (one-equation thermal model) and nonequilibrium thermal model (two-equation energy model) for a relatively small size matrix type HX [4]. A resistance-based model was implemented to predict the hydrodynamic and thermal performance of a carbon-foam–finned HX which combined many different correlations from the literature to predict the hydrodynamic and thermal resistances [25]. The success of the model strongly depends on the accuracy of the porous parameters. For this particular example, the model was proven to predict the hydrodynamic and the thermal performance within ±15% of the experimental data. A Compact Heat Exchanger Simulation Software (CHESS) has been developed [26–28] as a rating and design tool for industrial use based on the empirical correlation of the porous parameters to analyze the fin-and-tube part of a vehicle radiators (excluding inlet and outlet manifolds). It was demonstrated that by using CHESS, the thermal performance of different vehicle radiators was predicted within ±15% of the experimental values. Alternatively, a porous modeling-based CFD model for fluid flow and meso-scale ε-NTU-based modeling for thermal characteristic was utilized for an air-to-air cross-flow HX [29] to investigate the effect of the maldistribution on the thermal performance. A 3D CFD model coupled with porous medium approach has been developed to investigate the hydrodynamic performance of a plate-fin HX in which the porous parameters were also determined using a detailed CFD model on the unit cell [30].

A full-size 3D thermal modeling of a relatively small compact HX was conducted with different fin configurations and the heat transfer and friction factor parameters which can be used in conjunction with the LMTD or ε-NTU method [5]. Since the size of the HX was small, the meshing was not a problem and the computational model was utilized for the design of an inlet manifold for a better performance. Considering the size of the vehicle radiator, this approach is not an option. Thermal and structural analysis of a heavy-duty truck radiator which had finned structure both on the liquid and air side has been performed using Commercial CFD software, FLUENT® [31]. Forchheimer's relation was used for the porous modeling together with the experimental data. One-equation energy model was used together with the averaged equivalent thermal conductivity. The local heat transfer coefficients and pressure distribution gathered from the thermal analysis were used as a boundary condition for finite element structural analysis through which the thermal stresses and strains were obtained.

One alternative to all these approaches can be the modeling of the vehicle radiator with porous medium approach where the porous parameters are also deduced from a rigorous CFD modeling on a unit cell. Moreover, this procedure may be performed with a commercial CFD software which would have very strong meshing, solving and post-processing capabilities. However, implementation of the two-temperature energy equation is crucial for an accurate prediction especially for the vehicle radiators. This may not be straightforward with a commercial software. At this point, FLUENT® may be a viable solution since the two-temperature model capability has been included in version 14.5. More recently, a computational modeling of a fin-and-tube-type vehicle radiator has been conducted based on two-temperature model and the cooling capacity of a heavy-duty vehicle radiator has been estimated without any need for empirical and/or experimental data [32]. In the upcoming section, the computational methodology of such a computational model is outlined. This approach may allow CFD modeling to be an efficient rating and design tool for vehicle radiators. Although the proposed computational methodology is discussed for a vehicle radiator, it may also be implemented to any compact HX with repetitive fin structures which is an important problem for many industrial applications.

2. Computational modeling

The proposed computational methodology is implemented for a 4-row 39-column commercial available heavy-duty vehicle (more specifically tractor) radiator as shown in **Figure 1**. Tractor that uses the manufactured radiator has a 64 HP Perkins engine which requires a minimum cooling capacity of 55 kW according to the catalog data. The cooling capacity of this radiator was reported as 55.8 kW by the tractor company as a result of in-house experiments following the SAE-J1393 protocol [33]. Catalog data are tabulated in **Table 1**. The fin structure used on this radiator is a wavy fin (WF) structure which is a typical structure used in vehicle radiators due to its superior thermal performance. The selected wavy fin configuration is 84 mm in length.

Definition	Value
Rotational speed of engine [rpm]	2200
Inlet temperature [°C]	86.5
Outlet temperature [°C]	81
Ambient temperature [°C]	31
Inlet mass flow rate [kg/s]	2.41
Air velocity [m/s]	7
Heat rejection [kW]	55.8

Table 1. Catalog data for the four-row radiator.

The computational procedure starts with the determination of the porous parameters for a given mesh configuration. The geometric parameters are determined using the CAD model. On the other hand, to determine the flow-based parameters, a parametric study needs to be performed on a unit cell with high resolution which consists of one repeating section of the fin structure. For the determination of the porous medium coefficients, the flow field should be analyzed only for the section with the finned structure (physical fin simulations). To verify the extracted porous medium coefficients, the flow field within the unit cell together with included upstream and downstream fluid domain needs to be modeled both using actual fin geometry and porous modeling. This analysis needs to be performed only once for each mesh configuration of interest.

2.1. Determination of the porous parameters

Fin analysis is progressed under three main steps:

a. Simulating the unit cell straight fin model by using different air inlet velocities and obtaining the resultant pressure drop across the fin.

b. Fitting a second-order curve to the collected pressure versus velocity data gives the Darcy-Forchheimer's relation as:

$$\frac{\Delta p}{l} = aV + b\,V^2 = -\left(\frac{\mu}{\alpha}V + C_2\frac{1}{2}\rho V^2\right) \tag{8}$$

where a and b are the coefficients characterizing the flow.

c. Obtaining the inertial coefficient and viscous coefficient using the extracted coefficients in step (b) as:

$$I_c = \frac{2\,b}{\rho\,l} \tag{9}$$

$$V_c = \frac{a}{l\,\mu} \tag{10}$$

Obtaining the flow-based porous medium coefficients is followed by the determination of input parameters for heat transfer modeling. The necessary input parameters are the average heat transfer coefficient (HTC) and interfacial area density (IAD) for the two-equation energy model. Average heat transfer coefficient is obtained from FLUENT® post-processing which can be calculated by using the following relation:

$$HTC = \frac{Q}{T_w - T_{ref}} \tag{11}$$

The reference temperature in the above equation is the average temperature of the air between the inlet and outlet of the finned channel.

2.2. Physical fin simulations

Unit cell of a wavy fin model, Model-A shown in **Figure 2(a)**, is analyzed in order to obtain the porous medium parameters. Flow parameters are obtained by using the Forchheimer's relation. Model-A is simulated using different Reynolds numbers to obtain Forchheimer's curve. Once the parameters are obtained, Model-B, which is a unit cell of the wavy fin with additional upstream and downstream domains as shown in **Figure 2(b)**, is analyzed. Since the air domains (without fins) are attached at the inlet and the exit of the porous domain, the flow area contracts (at the inlet of the porous domain) and expands (at the exit of the porous domain) at the interfaces of these domains. To capture the physics, porous-jump boundary conditions are introduced to match the results of the two models [8]. Boundary conditions for Model-A are set as follows: the velocity inlet and pressure outlet boundary conditions are assigned for the fin inlet and outlet, respectively. Wall boundary condition is applied for the upper and lower walls. Constant wall temperature boundary condition is assigned to the walls as the thermal boundary condition (which is close to the real situation, since the temperature variation in the *z*-direction is small). Periodic boundary condition is used for the right and left sides. For Model-B, additional upstream symmetry and downstream symmetry are assigned for upstream and downstream domains. For both simulations, SIMPLE method is used with a least square-based approach for gradient reconstruction. In addition, standard scheme for pressure and the second-order up-winding schemes for momentum, turbulent kinetic energy and turbulent dissipation rate are employed. Relaxation factors are set to their default values. For both simulations, a minimum convergence of 1×10^{-5} is obtained for all residuals. One important step is the determination of the appropriate turbulence model. At this point, some benchmark solutions, empirical/experimental results can be employed for the determination of the appropriate turbulence model. For fin structure under consideration, k-ε realizable turbulence model with standard wall function is used (the detailed discussion on the justification of the use of this turbulence model can be found elsewhere [32]).

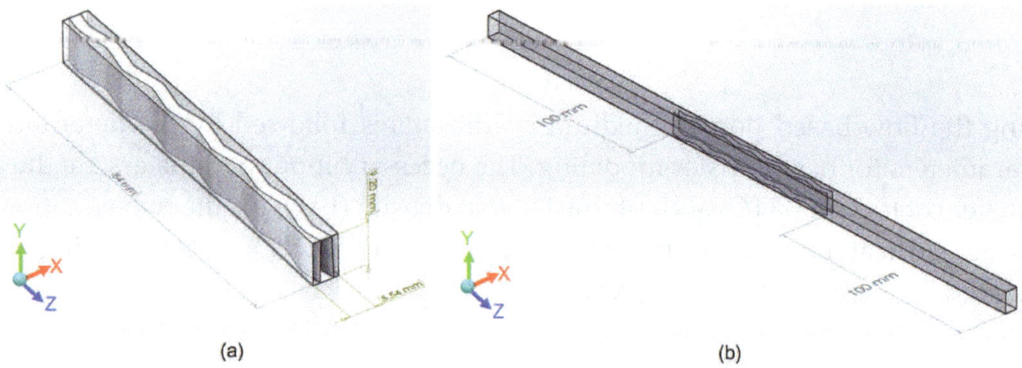

Figure 2. (a) Model-A: WF unit cell domain and (b) Model-B: WF unit cell with inlet and exit domains.

	Description	Unit
Domain length	84	mm
Number of cells	4,900,713	
Skewness (average)	0.241	
Turbulence modeling	k-ε-realizable	
Fin volume	2.2567×10^{-7}	m^3
Total volume	4.28131×10^{-6}	m^3
Porosity	0.9473	
Hydraulic diameter	0.00241	m
Turbulence intensity	0.058	
Turbulence length	0.000169	m
Solution method	SIMPLE	
Computation time/per simulation	30	min

Table 2. Input parameters for unit cell WF simulations.

Afterward, mesh independence analysis needs to be performed to ensure the mesh-independent solutions. It is observed that approximately 4,900,000 number of cells with 30 layers of boundary mesh for the fins generates a mesh-independent result for this particular fin configuration. **Table 2** contains the input parameters for the Model-A. In **Figure 3**, the pressure drop across the fin structure is plotted against velocity and a second-order curve is fitted to the simulation data. The corresponding inertial and viscous coefficients are determined as 17.3 and 4.01×10^6, respectively. Heat transfer parameters are obtained from the simulations of Model-B. For the simulation of Model-B, the input parameters are defined as 7.0 m/s for the inlet velocity, 304.2 K for the inlet temperature and 359.7 K for the temperature of fin walls in accordance the tabulated catalog data. Average surface heat transfer coefficient and tuned porous jump coefficients for the unit cell of a wavy fin are presented in **Tables 3** and **4**, respectively.

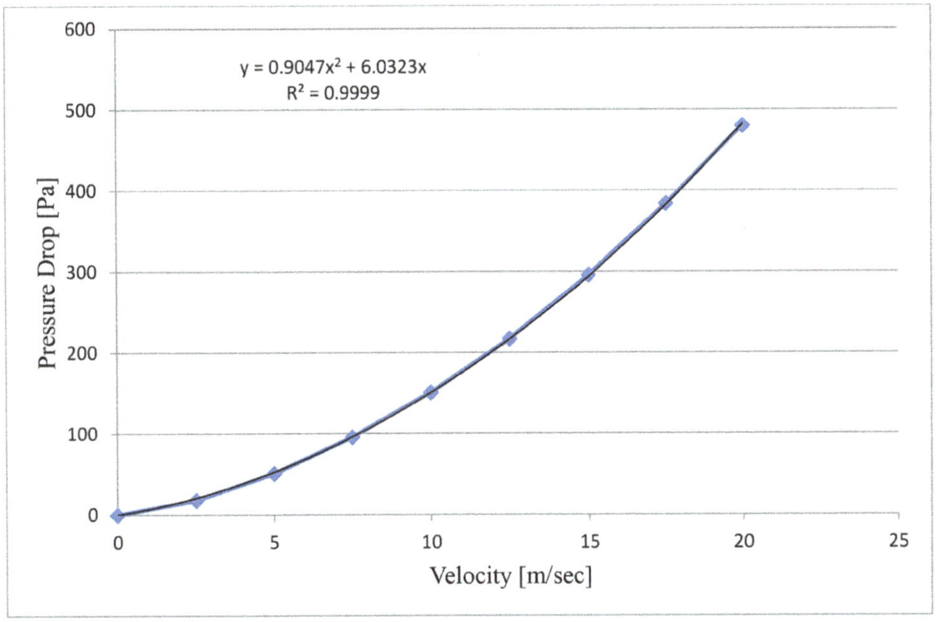

Figure 3. Unit cell physical WF simulation pressure versus velocity.

Interfacial area [m²]	Porous volume [m³]	IAD [1/m]	HTC [W/m² K]	Tref [K]
0.003957696	4.28131×10^{-6}	810	170	336

Table 3. Porous parameters of a WF.

	Face permeability [1/m²]	Thickness [m]	Inertial coefficient [1/m]
Inlet	4.01×10^{6}	0.1	3.42
Outlet	4.01×10^{6}	0.1	−5.2

Table 4. Porous jump coefficients for a unit cell of a WF.

2.3. Fin simulations with porous modeling

Once the porous coefficients are obtained, the flow field of the air can be modeled using the porous modeling. Upstream and downstream domains are attached for this analysis as shown in **Figure 4(a)**. To verify the porous modeling, the results are compared with the physical fin simulations. For the porous fin model, hexa-sweep meshing is used. The mesh of the porous model (**Figure 4(b)**) consists of 5320 cells. After completing the meshing process, boundary conditions are assigned. Besides the physical fin boundary condition configurations, additional porous-jump boundary conditions are introduced to match the physical fin simulations. All solver settings are taken to be the same as the physical fin simulations. After porous medium flow coefficients, porous-jump coefficients and heat transfer parameters are obtained from the simulation of a unit cell of a wavy physical fin and porous medium simulations are

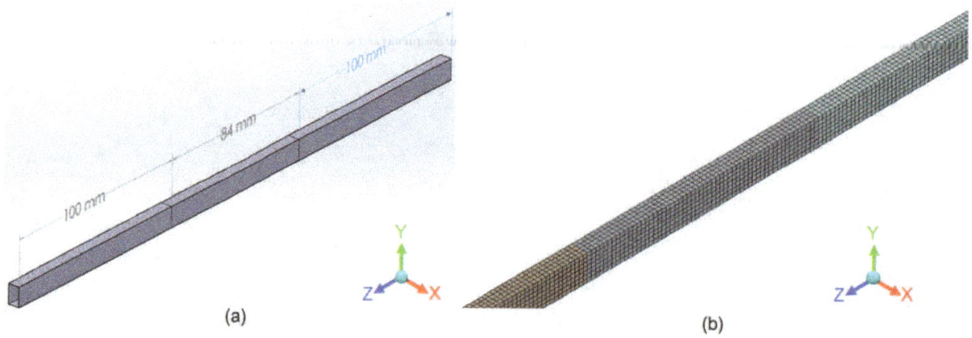

Figure 4. (a) Unit cell porous model with inlet and exit domains and (b) mesh configuration.

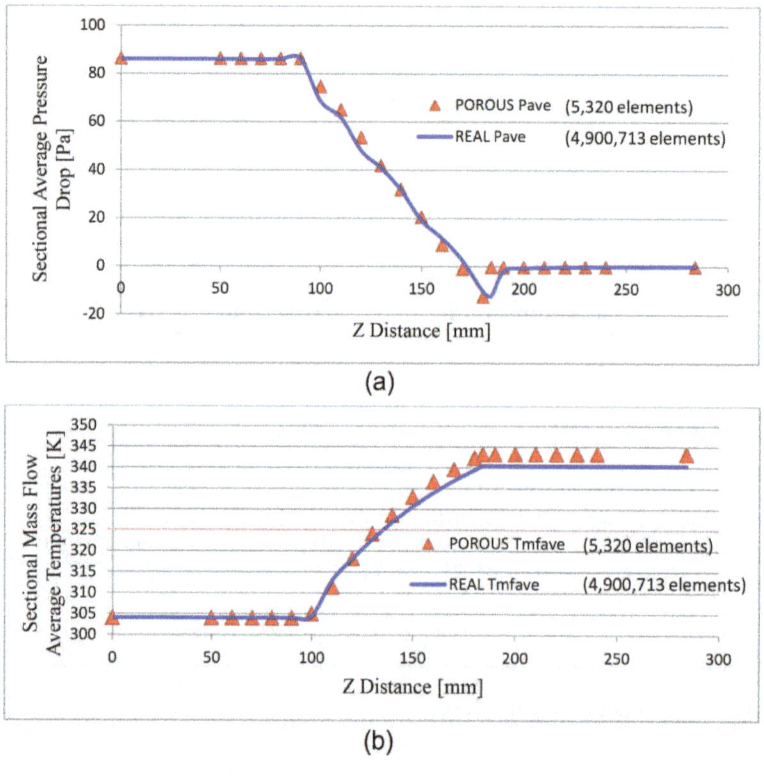

Figure 5. Comparison of physical WF and porous model: (a) sectional average pressure drop and (b) sectional mass-flow averaged temperature drop.

carried out with the same input parameters to verify the porous modeling. **Figure 5(a)** compares the sectional-averaged pressure drop for the physical fin and porous medium simulations. **Figure 5(b)** shows the same comparison for the sectional mass-flow-averaged temperature drop. As seen from **Figure 5**, an acceptable consistency is achieved with the porous modeling. One should note that the porous medium requires only 5320 cells per unit cell mesh; on the other hand, physical fin requires 4,900,713 cells. If a full-sized radiator is modeled with physical fins, the required cell number will be approximately 20 billion which is not feasible to analyze even with today's computing technology; therefore, by using porous modeling approach, full-sized model can be analyzed within reasonable computing time together with a reasonable accuracy. According to the presented results, pressure and temperature drop characteristics are coherent between the physical fin and porous medium. Contour representations for *y*+, velocity and temperature distribution across the fin are presented in **Figure 6**. It is seen from the Model-B results that *y*+ values are acceptable with respect to analysis results (for SST turbulence model maximum *y*+ value should be smaller than 1.0) [8] and velocity and temperature distributions have convenient characteristics.

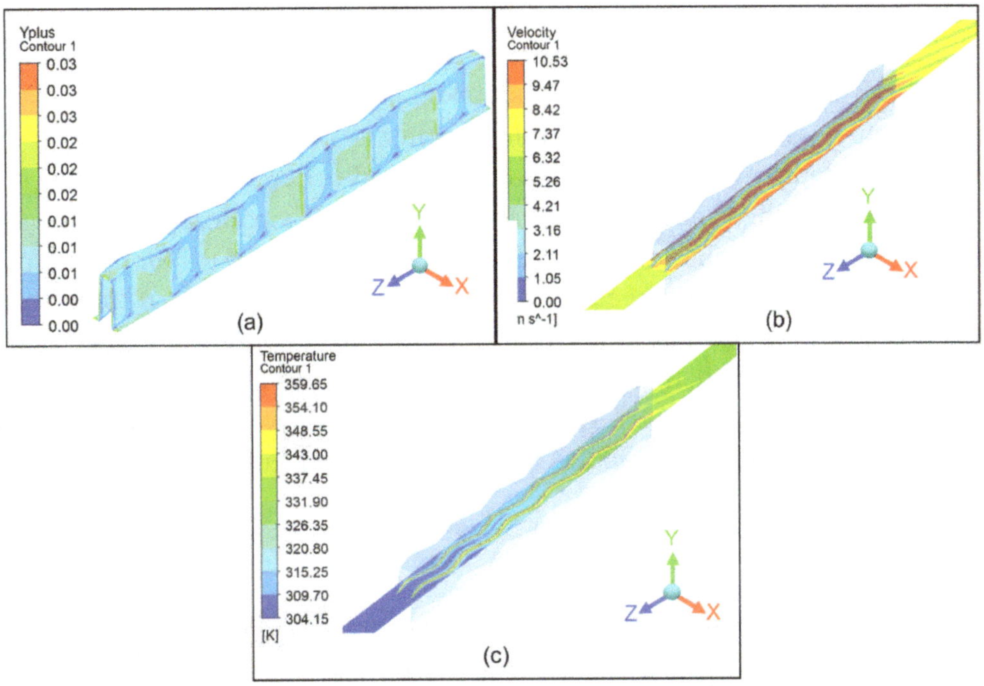

Figure 6. (a) y^+ contour, (b) velocity distribution and (c) temperature distribution across WF.

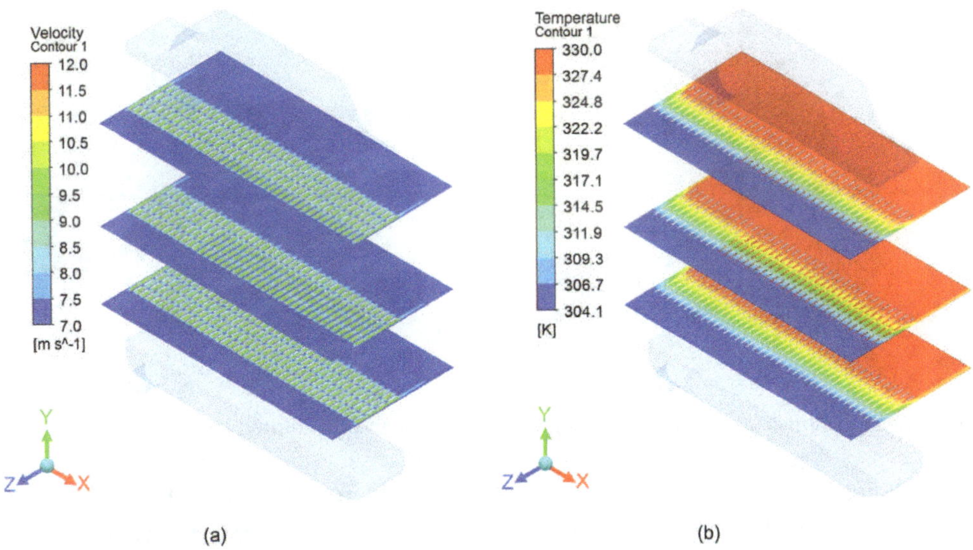

Figure 7. Air-side (a) velocity distribution and (b) temperature distribution.

2.4. Radiator modeling

3-D CAD model of the 4-row 39-column radiator is prepared by using CAD software. After forming the 3-D model, the meshing process is progressed. Fin, upstream, downstream and tube domains are meshed with hexa-type elements, while the upper and lower tanks are meshed with tetra elements. Tubes are meshed with a boundary layer mesh having two layers with 0.1 mm first layer height. The generated mesh consists of 53,355,356 cells with an average skewness value of 0.178. Mass flow inlet, pressure outlet, velocity inlet, pressure outlet, upstream wall and downstream wall boundary conditions are assigned for water inlet, water outlet, air inlet, air outlet and the outer surface boundary of the upstream and the downstream domains, respectively. The air inlet velocity is taken as 7.0 m/s with an inlet temperature of 304.2 K temperature, while the mass flow rate of water is 2.41 kg/s with an inlet temperature of 359.7 K in accordance with the catalog data. Second-order upwind scheme is used for momentum, turbulent kinetic energy (TKE) and turbulent dissipation rate (TDR). Relaxation factors are selected as 0.05 for momentum, 0.3 for TKE and TDR and 0.4 for turbulent viscosity in order to obtain optimized convergence rate and solution time. The heat transfer coefficient between the fins and air is taken as 170 W/m^2 K referring to the previous unit cell simulations. A converged solution is obtained after 472 iterations when the minimum residual is smaller than 1×10^{-4}. The simulations are performed on a DELL T5600 Workstation (Intel® Xeon®, 3.30 GHz, 2 processors, 16 cores, 128 GB RAM). The overall solution time is observed to be approximately 12 h and 40 min.

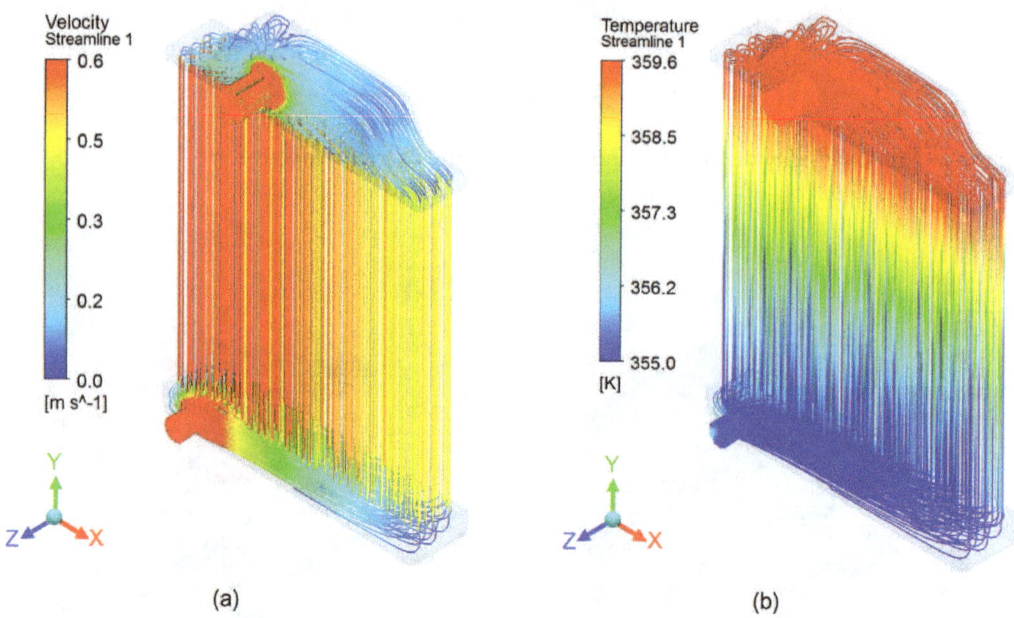

Figure 8. Water-side streamlines (a) colored according to the velocity and (b) colored according to the temperature.

Cross-sectional velocity and temperature distributions for the air-side and the water-side streamlines are presented in **Figures 7** and **8**, respectively. Temperature gradients are successfully achieved in *z*- and *y*-directions as expected. Air-side temperature is increasing in the flow direction as a result of the heat transfer from the water-side, while the water-side temperature is decreasing in the flow direction. The flow is not distributed uniformly among the tubes as shown in **Figure 8(a)**. However, to improve the performance of a radiator, the maldistribution of the flow at the header needs to be reduced [22–24]. Therefore, one can clearly state that there is a room for improvement for the design at hand. This nonuniform distribution of the flow among the tubes contributes also to the temperature in the *x*-direction. According to the simulation, the average outlet water temperature is found to be 354.3 K and total temperature drop of water through the radiator is calculated as 5.4 K which leads to a total heat capacity of:

$$Q = \dot{m}C_p \, \Delta T = 2.41 \, x4208 \, x5.36 = 54.4 \text{ kW} \tag{12}$$

The pressure drop for water which is also an important performance parameter for radiators is found to be 6.5 kPa. According to the catalog data, the outlet water temperature, temperature drop across the radiator and the cooling capacity are 354.2 K, 5.5 K and 55.8 kW. The same parameters are found to be 354.3 K, 5.4 K and 54.4 kW with the proposed CFD analysis. The deviation of the CFD results with the catalog is within 2.5% which is quite acceptable for a thermal analysis. Moreover, the proposed model solves the problem within a reasonable computational time. Considering the accuracy of the result and computational cost, the proposed methodology can be used as a rating and design tool for vehicle radiators.

3. Concluding remarks

Although the repetitive fin structures introduce a challenge for the computational modeling of a radiator, the repetitive nature also enables an efficient porous medium modeling. Moreover, again due to the repetitive nature, the porous parameters can be obtained by CFD modeling of a representative unit-cell with high resolution. A successful implementation of porous modeling can lead a dramatic reduction in computational cost and time. The implementation of the computational methodology through a commercial software also benefits from the powerful meshing, solving and post-processing capabilities. As demonstrated, CFD analysis of a radiator by using porous medium approach gives reasonable and reliable results. By using CFD analysis, design cost may be decreased dramatically by easing the experimental testing process. The porous parameters of a given fin geometry can be obtained within a couple of hours which may enable the hydrodynamic and thermal optimization of a radiator.

Optimization of radiators in terms of size and weight is desired to keep up with the constraints within competitive automotive industry. An efficient computational model enables the optimization process to be performed computationally for a range of different design parameters. Furthermore, more realistic computational models may be developed such as the inclusion of the radiator fan into model or the inclusion of the under hood equipment together with the increasing computational power of the computers. On top of these, the coupling of the flow and temperature field with the structural analysis may lead to far more efficient and robust radiator designs.

Acknowledgements

Financial support from the Turkish Scientific and Technical Research Council (Project No: 7130643) through YETSAN Radiator Co. Inc. is greatly appreciated.

Nomenclature

A_{fs}	interfacial area density
B_f	body force
C_2	inertial coefficient of porous domain
C_F	form-drag coefficient
E	total energy
h	enthalpy
h_{fs}	heat transfer coefficient for the fluid/solid interface
HTC	heat transfer coefficient
IAD	interfacial area density
I_c	inertial coefficient
J	diffusion flux
k	thermal conductivity
l	pipe length
Q	heat transfer
S^h	enthalpy source term
t	time
T	temperature
V	average velocity
V_c	viscous coefficient
Δ_p	pressure drop
α	permeability of porous domain
γ	porosity
μ	dynamic viscosity
$\tilde{\mu}$	effective viscosity
ϱ	density
τ	shear stress
$\bar{\bar{\tau}}$	stress tensor
l	pipe length

Q	heat transfer
S^h	enthalpy source term
t	time
T	temperature
V	average velocity
V_c	viscous coefficient
Δp	pressure drop
α	permeability of porous domain
γ	porosity
μ	dynamic viscosity
$\tilde{\mu}$	effective viscosity
ρ	density
τ	shear stress
$\bar{\bar{\tau}}$	stress tensor

Subscripts

f	fluid
s	solid
ref	reference
w	wall

Author details

Barbaros Çetin[1*], Kadir G. Güler[2,3] and Mehmet Haluk Aksel[4]

*Address all correspondence to: barbaros.cetin@bilkent.edu.tr

1 Department of Mechanical Engineering, Bilkent University, Ankara, Turkey

2 YETSAN Auto Radiator Co. Inc., Çorum, Turkey

3 General Electric Aviation Marmara Technology Center, Gebze, Istanbul, Turkey

4 Department of Mechanical Engineering, Middle East Technical University, Ankara, Turkey

References

[1] Incorpera F.P., DeWitt D.P. Fundamentals of Heat and Mass Transfer. 4th ed. John Wiley and Sons; New York 1996.

[2] Pavel B.I., Mohamad A.A. An experimental and numerical study on heat transfer enhancement for gas heat exchangers fitted with porous media. Int. J. Heat Mass Transfer. 2004;47:4939–4952.

[3] Taler D. Experimental determination of correlations for average heat transfer coefficients in heat exchangers on both fluid sides. Heat Mass Transfer. 2013;49:1125–1139.

[4] Hayes A.M., Khan J.A., Shaaban A.H., Spearing I.G. The thermal modelling of a matrix heat exchanger using a porous medium and the thermal non-equilibrium model. Int. J. Therm. Sci. 2008;47(10):1306–1315

[5] Ismail L.S., Ranganayakulu C., Shah R.K. Numerical study of flow patterns of compact plate-fin heat exchangers. Int. J. Heat Mass Transfer. 2009;52:3972–3983.

[6] Carluccio E., Starace G., Ficarella A., Laforgia D. Numerical analysis of a cross-flow compact heat exchanger for vehicle applications. Appl. Therm. Eng. 2005;25:1995-2013.

[7] Bejan A., Nield D.A. Convection in Porous Media. 3rd ed. Springer; New York 2006.

[8] ANSYS Inc. FLUENT® 17 User's Guide. 2016 (Section 6.2.3.8).

[9] Kim D., Kim S.J. Compact modelling of fluid flow and heat transfer in straight fin heat sinks. ASME.J. Electronic Packaging. 2004;126:247–255.

[10] Do K.H., Min J.Y., Kim S.J. Thermal optimization of an internally finned tube using analytical solutions based on a porous medium approach. ASME J. Heat Transfer. 2007;129:1408–1416.

[11] Kim S.J., Kim D. Forced convection in microstructures for electronic equipment cooling. ASME. J. Heat Transfer. 1999;121:639–645.

[12] Jeng T., Tzeng S., Hung Y. An analytical study of local thermal equilibrium in porous heat sinks using fin theory. Int. J. Heat Mass Transfer. 2006; 49:1907–1914.

[13] Jeng T.-M., Tzeng S.-C. A semi-empirical model for estimating permeability and inertial coefficient of pin-fin heat sinks. Int. J. Heat Mass Transfer. 2005;48:3140–3150.

[14] Kulasekhara N., Purushotham H.R., Junjanna G.C. Performance improvement of a louver-finned automobile radiator using conjugate thermal CFD analysis. Int. J. Eng. Res. Technol. 2012;1(8):1–13.

[15] Wen M.Y., Ho C.Y. Heat transfer enhancement in fin and tube heat exchanger with improved fin design. Appl. Therm. Eng. 2009;29:1050–1057.

[16] Yan W.M., Sheen P.J. Heat transfer and friction characteristics of fin and tube heat exchangers. Int. J. Heat Mass Transfer. 2000;43:1651–1659.

[17] DeJong N.C., Zhang L.W., Jacobi A.M., Balachandar S., Tafti D.K. A complementary experimental and numerical study of the flow and heat transfer in offset strip-fin heat exchangers. ASME J. Heat Transfer. 1998;120:690–698.

[18] You H., Chang C.H. Determination of flow properties in non-Darcian flow. ASME. J. Heat Transfer. 1997;119:190–192.

[19] Yang J., Zeng M., Wang Q. Forced convection heat transfer by porous pin fins in rectangular channels. ASME J. Heat Transfer. 2010;132:1–8.

[20] Baliga B.R., Azrak R.R. Laminar fully developed flow and heat transfer in triangular plate-fin ducts. ASME J. Heat Transfer. 1986;108:24–32.

[21] Zhang L.Z. Laminar flow and heat transfer in plate-fin triangular ducts in thermally developing entry region. Int. J. Heat Mass Tran. 2007;50:1637–1640.

[22] Jain S., Deshpande Y. CFD modelling of a radiator axial fan for air flow distribution. World Academy Sci. Eng. Technol. 2012;71:1085–1090.

[23] Zhang Z., Li Y. CFD simulation on inlet configuration of plate-fin heat exchangers. Cryogenics. 2003;43:673–678.

[24] Wasewar K.L., Hargunami S., Atluri P., Kumar N. CFD simulation of flow distribution in the header of plate-fin heat exchangers. Chem. Eng. Technol. 2007;30(10):1340–1346.

[25] Yu Q., Straatman A.G., Thompson, B.E. Carbon-foam finned tubes in air-water heat exchangers. Appl. Therm. Eng. 2006;63:63–71.

[26] Oliet, C., Castro, J., Perez-Segarra, C.D. Parametric studies on automotive radiators. Appl. Therm. Eng. 2009;27:2033–2043.

[27] Perez-Segarra C.D., Oliet C., Oliva A. Thermal and fluid dynamic simulation of automotive fin-and-tube heat exchangers, part 1: mathematical model. Heat Transfer. Eng. 2008;29(5):484–494.

[28] Oliet C., Perez-Segarra C.D., Oliva A. Thermal and fluid dynamic simulation of automotive fin-and-tube heat exchangers, part 2: experimental comparison. Heat Transfer. Eng. 2008;29(5):495–502.

[29] Zhang L.Z. Flow Maldistribution and thermal performance deterioration in a cross-flow air to air heat exchanger with plate-fin cores. Int. J. Heat Mass Transfer. 2009;52:4500–4509.

[30] Wang W., Guo J., Zhang S., Yang J., Ding X., Zhan X. Numerical study on hydrodynamic characteristics of plate-fin heat exchanger using porous media approach. Comput. Chem. Eng. 2014;61:30–37.

[31] Mao S., Cheng C., Li X., Michaelides E.E. Thermal/structural analysis of radiators for heavy-duty trucks. Appl. Therm. Eng. 2010;30:1438-1446.

[32] Guler K.G. Computational modelling of fin-and-tube type vehicle radiators based on porous medium approach [thesis]. Middle East Technical University, Ankara, Turkey; 2014.

[33] Society of Automotive Engineers International. Heavy Duty Vehicle Cooling Test Code (J1393) [Internet]. September 17, 2012. Available from: http://standards.sae.org/j1393_201209/ [Accessed: August 10, 2016]

6

Heat Exchanger Design with Topology Optimization

Mark Christian E. Manuel and Po Ting Lin

Abstract

Topology optimization is proving to be a valuable design tool for physical systems, especially for structural systems. However, its application in the field of heat transfer is less evident but is constantly progressing. In this chapter, we would like to introduce topology optimization in the context of heat exchanger design to the general reader. We also provide a chronological review of available literature to see the current progress of topology optimization in the field of heat transfer and heat exchanger design. We expect that topology optimization will prove to be a valuable tool in heat exchanger design for the coming years.

Keywords: topology optimization, heat transfer, heat exchanger

1. Introduction

The need for high-performance heat-dissipating devices is highly needed in today's rapidly changing power device and electronics markets [1, 2]. With worldwide movements on the implementation of Industry 4.0, we will see more radical changes in the way tangible products are manufactured [3]. At the same time, rapid product design cycles are becoming more of a standard rather than a demand. Thus, the need for automated design processes carried out with the use of computer as tools has never been so imperative. Computational design procedures have been more widely accepted during the past decades due to the improvements in computing technologies [4]. Together with this, rapid advancements in the algorithms and automated design procedures have flourished. Topology optimization can be viewed as one of the most promising automated design procedures, which has been an active topic of research for almost three decades.

Topology optimization is an automated, 'best material layout' process, which follows the governing equations of one or more physics taken into consideration under a user-defined set of conditions and limitations. Several methods and techniques are already well developed especially for the field of structural engineering. Topology optimization is slowly being used in mainstream design processes of tangible products due to the advancements in computational power of computers, the optimization methods, and techniques used in topology optimization itself.

Computational tools have been developed to aid and answer some of the engineering queries, but the main design of the structure is usually left to experienced and specialized professionals. Commonly applied modern-day topology optimization methods utilize finite element analyses (FEA) where each discretization is treated as a design variable. By choosing and varying the adequate material property related to the investigated case, we would iteratively investigate which element is helpful, thus material is 'allocated', and which ones are not, thus can be left as 'void', from the design space. We can also set areas that must be filled with material or areas where materials should not be placed. There are a number of learning materials for topology optimization, most are from one research group from Denmark. Among their developments are a free mobile app, TopOpt [5] and TopOpt3D [6], which can execute structural topology optimization and output. STL files ready for three-dimensional (3D) printing. The interface, some common definitions for structural topology optimization and an example are presented in **Figure 1**.

The earliest work related to topology optimization can be traced back to the ingenious Australian inventor who formulated Michell's truss theory [7] (named after inventor George Michell). The said theory dealt with the least-volume topology of trusses with a single load condition and a stress constraint. Not only was this imaginatively ingenious, it was also ahead of his time where almost nothing was known about the techniques of structural optimization.

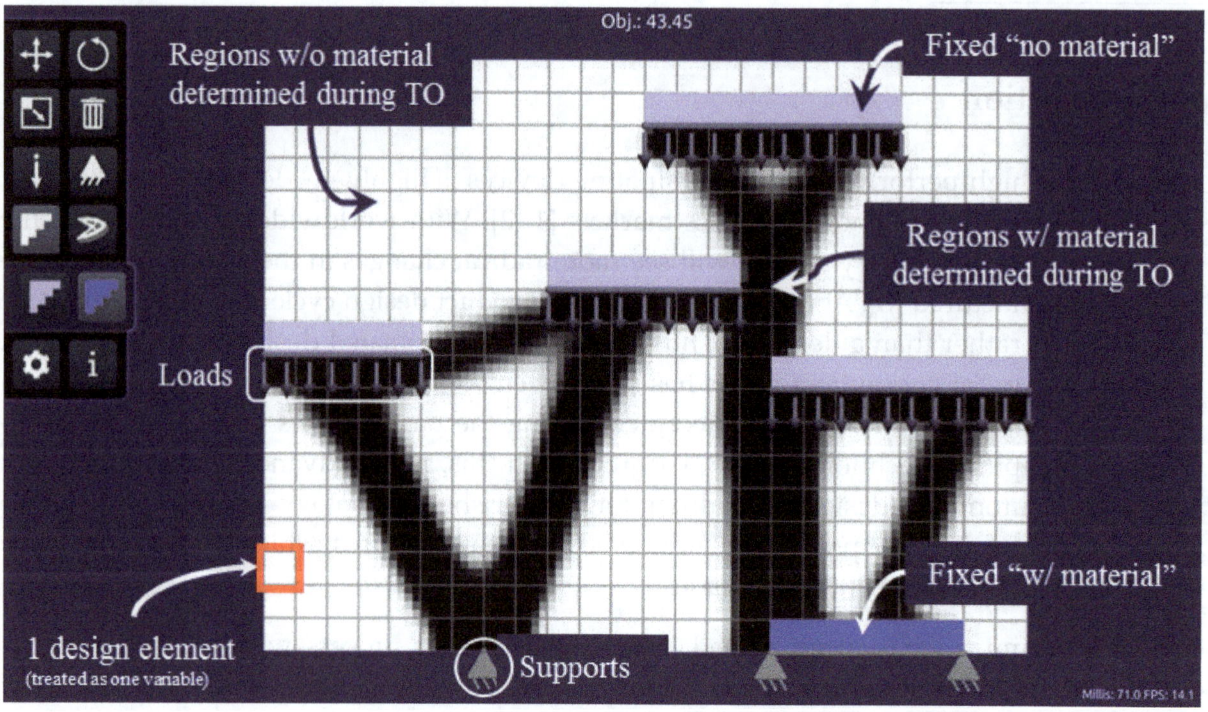

Figure 1. TopOpt app [5] developed by DTU reflecting the essential elements of topology optimization.

His works were ignored for almost half a decade where it was rediscovered by Cox [8, 9] and Owen [10] in the 1960s, the same time when computers were acknowledged as automation tools. It was Hemp [11] and his co-workers who had spent most of their professional lives and comprehensively studied Michell structures. Modern-day computer-aided topology optimization can probably be traced back to the works of Bendsoe and Kikuchi [12] on homogenization who had also coined and popularized the term topology optimization. For the following decades, their works had sparked the interest of many researchers and might not have necessarily had any product-related applications. In the next section, we first briefly discuss the main methods commonly used in topology optimization.

2. Topology optimization methods and learning codes

Different methods have been developed in finding solutions to the optimal layout problem. Since Bendsoe and Kikuchi's work in 1988 [12], focus has been more on finite element (FE)-based topology optimization of continuum structures. Different methods have been developed since. The differences in the different methodologies lay in the way the design space, and consequently, the design variable is parameterized and controlled. Some methods directly define the design variables on the finite element domain, while others define a separate function from which the generated structure is interpreted. In both cases, '0-1' designs or 'void/solid' designs are desired because they can be easily interpreted and physically realized. Here, 1 or solid means that material is allocated on the design element and 0 or void means that material is not present in the design element. In some methods, 'grey' or intermediate densities, which are values between 0 and 1, are encountered and observed. The following subsections outline some of the popular methods for topology optimization. We also list references at the end of each method where codes (usually written in MATLAB) are readily available for interested readers. These codes also contain more information on the mathematical background and rationale for each method. These codes, however, are usually written in the context of structural topology optimization but could be modified appropriately to solve heat-transfer problems. The detailed modifications needed are explicitly given in Appendix B of Ref. [13].

2.1. Homogenization method

Pioneering work by Bendsøe and Kikuchi [12] posed a structural layout problem within the context of homogenization theory. In their method, now known as the homogenization method, they treat each element as porous material whose microstructures can be modelled and controlled. By tuning these microstructures, macro-scale material properties are realized which are best suited for the stress experienced from each element. The periodic microstructures are defined for each discretized unit cell in the finite element domain. In their work, they had demonstrated two potential microstructures that could be generated on each unit cell. The first one being a perforated microstructure in the form of a square cell with a rectangular void with three control parameters (μ_1, μ_2 and θ). The second was a layered microstructure with two isotropic constituents with the same control parameters. These two types of microstructure definition are visualized in **Figure 2 (a)**. Under the assumption of infinitesimally small periodic unit cells and the adequate microstructure definition, it was deemed that any

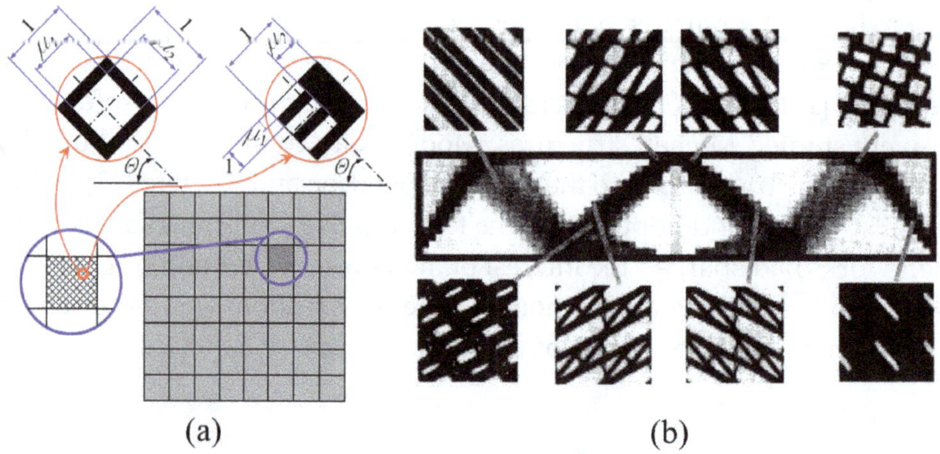

Figure 2. (a) Schematic of the treatment of design variables contained in elements showing two possible microstructures. (b) Result for a truss problem reflecting generated microstructures [14].

anisotropic macro-scale representation of the material can be achieved such as pure solid, pure void, composite and porous material.

A single set of variables corresponding to each microstructure can be used for each design element or can be extended to a sub-mesh to generate finer structures. Topology optimization, in this sense, becomes a problem to determine the optimal combination of these design variables, which corresponds to the optimal macro-scale distribution of properties which minimize a given objective function. This approach was investigated in the 1990s but has received less attention in the recent years due to the emergence of more efficient methods. Nevertheless, it gave the fundamental concepts and ideas in the other methods. Additionally, some methods that will be later mentioned apply alternative formulations to alleviate the common numerical issues found in explicit topology parameterization. Nowadays, it has found its application in finding ways of how to realize high-performing microstructures and is called 'inverse homogenization'.

2.2. 'Hard-kill' methods

'Hard-kill' methods are a generalization of methods that explicitly treat each element as material or void. Unlike other methods, they do not relax the '0-1' problem on topology optimization. These methods gradually remove (in some cases add) elements that represent absence (or presence) of a material into the design domain explicitly for each iteration step. A few of these methods utilize combinatorial techniques such as genetic algorithms and simulated annealing, to name a few. Another 'hard-kill' method that is based on sensitivity information is known as the concept of using topological derivatives (or topological sensitivity) [15]. The concept of topological derivatives is that undesired computational nodes are explicitly removed. The most well-known 'hard-kill' method in topology optimization is the evolutionary structural optimization (ESO) [16] and, more recently, the bi-directional evolutionary structural optimization (BESO) [17]. BESO is differentiated from ESO in a way that ESO only allows for the removal of elements while BESO allows for both the addition and removal of elements that represent the presence or absence of a material based on an 'optimization criterion', which is evaluated in each small domain or element. This is analogous to slowly evolving the shape of a structure towards the desired optimum result by removing (or adding) the elements that do not contribute to the improvement of the desired objective function. The choice of material to be removed (or added) is based on heuristic criteria, which is based on sensitivity information of the iteration steps. As a result of these heuristic features, the technicality of this method is often questioned for a robust theoretical basis does not exist [18, 19]. One of the most attractive features of these hard-kill methods is its simplicity with which they can be utilized with commercial finite element packages. It is claimed that the integration of algorithms based on hard-kill methods with finite element analysis (FEA) solvers requires only minor modification in the pre- or post-processing steps [19]. Also, structures generated are free from intermediate or 'grey' material representations due to the nature of its solution method of explicitly removing (or adding) material in the finite element system. More recently in [20], BESO has been relaxed to prevent the concerns given in [18, 19] and was termed 'soft-kill' method. An attempt to visually present the conceptual differences between the different 'hard-kill' methods is presented in **Figure 3**. In **Figure 3 (a)**, elements are essentially removed from the FEA routine as executed in the original ESO. In **Figure 3 (b)**, a void element is essentially allowed to 'roam' on neighbouring elements until such time it finds an optimal location and this is common for combinatorial techniques. In **Figure 3 (c)**, a node is essentially removed and creates an area of void elements, and this is the concept behind the topological derivatives. A MATLAB code of the relaxed BESO implementation is also given in Ref. [20].

2.3. Boundary variation methods

Boundary variation methods are among the most recent and noteworthy contributions that lead to advancements in structural topology optimization. Boundary variation methods have originated in shape optimization techniques and had been recently introduced to structural topology optimization. They are differentiated from the other methods from the fact that structure domain and boundaries are represented based on implicit functions rather than an explicit parameterization of the design domain. In most methods, the design variable in the domain is given explicitly as values from 0 to 1 where 0 would represent the absence of material and 1 represents the presence of material. For boundary variation methods, the structural

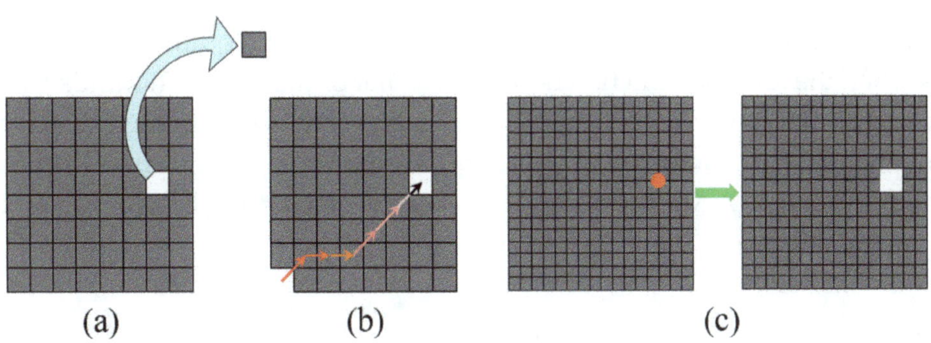

Figure 3. Conceptual differences of different 'hard-kill' methods (a) ESO, (b) combinatorial techniques and (c) topological derivative.

boundaries are implicitly defined as the contour line of a field which is a function of the design variable. Boundary variation methods are currently dominated by two methods: the level-set method and the phase-field methods. Both of these methods produce results in the design domain with crisp and smooth edges that require little post-processing effort to realize the relevant structural features. Additionally, these methods are fundamentally different from shape optimization techniques because they allow both the movement of the structural boundary and topological changes (e.g. formation, disappearance and merging of void regions).

2.3.1. *Level-set method*

Level sets for moving interface problems in physics were first developed by Osher and Sethian [21], with the fundamental goal of tracking the motion of curves and surfaces. This method has been applied in a wide variety of research areas [22, 23] including topology optimization. The level-set method was first applied to topology optimization in the early 2000s by Sethian and Wiegmann [24], where it was used to represent the free boundary of a structure for linearly elastic problems in structural design. In another direction, Osher and Santosa [25], in about the same time, combined level sets with a shape sensitivity analysis framework for the optimization of structural frequencies.

In the level-set method, the boundaries of the structure are represented on the zero-level curve (or contour) of the scalar function Φ which is consequently called the level-set function. Topological functions such as the changes in the boundary, merging of boundaries and formation of new voids are performed on the level-set function. The geometric boundary shape is modified by controlling the motion of the level set according to the physical problem and optimization conditions [26]. It is worth noting that most level-set formulations still rely on finite elements despite the smooth boundary representation. Thus, boundaries are still represented by discretized mesh which leads to some unsmooth results. Alternative techniques such as the extended finite elements (XFEMs) [27] have been utilized to represent the geometry in the analysis of the model which produces superior, smooth and continuous boundaries.

The level-set method does not exhibit intermediate material densities since the presence or absence of material on the domain is determined at the zero-level-set function. However, current level-set methods are known for their dependency on the initial design and locations of the level-set functions. This drawback poses a severe problem in the acceptability of solutions of level-set functions but new developments have been made to address and improve this deficiency [28]. Also, at some cases the level-set method might require re-initialization during the process when the level-set function becomes too flat or too steep. This adds computational complexity and additional tuning parameters to the algorithms which is undesirable especially for implementation with commercially available software. A visualization of this concept is presented in **Figure 4 (a)**. A MATLAB code for the level-set method is available in Ref. [29].

2.3.2. *Phase-field method*

The phase-field method originates from theories developed to track and represent phase transition and phase interface phenomena in surface dynamics [30]. This method has been utilized for solid-liquid transitions, diffusion, solidification, crack propagation, multiphase flow and

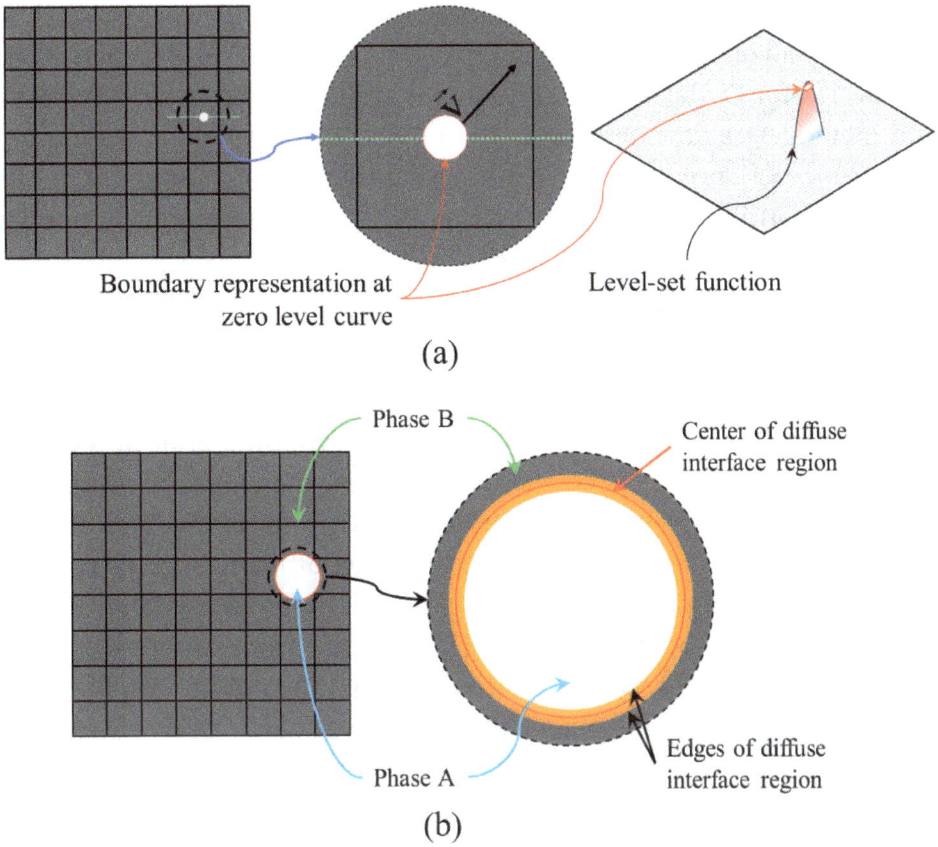

Figure 4. Conceptual difference between (a) level-set method and (b) phase-field method.

eventually in topology optimization [31]. In the application of these theories, a phase-field function is specified over the design domain that is composed of two phases (e.g. A and B), which are represented by two variables as a function of the phase-field function. The boundary region between phases is a continuously varying region of thin finite thickness.

In topology optimization utilizing the phase-field method [31–33], this interface region defines the structural boundary, thus separating material from void, and is modified via a dynamic evolution of the phase-field function. The primary difference between the level-set and phase-field methods is mainly due to the fact that in the phase-field method, the interface between the boundaries of the two distinct phases is not tracked throughout optimization. Whereas in the level-set method, the boundary is tracked as it moves during the optimization process. In other words, the governing equations of phase transition are solved over the complete design domain without the initial information of the phase interface location. Consequently, phase-field methods do not require the re-initialization step as do level-set functions. Its conceptual difference with the level-set method is presented in **Figure 4 (b)**. A MATLAB code for the phase-field method is available for download by visiting the website of Ref. [31].

2.4. Density-based methods

Currently, the most widely used methods for structural topology optimization are explicit parameterizations that are broadly classified as density-based methods. Variations of this method are termed 'material interpolation', 'artificial material', 'power law' and 'solid isotropic material with penalization (SIMP)' methods. Although SIMP is only one of the methods, its popularity has led for the term to be colloquially used in place of density-based methods. Density-based methods are an extension of the works on the homogenization method. This type of method has experienced much popularity in recent years in this community due to its conceptual simplicity and ease in implementation. Nearly all commercial topology optimization tools utilize a density-based method [34].

Similarly with the homogenization method, these density-based methods operate on fixed domain of finite elements. The main difference is that, rather than a set of microstructure properties, each finite element contains only a single design variable. This variable is often understood as the element material density, ρ_e. The relevant material property of each element concerned with the physics involved, for example, the elastic modulus for structural problems or thermal conductivity for heat-transfer problems, is made as a function of the density design variable. This is usually accomplished by utilizing an interpolation function. The topology generated in **Figure 1** was based on this method. Tremendous amount of literature is available for this method and the book [13] contains much discussion on this method as well as an '99-line code' for MATLAB which pioneered the publication of codes for educational purposes in topology optimization. It has been reworked by Andreassen et al. in [35] which shortened the code as well as greatly improving its efficiency. Another rework was made by Liu et al. in [36] which provides the code's extension to 3D problems in the MATLAB environment. More recently, Aage et al. [37] has released their code which utilized Portable, Extensible Toolkit for Scientific Computation (PETSc) and can handle problem scales which are not practical in MATLAB.

3. The heat-transfer problem in the context of topology optimization

The heat-transfer problem (as shown in **Figure 5**), in its weak form in the design domain, can be generalized as

$$\begin{aligned}
\rho c\, T_t &= k\Delta T + q_v && \text{in } \Omega \times [0, t] \\
T &= T_b && \text{on } \Gamma_1 \\
-k\frac{\partial T}{\partial n} &= q && \text{on } \Gamma_2 \\
-k\frac{\partial T}{\partial n} &= h(T - T_\infty) && \text{on } \Gamma_3 \\
T|_{t=0} &= T_0
\end{aligned} \tag{1}$$

where ρ is the material density, c is the specific heat of material, T_t is the temperature for a particular given time in transient cases, k is the thermal conductivity and q_v is an internal heat generation rate per unit volume. In general, three types of boundary conditions may exist and can be considered: a temperature condition on Γ_1, a heat flux conduction on Γ_2, and a convective condition on Γ_3. T_0 is the initial temperature at time $t = 0$, T_b is a temperature imposed on Γ_1, q is a heat flux boundary condition imposed on Γ_2, h is the convective heat-transfer coefficient

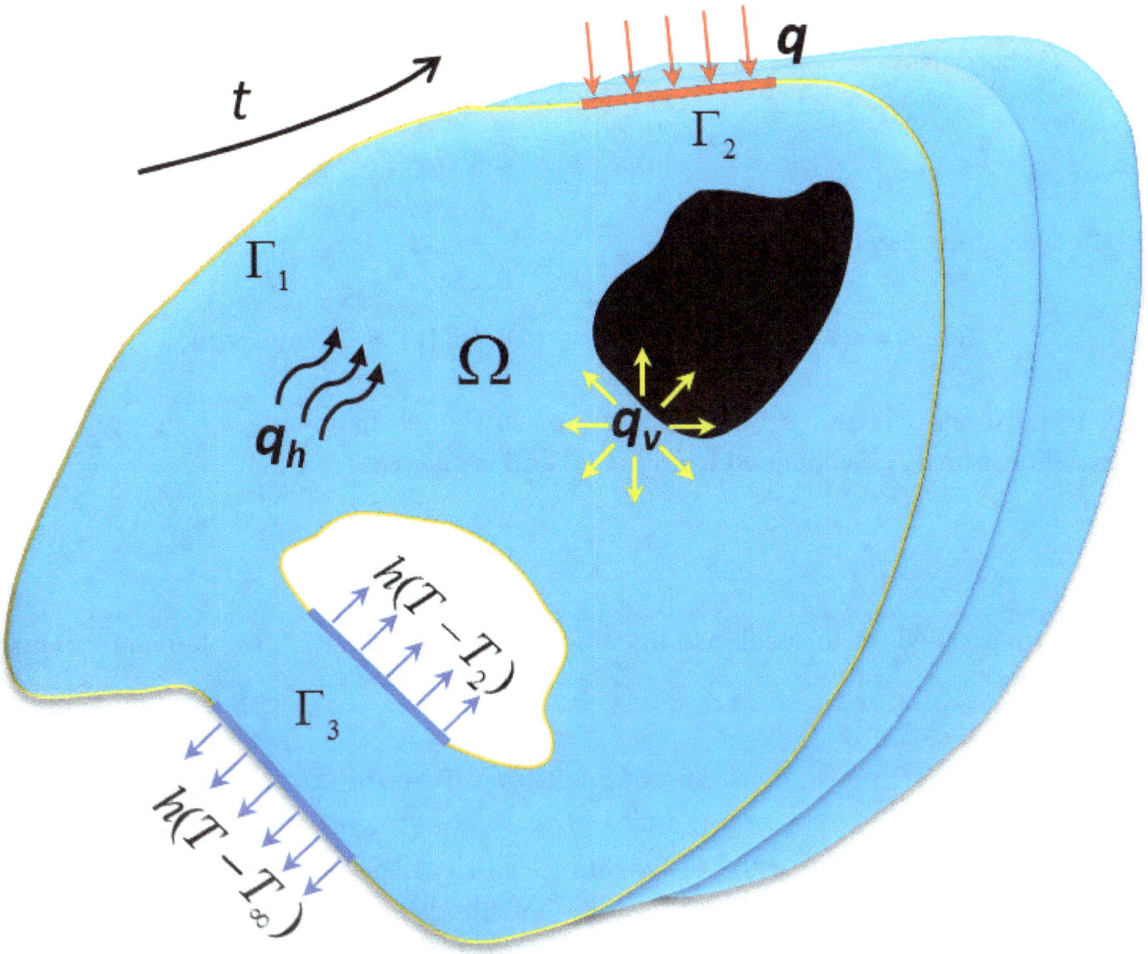

Figure 5. Generalized heat-transfer problem.

on Γ_3 and T_∞ is a fixed reference temperature, n is the boundary normal vector. Special treatment is needed for methods which produce intermediate densities for problems considering convective boundary conditions since boundaries are not well defined.

Simplifying to a steady-state heat pure heat conduction case with only temperature boundary conditions and heat flux boundary conditions considered, Eq. (**1**) is reduced to

$$\begin{aligned} q_v &= k\Delta T \quad \text{in } \Omega \\ T &= T_b \quad \text{on } \Gamma_1 \\ -k\frac{\partial T}{\partial n} &= q \quad \text{on } \Gamma_2 \end{aligned} \tag{2}$$

This form is often considered for the 'volume-to-point' problem commonly investigated in heat conduction problems. Using the virtual temperature field, v, the weak formulation of the heat conduction problem is given by

$$\int_{\Omega} (k\Delta T - q_v)v dx = 0 \tag{3}$$

After integration by parts has been carried out and applying the heat flux boundary condition, Eq. (3) becomes

$$\int_{\Omega} k \nabla T \cdot \nabla v - q_v v dx = \int_{\Gamma_1} k \nabla T \cdot n v d\Gamma = -\int_{\Gamma_2} q v d\Gamma \tag{4}$$

And the weak form can be written as

$$a(T, v) = \ell(v) \forall v \in \tilde{\mathbf{T}} \tilde{\mathbf{T}} = \{T \in [H^1(\Omega)] | T = T_b \text{ on } x \in \Gamma_1\} \tag{5}$$

where v is in $\tilde{\mathbf{T}}$ and $\tilde{\mathbf{T}}$ is a subset of a Sobolev space. The left-hand side $a(T, v)$ represents the energy bilinear form. It is obtained from Eq. (4) and is given as

$$a(T, v) = \int_{\Omega} k \nabla T \cdot \nabla v \tag{6}$$

The $\ell(v)$ term is called the thermal load linear form and can similarly be obtained from Eq. (4) and is given as

$$l(v) = \int_{\Omega} q_v v dx - \int_{\Gamma_2} q v d\Gamma \tag{7}$$

This is often used for deriving the propagating velocity of the material boundaries by the material derivative theory in boundary variation methods and the homogenization method. One design objective or thermal compliance measure, c, that is considered as the mean temperature could be expressed as

$$c(\Omega) = \int_{\Omega} k \nabla T \cdot \nabla T dx \tag{8}$$

And finally the topology optimization problem of the heat conduction problem is expressed as

$$\begin{aligned} \min_{\Omega \subseteq D} \quad & c(\Omega) \\ \text{s.t.} \quad & a(T, v) = \ell(v), \quad \text{for all } v \in \tilde{\mathbf{T}} \\ & \int_D dx \leq V_{\max} \end{aligned} \tag{9}$$

where D contains the material distributed in the design domain Ω and $V_{\max}$ is a volumetric constraint. In the context of density-based topology optimization, we introduce the element density, ρ_e, and applying a discretized optimization model, for example, FEA, the heat conduction problem is defined as

$$\begin{aligned} \min_{\rho_e} \quad & c(\rho_e) = \mathbf{T}^T \mathbf{K} \mathbf{T} = \mathbf{Q}^{\mathrm{T}} \mathbf{T} \\ \text{s.t.} \quad & \mathbf{K} \mathbf{T} = \mathbf{Q} \\ & V_f = V(\rho_e) / V_o \\ & 0 < \rho_e \leq 1 \end{aligned} \tag{10}$$

where K is the global thermal conductivity matrix, **T** is the node temperature vector and **Q** is the applied thermal load. It is to be noted that the global thermal conductivity matrix if formed from the assembly of individual element thermal conductivity matrix, $\mathbf{k}_e$, and the material interpolation schemes is applied here as is formally given as

$$\mathbf{K}(\rho_e) = \sum_{e=1}^{N} k_{eff}(\rho_e)\,\mathbf{k}_e \tag{11}$$

where k_{eff} is the material interpolation scheme. The objective function could then be expressed as

$$c(\rho_e) = \mathbf{Q}^T\mathbf{T}, \quad \text{where } \mathbf{T} \text{ solves: } \sum_{e=1}^{N} k_{eff}(\rho_e)\,\mathbf{k}_e\mathbf{T} = \mathbf{Q} \tag{12}$$

One simple form of the interpolation scheme was presented when SIMP was introduced and is given as

$$k_{eff}(\rho_e) = \left(k_{\max} - k_{\min}\right)\rho_e^{\,p} \tag{13}$$

Gradients are usually required by the optimization algorithms needed for the update process in topology optimization. These are easily derived for the objective and constraints involving only ρ_e. For functions that depend also on temperatures, derivative can be obtained using the chain rule. These expressions will then contain derivatives of temperature, which in turn can be obtained by taking the derivative of the equilibrium equation, **KT** = **Q**. The most effective method for calculating the derivatives is to use the adjoint method, where derivatives of the temperature are not calculated explicitly. For the thermal compliance problem given above, we rewrite the objective function by adding a zero function:

$$c(\rho_e) = \mathbf{Q}^T\mathbf{T} - \lambda^T(\mathbf{KT} - \mathbf{Q}) \tag{14}$$

where λ is called the Lagrangian multiplier which is an arbitrary, but fixed real vector. We then obtain the derivative as

$$\frac{\partial c}{\partial \rho_e} = (\mathbf{Q}^T - \lambda^T\mathbf{K})\frac{\partial \mathbf{T}}{\partial \rho_e} - \lambda^T\frac{\partial \mathbf{K}}{\partial \rho_e}\mathbf{T} \tag{15}$$

which can be re-written as

$$\frac{\partial c}{\partial \rho_e} = -\lambda^T\frac{\partial \mathbf{K}}{\partial \rho_e}\mathbf{T} \tag{16}$$

When λ satisfies the adjoint equation:

$$\mathbf{Q}^T - \lambda^T\mathbf{K} = 0 \tag{17}$$

This equation is in the form of an equilibrium equation and for thermal compliance we see that we obtain directly that $\lambda = \mathbf{T}$. Moreover, the form of the stiffness matrix means that the derivatives of the thermal compliance $c(\rho_e)$ for the main problem in Eq. (**10**), considering the SIMP interpolation as presented in Eq. (**13**), are

$$\frac{\partial c}{\partial \rho_e} = -p(k_{\max} - k_{\min})\, \rho_e^{\,p-1}\, \mathbf{T}^T \mathbf{K} \mathbf{T} \tag{18}$$

Thus, the derivative for the thermal compliance problem becomes easier to compute. It is also worth noting that the derivative is 'localized' to the element level; however, there is an effect from other design variables hidden in the temperature, **T**. The sensitivity is negative for all elements, so intuitively, additional material in any element decreases compliance, and makes the overall objective go lower. Using this sensitivity information, the material is redistributed and the process is repeated until a convergence criterion for the topology optimization process is attained. Each of the paper in the following chapter discusses the complete topology optimization process with more depth and varies depending on the method they have utilized.

4. Chronology

Interests in topology optimization can be represented by the recent amount of publications and citations over the years as presented in **Figure 6(a)** and **(b)**respectively. Although this figure might not accurately represent the exact number of papers, we can still see that the contribution of the papers related to 'heat' is roughly around 1/20th of the total contributions for topology optimization. It has also been increasing especially within the past decade. The amount of papers that are directly related to heat exchangers is arguably much less in number. The following subsections present a number of papers related to the interest of this chapter in its chronological order. For the completeness of the review, some papers at the end of each year are cited but no further elaborations are made due to access restrictions.

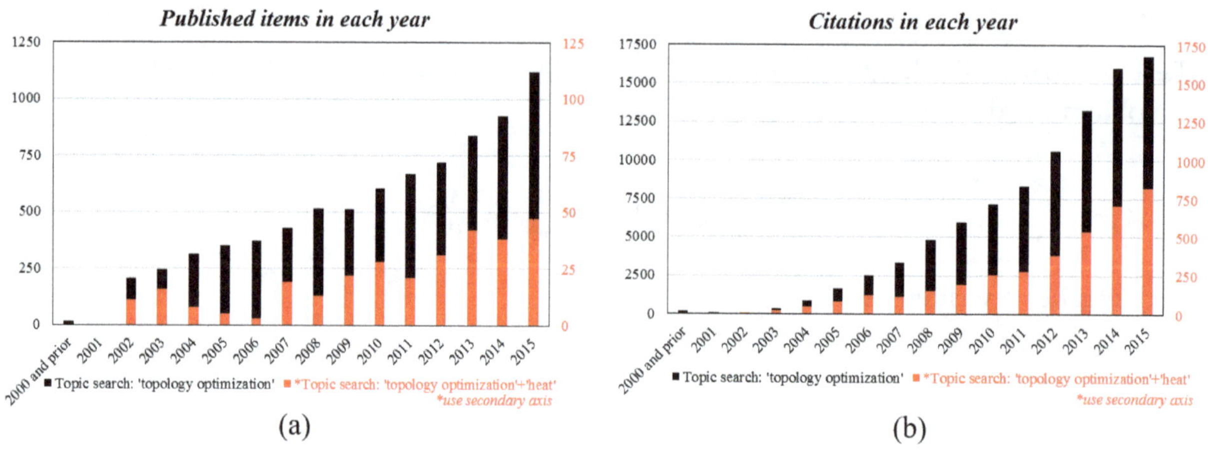

Figure 6. Some scholarly metrics [38] for topology optimization: (a) publication count and (b) citation count. Highlighted in red are search results for 'topology optimization' + 'heat'. Note that results for this search are reflected on the red axis at the right-hand side and is scaled 1:10.

4.1. Prior to 2005

Rodrigues and Fernandez [39, 40] and Jog [41] utilized topology optimization for designing thermoelastic structures. Heat transfer was treated as one of the involved physics and as an extension to structural mechanics problems. It is worth noting that this was the beginning of the consideration for heat-transfer applications for topology optimization. However, in this chapter, we restrict ourselves to papers that focus more on heat conduction topology optimization (and a few convection cases) that is more directly related to cooling applications, such as the case for heat exchanger design.

Bejan [42, 43] introduced constructal theory in the context of heat transfer. Although it is not directly categorized as topology optimization due to restrictions on size and orientation of each building block, it has provided interesting discussions and has formulated a fundamental problem for the heat-transfer community. The problem is now known as 'volume-to-point (VP) problem' or 'access problem' and discusses the need to layout a fixed amount of material in a heat-generating domain (such as a CPU).

Xie et al. [44] used ESO explicitly for conduction problems. Several generalized claims were given regarding topology optimization, which might not necessarily be true on other methods. The paper is recognized as the first topology optimization paper presented directly solving pure conduction problems. In this paper, an element's rejection is based on the integral of different thermal parameters, more specifically, integral of the temperature surrounding the element. They have highlighted the simplicity of the ESO method to generate novel structures and had considered anisotropic cases in one of the examples. He had also multiple loading cases and had presented two ways to introduce the loading cases, which generated distinct designs.

Turteltaub [45] used SIMP for finding optimal material properties for transient heat conduction problems. Although the generated final designs were rich in intermediate densities due to the lack of penalization, this paper had first offered the possibility to extend topology optimization for transient problems. It was also mentioned that in the heat-transfer problems, special care should be given especially for convective boundary conditions. Though he did not use any explicit boundary-tracking scheme, it was already recognized that difficulty in convective boundaries are present.

Haslinger et al. [46] applied the original homogenization method for conducting structures. Although the paper had focused more on convergence analysis and approximation strategies, it has utilized rank-two laminated structures to demonstrate the optimal heat conductor configurations for its test problems. The effective conductivity of the rank-one laminates was assumed to consist of harmonic and arithmetic means. Numerical minimization was performed by a subroutine from the Numerical Algorithms Group (NAG) Numerical Library which implemented a sequential quadratic programming (SQP) approach.

Cheng et al. [47] introduced bionic optimization strategy for constructing better performing conductive paths. This was directly addressing and comparing results with Bejan's original work. There is not much detail regarding their implementation but it can be viewed as a heuristic 'hard-kill' method. In the same year, Novotny et al. [15] introduced the concept of topological derivative. Although it is viewed as a 'hard-kill' method since it explicitly creates holes in the design domain, the concept is very far from ESO since it utilizes concepts from shape sensitivity analysis to evaluate the topological derivatives. Several theorems were presented in how it can be utilized for the design of conducting structures and has considered Robin boundary conditions in the formation of new holes. In a seemingly unrelated development, Borvall and Petersson [48] introduced the use of topology optimization for fluidic systems governed by Stokes flow. This field of topology optimization has its own unique developments and only a few papers which are relevant to the context of this chapter are mentioned. In this year, Bendsoe and Sigmund [49] also published their book on topology optimization which had some mention of heat-transfer topology as well as instructions for converting the learning code to conduction heat-transfer topology optimization. Guo et al. [50] presented the least dissipation principle.

Xie [51] presented some changes in the ESO method for heat conduction applications which includes some revision for the criterion for rejection through some sensitivity measure. It is also mentionable that this paper had contained a good compilation of literature for shape sensitivity analysis in the field of heat transfer. It was not explicitly stated but the methods implemented were not as aggressive to the original ESO paper where degeneration was considered. This is more properly termed nowadays as 'soft-kill' ESO. Also, the design variable was constructed in terms of the element's thermal conductivity. New interesting problems are given in the context of proper insulation design. Alberto and Sigmund [52] also published on multiphysics problems governed by Poisson's equation, which includes conduction heat transfer. Ha et al. [53] presented non-linear heat conduction problems. Moon et al. [54] presented reliability-based topology optimization considering convection heat transfer.

In this transitory stage, we can see that most of the existing methods are directly being migrated from structural topology optimization into the context of heat transfer. Here, we see ESO, homogenization method and SIMP which is more complex and harder to understand compared to papers in the next years. It is also worth noting that SIMP has already considered transient problems. The topological derivative is also introduced first in the context of heat-transfer problems which will later be a very powerful addition to level-set methods. It can also be said that in this year, fluid flow topology optimization has just started.

4.2. 2005–2010

Yoon and Kim [55] introduced an element connectivity parameterization (ECP) to alleviate problems in applying SIMP to multiphysics problems. A more specific problem of temperature undershooting was emphasized as a numerical instability when applying SIMP to include heat convection formulations on the generated structure boundaries (termed as 'side convection' in the paper). These undershootings in temperatures were deemed to be impossible and infeasible solutions which needed to be strongly addressed. Thus, their paper had given special attention to heat transfer utilizing the zero-length heat conductors as element connectivity measures in ECP. Good results were obtained using the method which was again extended to heat-dissipating structures and electro-thermal actuators. In the same year, Ha and Cho [56] introduced the level-set approach explicitly for heat conduction problems.

Their paper contains a detailed yet understandable introduction for level-set methods in the context of heat-conducting structures. It is also worth noting that due to the nature of the level-set method of clear and well-defined boundaries, convection heat transfer was already considered. However, it was not directly applied to the evolving boundaries. Also, it was well reported in this pioneering paper for level-set method for heat-transfer topology optimization that the generated structures were highly dependent on the initial distribution of holes in the design domain since this implementation cannot create new holes during the optimization process. It was also mentioned that density-based method (SIMP) yielded better results for most cases, in terms of the number of iterations needed to achieve the converged results. Thermal compliance values for both methods were in very close agreements.

Gersborg-Hansen et al. [57] introduced the use of the finite volume method (FVM) for heat conduction problems. It is worth noting that all previous papers were utilizing finite element methods and formulations. Their justification for utilizing FVM was made in the context of guaranteeing element-wise conservation of the physical quantities and to give access to FVM users to topology optimization. Element interface heat fluxes were calculated using the value of thermal conductivities based on the arithmetic and harmonic means of the surrounding nodes. The SIMP method was utilized for their implementations. Two unique compliance measures were investigated. It was mentioned that the results from FEM and FVM were qualitatively similar and the designs suffered high-mesh dependence when the compliance measure for arithmetic average was used even though the penalty value, p, in SIMP was increased up to 5 using the continuation approach. Using the harmonic average in the FVM formulation also reduces checkerboard formation up in their test cases. Donoso [58] revisited the VP problem in 3D space and used the optimality criteria (OC) method to find the solution.

Zhuang et al. [59] utilized the concept of topological derivative in conjunction with the level-set method. The topological derivative was used to create new holes during the topology optimization process and thus eliminating the dependence on the initial hole distribution. A fixed cutting ratio was set for the topological derivative for generating new holes. Multiple load cases were also one of the highlights of their paper and highly consistent results and convergence curves were presented. Xu et al. [60], on the other hand, presented a combinatorial approach for optimizing the heat conduction paths. In their paper, they tried to solve the volume-to-point problem using simulated annealing and genetic algorithm (hard-kill approaches). Their paper had clearly presented their implementation scheme and had made comparisons with the results of bionic optimization. The optimal results were generalized as all high conductivity materials are continuous, no holes are present. For cases in which the thermal conductivity ratio is relatively small, shapes are thick and short surrounding the heat sink. When the thermal conductivity ratio is increased, the shape becomes more slender. Mathieu-Potvin and Gosselin [61] developed an evolutionary algorithm which tries to solve the VP problem. Their evolutionary algorithm aimed to minimize the hotspot temperature by displacing elements. Displacements were either by swapping of a heat-generating cell with a void cell and swapping a heat-generating cell with a conductive cell based on heat flux or by element-averaged heat flux and temperature. It is worth noting that in their implementation, an extended domain was utilized and during the evolution process, the cell elements can rearrange themselves in the extended domain. Due to the nature of the algorithm, exact repeatability of results is most unlikely but measures were adopted to find approximate performances and determine the algorithm's robustness. Results were also compared to constructal theory in terms of the temperature objective function, $k\varphi$, dimensionless distance measure, uniformity distribution measure and fractal dimension. Good discussions were made regarding each of these performance measures. Also in the same year, Bruns [62] clarified and resolved the problems presented by Yoon and Kim in their 2005 paper for convection-dominated heat-transfer problems in the context of density-based topology optimization. He has discussed the necessary techniques to prevent the 'undershoot' in temperatures mentioned by ensuring that the convection term contributions are treated as lumped matrix. Side convection terms are weighed by a density difference interpolation scheme and half of the total contribution is associated with two elements connected along the same edge. He has also used the SINH (pronounced as 'cinch') [63] method. He has concluded that poor convection modelling can greatly influence the design process. Kim et al. [64] reconsidered the printed circuit board (PCB) cooling problem but had included mechanical constraints. Zhuang et al. [65] minimized the quadratic mean temperature using the level-set method. Yoo and Kim [66] considered three-dimensional cooling fins using the ECP method.

He and Liu [67] used the bi-directional evolutionary structural optimization (BESO). BESO is differentiated from ESO since BESO allows element addition which is not allowed in ESO. Using a uniform heat distribution problem, he compared the results with SIMP-based solutions. Special attention was given to the lack of intermediate elements in the ESO results, thus, easier manufacturability. Gao et al. [68] published another BESO paper and have considered both design-independent and -dependent loading. Design-dependent loading in this paper was defined as heat loads that vary whether or not material is present. In other words, without the presence of material heat cannot be generated. One case was presented to elaborate the difference and the effect of this assumption. Zhang and Liu [69] mentioned a new method for designing heat-conducting paths based on SIMP. This is related to a later publication mentioned in 2011. Yamasaki et al. [70] presented level-set method for both vibration and heat conduction problems.

Iga et al. [71] introduced convection and heat generation design-dependent effects. He has used a different homogenization approach (termed as the homogenization design method in their paper) and defined a hat function in which the convection boundary conditions are easily applied. The hat function serves as the boundaries between the solid and the void regions. Interest in this paper is given for the utilization of a surrogate model for several fin models for including a better representation of the convection condition. They have also utilized sequential linear programming (SLP) for the update process during topology optimization. They have presented several examples which exhibit the adverse effects if an inappropriate convection modelling is used. Marczak and Anflor [72] introduced the boundary element method (BEM) as an alternative to FEA and FVM. In their paper, topological derivative was used as the means to generate the optimal topologies. BEM is differentiated from FEA and FVM since it does not directly compute based on cells or elements. BEM is considered as 'mesh-free' methods. Although nodes are still present inside the modelled domain, they are treated more as 'recovery points'. Their examples and results were compared to the first ESO paper by Xie et al. in 1999. Dede [73] presented the use of COMSOL Multiphysics coupled with MATLAB for multiphysics topology optimization of heat and flow systems. Kim et al. [74] considered non-linear heat conduction and had designed structures based on the level set with topological derivatives. Pingen and Meyer [75] presented topology optimization for thermal transport.

Yoon [76] considered a sequential computational procedure to design heat-dissipating structures that considers forced convective heat transfer. A staggered approach was used where the flow field was solved first. Artificial damping force was introduced to the Navier-Stokes equation, which was similar to techniques used in immersed boundary methods (IBMs). A total of four material properties were interpolated in his implementation. He had utilized density-based approach and SIMP interpolations for the material properties. Kim et al. [77] compared results for different sensitivity analyses formulations. They have reported the computational time for the finite difference method and two different design sensitivity analyses (DSA). It was reported that a large difference in terms of performance was present between the direct differentiation method and the adjoint variable method (a factor of about 142). The SIMP method was utilized but was not mentioned explicitly. A 3D example was also provided in one of their examples which considered a single convection boundary condition. Dede [78] performed investigations on topology-optimized designs for impinging jets. Single-jet geometry was investigated from coupled thermal-fluidic simulations in a commercial software package. SIMP-based topology optimization was then performed in MATLAB with MMA. The result from the single impinging jet was made as basis for a textured surface geometry for a 3D slot jet. It is worth noting that the two-dimensional (2D) model was made under the assumptions of laminar flow and the 3D multi-jet structure is expected to fall within the turbulent regime. Zhuang et al. [79] utilized level-set method for the design of multi-material heat-conducting structures.

It can be said that the interest in heat-transfer topology optimization started in this time period. The papers presented in this time period were mostly dedicated and developed for heat transfer and design of heat exchangers. Investigations to include convection heat transfer as well as other design-dependent effects are also evident. Level-set method that is combined with the concept of topological derivative can be treated as state of the art during this time period. Also, 'mesh-less' topology optimization was introduced. SIMP has remained as a key method and its integration for FVM users has been mostly utilized.

4.3. 2011–2015

Yamada et al. [80] utilized the level-set method to include design-dependent effects such as convection boundary loading. A fictitious interface energy term was introduced for the design-dependent boundary conditions. Three-dimensional examples were given which clearly demonstrates clear and smooth optimal configurations. A regularization parameter was also utilized to tune the complexity of the optimal results. Convection loading was based on a fixed value. Zhang et al. [81] emphasized on the objective functional in topology optimization. It was highlighted that the cooling problem, as given by Bejan, needs to minimize the maximum temperature but most problems minimize the heat dissipation efficiency (termed as dissipation of heat potential capacity, DHTPC). A one-dimensional problem was revisited and a new objective of minimizing the geometric average temperature is presented in the context of a topology optimization problem. It was not explicitly mentioned what method was used and it is hypothesized that ESO was used due to the chosen problem and the well-defined boundaries of the optimal results (they have mentioned feasible direction method). Li et al. [82] had used the rational approximation of material property (RAMP) material interpolation scheme, OC based on density approach and a density filter was explored. Papoutsis-Kiachagias et al. [83] presented a constrained topology optimization for laminar and turbulent flows, including heat transfer.

Marck et al. [84] performed multi-objective optimization (MOO) using the SIMP method. The MOO was carried out with the two separate goals of minimizing the average temperature and minimizing the variance in the temperature. A very detailed and elaborate description of the FVM-based topology optimization was given. Tests regarding the mesh dependence, sensitivity and density filters as well as the heat transfer in the domain of the VP problem were carried out. Interestingly, this paper had obtained results which had discontinuity in the structure. Dede [85] optimized and designed multi-pass-branching microchannels with topology optimization as a tool. Gregersen et al. [86] considered finite volume-based topology optimization of coupled fluid dynamic and thermal conduction systems. Lee [87] completed his dissertation for topology optimization of convective cooling systems.

Koga et al. [88] demonstrated the complete product development cycle of a topology-optimized water-cooled heat exchanger. A fully coupled problem was solved using finite element method with some modifications to avoid numerical instabilities. A weighted logarithmic multi-objective function was used which contained a function to represent the power dissipation for the fluid flow and the heat dissipation for the heat-transfer problem. SINH was used in their implementation together with a weighed density filter. The heat exchanger was manufactured through electrical discharge machining and precision CNC milling. The experiments have matched well with the numerical simulations. It is worth noting that although the heat exchanger is a three-dimensional device, 2D modelling was employed for the topology optimization process. Burger et al. [89] explored the 3D solution for the volume to point (called volume to surface in this case) utilizing SIMP with method of moving asymptotes (MMAs) implementation. Full and partial Dirichlet boundary conditions were considered. In the partial Dirichlet boundary condition, only a square surface was given a fixed temperature. In the full Dirichlet boundary condition consideration, a volume of non-designable domain was set and the temperature conditions were set at the surfaces of a small volume before the allowable design domain. Different conductivity ratios varying volumetric constraints were explored as well as multiple boundary condition locations. Tree-like structures with four main branches extending to the corners of the design domain were the dominant optimal design features. Zhuang et al. [90] utilized triangular meshes on a transient heat conduction problem. Level-set method with topological derivative was used for the topology optimization process. Radial basis functions were used for defining the boundaries. A narrow band algorithm on the triangular mesh further improves the numerical efficiency. Dirker and Meyer [91] have performed performance tests for SIMP with MMA in an FVM setting. The VP problem was considered. It is worth noting that this work did not utilize any filtering techniques. A total of seven implementation cases were investigated. Six of the cases used predefined penalization parameters ranging from 1 to 5 in 0.5 intervals. Two volumetric constraints as well as three conductivity ratios were considered. Marck et al. [92] discussed topology optimization for heat and mass transfer problems in great detail for laminar flows. Jing et al. [93] has used BEM and level-set method for 2D heat conduction problems. Matsumori et al. [94] published fluid-thermal interaction problems under constant input power. Kontoleontos et al. [95] published an adjoint-based constrained topology optimization for viscous flows, including heat transfer.

Zhuang and Xiong [96] proposed a new compliance measure for transient heat conduction problems. They have suggested that the peak values of the given compliance during the time iterations are to be minimized. SIMP with MMA utilized for this study. The equivalent static load-based topology optimization for transient problems was deemed to be more practical and computationally efficient. Cheng and Chen [97] introduced a non-constrained formulation with a volume-of-solid (VOS) function to represent the bounds of the domain. This work is interesting since the objective function was defined as the heat-transfer index ($\dot{Q}/m$). Oevelen and Baelsmans [98] demonstrated solutions to a conjugate heat-transfer problem using a two-layer-reduced model to represent a full-3D solution. A test case considered Stokes flow and a highly branching flow network was obtained. SIMP interpolation scheme was used. They have acknowledged the artificial flow through a solid network if the penalization for the flow equations is not sufficient. They have further explored the effects of target temperature and bottom-layer thickness. Dede et al. [99] utilized topology optimization in the design and justification of novel structures that can shield, focus or reverse heat flux over a target domain. Anisotropic material constituents utilizing two-phase material microstructure descriptions for non-symmetric inclusions embedded in a matrix medium were manipulated to obtain the desired performance of the structures. Results were compared with a test case and extensions to arbitrary geometries were explored. Alexandersen et al. [100] made tremendous efforts for investigating heat topology optimization considering buoyancy forces. In this situation, the strongly coupled physics phenomena are very hard to model and thus making topology optimization for this kind of systems even more cumbersome. They have utilized SIMP interpolations for some of the material properties and density-based methods were sought for their implementations. They have demonstrated that effects of buoyancy affect the generated design significantly and have presented a natural convection heat exchanger as well as a buoyancy-driven micro-pump. In their paper, elements of large-scale simulations are already evident and they mentioned the difficulties they have encountered as well as their proposed solutions to overcome them. Lee [101] presented a multi-material heat conduction problem using a multiphase level set. Jing et al. [102] presented the topological sensitivity of the objective function on morphing boundaries.

Yaji et al. [103] utilized the level-set method to obtain the optimal design for a fully coupled thermo-fluidics problem. Tikhonov-based regularization scheme enabled the qualitative control for the geometric complexity of the generated structures. An optimization algorithm together with a smoothed Heaviside function was needed for the stabilization of the numerical computations. In this paper, both 2D and 3D examples were demonstrated with smooth and well-defined boundaries. Zhuang and Xiong [104] introduced additional temperature constraints on a defined region in the design domain. Their work had considered transient problems and had utilized the equivalent temperature field as a more effective means to solve the time-dependent finite element problem. In addition, they have utilized three materials in some of their examples using SIMP method. Jing et al. [105] utilized the BEM for the implementation of level-set method and considers design-dependent boundary conditions. The level-set method is used to represent the structural boundary and the boundary mesh for the BEM analysis is constructed on the iso-surface of the level-set function. Topological derivative is also utilized to make new holes. Cheng and Chen [106] utilized their volume-of-solid method for the topological design of the laminated metallic composite materials arranged in two predefined configurations. Similar to the previous paper, they have presented two new very interesting objective functions ($\dot{Q}/V$ and $\dot{Q}/USD$). Dede et al. [107] recently demonstrated a complete product cycle development for developing a forced air-cooled heat-sink-made additive manufacturing (AM). They have applied SIMP-based topology optimization and had utilized a modified hat function to define the heat convection loading surface for their problem. A parabolic distribution of the heat-transfer coefficient was assumed in relation to the forced air cooling. Two-dimensional models are first tested and compared to some common heat-sink geometries found in the market. A quarter of a 3D model was then implemented and volume reconstruction was also mentioned to obtain a water-tight design suitable for additive manufacturing. Experiments were then conducted and the topology-optimized structures are compared with the commercially available design. Results showed that the designed heat sink performed better compared to other heat sinks but due to the inferior material properties and porous structure of the AM-produced design, it was not performing as to its numerical design specifications. Alexandersen et al. [108] recently published the culmination of their buoyancy flow works by implementing a large-scale three-dimensional model of designed heat sinks. A total of 16.38 million design elements with 83.08 million degrees of freedom were solved in one of their examples for a passive heat-sink cooler for light-emitting diodes (LEDs). Lohan et al. [109] presented generative design algorithms for heat conduction. A dissertation study utilizing boundary element method was recently finished by Jing [109]. Dede [110] designed and fabricated a multi-device single-phase-branching microchannel cold plate.

In this time period, highlight is given to product design cycles and actual realization of topology-optimized designs. It is also evident that trends are going for incorporation of fluid flow either directly (through coupled analysis of both the fluid and heat-transfer domains) or indirectly (through convection boundary conditions). Interests for transient problems have also re-emerged with techniques such as the equivalent temperature field being utilized to reduce the burden of the finite analysis for the governing equations of the system. Level-set method is also evolving rapidly by utilizing other techniques such as topological derivative and BEM to make up for their weak points. Density-based methods, especially SIMP, are still staple with most of the works for 3D modelling and thermo-fluidic systems. Massive implementations with millions of DOFs are also slowly being realized, mostly utilizing density-based methods. As an additional foresight, it can be mentioned that none of the above works have considered radiation effects, though some problem formulations can accommodate radiation by utilizing the convection form of radiation. In the future, this work could be sought but would pose the problem for the discretized method of properly identifying cavities and formations inside the evolving domain. View factor computation is also one complication which would be very expensive to perform since radiating boundaries would change in each iteration.

5. Conclusion

In this chapter, we have re-introduced topology optimization with special focus on the progress of heat exchanger design over the past two decades. We have first given an overview of its historical background in terms of structural topology optimization. We have then conceptually introduced the different methods developed over the years in topology optimization. Learning references for each of the methods mentioned, together with MATLAB codes, were cited and is expected to help those who are interested in further learning and investigating topology optimization. A chronological review highlighting the different progress over the years related to heat exchanger design was also given.

Novel heat-transfer structures are still being realized to further drive design performance to its limits. Topology optimization, as a physics-based and automated layout optimization method, will indeed serve as a valuable design tool for heat-transfer systems. Heat exchanger designs arising from topology optimization has now been realized and continuous efforts are still being made to further improve both methods and implementation. Topology optimization is expected to play a bigger role in the coming years for heat exchanger design.

Nomenclature

Abbreviations

2D/3D	Two-dimensional/three-dimensional
BEM	Boundary element method
BESO	Bi-directional evolutionary structural optimization
DHTPC	Dissipation of heat-transfer potential capacity
DOF	Degrees of freedom
DSA	Design sensitivity analysis
ECP	Element connectivity parameterization
ESO	Evolutionary structural optimization
FEA	Finite element analysis
FEM	Finite element method
FVM	Finite volume method
IBM	Immersed boundary method
MMA	Method of moving asymptotes

MOO	Multi-objective optimization
OC	Optimality criteria
RAMP	Rational approximation of material properties
SIMP	Solid isotropic material with penalization
SLP	Sequential linear programming
SQP	Sequential quadratic programming
TO	Topology optimization
VP	Volume-to-point problem

Symbols and variables (in the order of appearance)

μ	Dimension control parameter in homogenization method
θ	Orientation control parameter in homogenization method
ρ	Density variable for density-based topology optimization, material density
c	Standard variable for the design objective or compliance, material-specific heat
T	Temperature
k	Thermal conductivity
q	Heat flux
n	Boundary normal vector
h	Convection heat-transfer rate
Ω	Domain
t	Time
Γ	Boundaries
v	Virtual temperature field
a	Energy bi-linear form
l	Thermal load linear form
x	Design variable
V	Volume

$\tilde{T}$	Test temperature vector
H	Sobolev space
$\mathbf{T}$	Temperature vector in heat-transfer TO
$\mathbf{K}$	Global stiffness matrix for FEA
$\mathbf{Q}$	Applied thermal load in heat-transfer TO
$\mathbf{k}$	Global stiffness matrix for FEA
λ	Langrangian multiplier

Subscripts

$0, 1, 2, .., i$	Standard discrete numerical counter
e	Element in discretization
mat	Material
min	Minimize/minimummax
t	Time, to imply transient case in derivation
v	Per unit volume, in derivation
b	Imposed boundary condition, w/temperature in derivation
eff	Effective, used with thermal conductivity, k
p	Penalty parameter for SIMP

Author details

Mark Christian E. Manue[1,*] and Po Ting Lin[2]

*Address all correspondence to: marchm.090407@gmail.com

1 School of Mechanical and Manufacturing Engineering, Mapua Institute of Technology, Manila, Philippines

2 Mechanical Engineering Department, Chung Yuan Christian University, Chungli City, Taoyuan, Taiwan

References

[1] Bourdin, B. and A. Chambolle, *The Phase-Field Method in Optimal Design*, in IUTAM Symposium on Topological Design Optimization of Structures, Machines and Materials:

Status and Perspectives, M.P. Bendsøe, N. Olhoff, and O. Sigmund, Editors. 2006, Springer Netherlands: Dordrecht. p. 207–215.

[2] Pedersen, C.B.W. and P. Allinger, Industrial Implementation and Applications of Topology Optimization and Future Needs, in IUTAM Symposium on Topological Design Optimization of Structures, Machines and Materials: Status and Perspectives, M.P. Bendsøe, N. Olhoff, and O. Sigmund, Editors. 2006, Springer Netherlands: Dordrecht. p. 229–238.

[3] Hsu, K.S., Manuel, M.C.E. and Lin, P.T., *3D printed swirled liquid impinging cooler*, in *2014 Innovative Design Contest, IDC'14*. 2014: Taipei, Taiwan.

[4] Gross, M., *Now more than ever: computational thinking and a science of design (< Special Issue> What is "What's the Design"?)*. デザイン学研究. 特集号, 2009. **16**(2): p. 50–54.

[5] Aage, N., et al., *Interactive topology optimization on hand-held devices*. Structural and Multidisciplinary Optimization, 2013. **47**(1): p. 1–6.

[6] Aage, N., *TopOpt 3D UserGuide*. 2014.

[7] Michell, A.G.M., *LVIII. The limits of economy of material in frame-structures*. The London, Edinburgh, and Dublin Philosophical Magazine and Journal of Science, 1904. **8**(47): p. 589–597.

[8] Cox, H.L., *The theory of design*. Aeronautical Research Council Report No. 19791, 1958, Great Britain.

[9] Cox, H.L., *The design of structures of least weight*. 1965, New York, NY: Pergamon Press.

[10] Owen, J. and Brynmor, B., *The analysis and design of light structures*. 1965, New York, NY: Elsevier.

[11] Hemp, W.S., *Optimum structures*. 1973, Oxford: Clarendon Press.

[12] Bendsøe, M.P. and Kikuchi, N., *Generating optimal topologies in structural design using a homogenization method*. Computer Methods in Applied Mechanics and Engineering, 1988. **71**(2): p. 197–224.

[13] Bendsoe, M.P. and Sigmund, O., *Topology optimization: theory, methods, and applications*. 2013, Berlin: Springer-Verlag.

[14] Eschenauer, H.A. and Olhoff, N., *Topology optimization of continuum structures: A review*. Applied Mechanics Reviews, 2001. **54**(4): p. 331–390.

[15] Novotny, A.A., et al., *Topological sensitivity analysis*. Computer Methods in Applied Mechanics and Engineering, 2003. **192**(7): p. 803–829.

[16] Xie, Y.-M. and Steven, G.P., *Basic evolutionary structural optimization*. 1997, London: Springer-Verlag.

[17] Huang, X. and Xie, M., *Evolutionary topology optimization of continuum structures: methods and applications*. 2010, United Kingdom: John Wiley & Sons.

[18] Zhou, M. and Rozvany, G., *On the validity of ESO type methods in topology optimization*. Structural and Multidisciplinary Optimization, 2001. **21**(1): p. 80–83.

[19] Rozvany, G., *A critical review of established methods of structural topology optimization*. Structural and Multidisciplinary Optimization, 2009. **37**(3): p. 217–237.

[20] Huang, X. and Xie, Y.-M., *A further review of ESO type methods for topology optimization*. Structural and Multidisciplinary Optimization, 2010. **41**(5): p. 671–683.

[21] Osher, S. and Sethian, J.A., *Fronts propagating with curvature-dependent speed: algorithms based on Hamilton-Jacobi formulations*. Journal of Computational Physics, 1988. **79**(1): p. 12–49.

[22] Sethian, J.A., *Level set methods and fast marching methods: evolving interfaces in computational geometry, fluid mechanics, computer vision, and materials science*. Vol. 3. 1999, United Kingdom: Cambridge University Press.

[23] Osher, S. and Fedkiw, R., *Level set methods and dynamic implicit surfaces*. Vol. 153. 2006, New York: Springer Science.

[24] Sethian, J.A. and Wiegmann, A., *Structural boundary design via level set and immersed interface methods*. Journal of Computational Physics, 2000. **163**(2): p. 489–528.

[25] Osher, S.J. and Santosa, F., *Level set methods for optimization problems involving geometry and constraints: I. frequencies of a two-density inhomogeneous drum*. Journal of Computational Physics, 2001. **171**(1): p. 272–288.

[26] van Dijk, N.P., et al., *Level-set methods for structural topology optimization: a review*. Structural and Multidisciplinary Optimization, 2013. **48**(3): p. 437–472.

[27] Makhija, D. and Maute, K., *Numerical instabilities in level set topology optimization with the extended finite element method*. Structural and Multidisciplinary Optimization, 2014. **49**(2): p. 185–197.

[28] Dunning, P.D. and Kim, A.H., *A new hole insertion method for level set based structural topology optimization*. International Journal for Numerical Methods in Engineering, 2013. **93**(1): p. 118–134.

[29] Challis, V.J., *A discrete level-set topology optimization code written in Matlab*. Structural and multidisciplinary optimization, 2010. **41**(3): p. 453–464.

[30] Chen, L.-Q., *Phase-field models for microstructure evolution*. Annual Review of Materials Research, 2002. **32**(1): p. 113–140.

[31] Takezawa, A., Nishiwaki, S. and Kitamura, M., *Shape and topology optimization based on the phase field method and sensitivity analysis*. Journal of Computational Physics, 2010. **229**(7): p. 2697–2718.

[32] Wang, M.Y. and Zhou, S., *Phase field: a variational method for structural topology optimization*. Computer Modelling in Engineering & Sciences, 2004. **6**(6): p. 547–566.

[33] Bourdin, B. and Chambolle, A., *The phase-field method in optimal design. in IUTAM Symposium on Topological Design Optimization of Structures, Machines and Materials*. 2006, Springer Netherlands: Dordrecht. p. 207–215.

[34] Pedersen, C.B.W. and P. Allinger, *Industrial Implementation and Applications of Topology Optimization and Future Needs, in IUTAM Symposium on Topological Design Optimization of Structures, Machines and Materials*. 2006, Springer Netherlands: Dordrecht. p. 229–238.

[35] Andreassen, E., et al., *Efficient topology optimization in MATLAB using 88 lines of code*. Structural and Multidisciplinary Optimization, 2011. **43**(1): p. 1–16.

[36] Liu, K. and Tovar, A., *An efficient 3D topology optimization code written in Matlab*. Structural and Multidisciplinary Optimization, 2014. **50**(6): p. 1175–1196.

[37] Aage, N., Andreassen, E. and Lazarov, B.S., *Topology optimization using PETSc: An easy-to-use, fully parallel, open source topology optimization framework*. Structural and Multidisciplinary Optimization, 2015. **51**(3): p. 565–572.

[38] Thomson Reuters. *Citation Reports*. Dec. 2015; Available from: http://apps.webofknowledge.com/.

[39] Rodrigues, H. and Fernandes, P., *Topology optimal design of thermoelastic structures using a homogenization method*. Control and Cybernetics, 1994. **23**(3).

[40] Rodrigues, H. and Fernandes, P., *A material based model for topology optimization of thermoelastic structures*. International Journal for Numerical Methods in Engineering, 1995. **38**(12): p. 1951–1965.

[41] Jog, C., *Distributed-parameter optimization and topology design for non-linear thermoelasticity*. Computer Methods in Applied Mechanics and Engineering, 1996. **132**(1): p. 117–134.

[42] Bejan, A., *Constructal-theory network of conducting paths for cooling a heat generating volume*. International Journal of Heat and Mass Transfer, 1997. **40**(4): p. 799–816.

[43] Ledezma, G.A., Bejan, A. and Errera, M.R., *Constructal tree networks for heat transfer*. Journal of Applied Physics, 1997. **82**(1): p. 89–100.

[44] Li, Q., et al., *Shape and topology design for heat conduction by evolutionary structural optimization*. International Journal of Heat and Mass Transfer, 1999. **42**(17): p. 3361–3371.

[45] Turteltaub, S., *Optimal material properties for transient problems*. Structural and Multidisciplinary Optimization, 2001. **22**(2): p. 157–166.

[46] Haslinger, J., et al., *Optimization of conducting structures by using the homogenization method*. Structural and Multidisciplinary Optimization, 2002. **24**(2): p. 125–140.

[47] Cheng, X., Li, Z. and Guo, Z., *Constructs of highly effective heat transport paths by bionic optimization*. Science in China Series E: Technological Sciences, 2003. **46**(3): p. 296–302.

[48] Borrvall, T. and Petersson, J., *Topology optimization of fluids in Stokes flow*. International Journal for Numerical Methods in Fluids, 2003. **41**(1): p. 77–107.

[49] Bendsoe, M.P. and Sigmund, O., *Topology optimization: theory, methods and applications*. 2003, Springer: Berlin.

[50] Guo, Z., Cheng, X. and Xia, Z., *Least dissipation principle of heat transport potential capacity and its application in heat conduction optimization*. Chinese Science Bulletin, 2003. **48**(4): p. 406–410.

[51] Li, Q., et al., *Evolutionary topology optimization for temperature reduction of heat conducting fields*. International Journal of Heat and Mass Transfer, 2004. **47**(23): p. 5071–5083.

[52] Alberto, E. and Sigmund, O., *Topology optimization of multiple physics problems modelled by Poisson's equation*. Latin American Journal of Solids and Structures, 2004. **1**(2): p. 169–184.

[53] Ha, Y., Kim, M.-G. and Cho, S., *Topology optimization of nonlinear heat conduction problems using adjoint design sensitivity analysis method in Proceedings of the 10th AIAA/ISSMO Multidisciplinary Analysis and Optimization Conference*. 2004, Albany, New York.

[54] Moon, H., Kim, C. and Wang, S., *Reliability-based topology optimization of thermal systems considering convection heat transfer*. in *10th AIAA/ISSMO multidisciplinary analysis and optimization conference*. 2004. New York.

[55] Ho Yoon, G. and Young Kim, Y., *The element connectivity parameterization formulation for the topology design optimization of multiphysics systems*. International Journal for Numerical Methods in Engineering, 2005. **64**(12): p. 1649–1677.

[56] Ha, S.-H. and Cho, S., *Topological shape optimization of heat conduction problems using level set approach*. Numerical Heat Transfer, Part B: Fundamentals, 2005. **48**(1): p. 67–88.

[57] Gersborg-Hansen, A., Bendsøe, M.P. and Sigmund, O., *Topology optimization of heat conduction problems using the finite volume method*. Structural and Multidisciplinary Optimization, 2006. **31**(4): p. 251–259.

[58] Donoso, A., *Numerical simulations in 3D heat conduction: minimizing the quadratic mean temperature gradient by an optimality criteria method*. SIAM Journal on Scientific Computing, 2006. **28**(3): p. 929–941.

[59] Zhuang, C., Xiong, Z. and Ding, H., *A level set method for topology optimization of heat conduction problem under multiple load cases*. Computer Methods in Applied Mechanics and Engineering, 2007. **196**(4): p. 1074–1084.

[60] Xu, X., Liang, X. and Ren, J., *Optimization of heat conduction using combinatorial optimization algorithms*. International Journal of Heat and Mass Transfer, 2007. **50**(9): p. 1675–1682.

[61] Mathieu-Potvin, F. and Gosselin, L., *Optimal conduction pathways for cooling a heat-generating body: A comparison exercise*. International Journal of Heat and Mass Transfer, 2007. **50**(15): p. 2996–3006.

[62] Bruns, T.E., *Topology optimization of convection-dominated, steady-state heat transfer problems*. International Journal of Heat and Mass Transfer, 2007. **50**(15): p. 2859–2873.

[63] Bruns, T.E., *A reevaluation of the SIMP method with filtering and an alternative formulation for solid–void topology optimization*. Structural and Multidisciplinary Optimization, 2005. **30**(6): p. 428–436.

[64] Kim, M.S., et al., *Topology optimization of a PCB substrate considering mechanical constraints and heat conductivity*. Journal of Mechanical Science and Technology, 2007. **21**(12): p. 2041–2047.

[65] Zhuang, C., Xiong, Z. and Ding, H., *Minimizing the quadratic mean temperature gradient for the heat-conduction problem using the level-set method*. in Institution of Mechanical Engineers, Part C: Journal of Mechanical Engineering Science. 2007. **221**(2): p. 235–248.

[66] Yoo, S.M. and Kim, Y.Y., *The topology optimization of three-dimensional cooling fins by the internal element connectivity parameterization method in Proceedings of the KSME 2007 spring annual meeting*. 2007, Republic of Korea: KSME.

[67] He, D. and Liu, S., *BESO method for topology optimization of structures with high efficiency of heat dissipation*. International Journal for Simulation and Multidisciplinary Design Optimization, 2008. **2**(1): p. 43–48.

[68] Gao, T., et al., *Topology optimization of heat conduction problem involving design-dependent heat load effect*. Finite Elements in Analysis and Design, 2008. **44**(14): p. 805–813.

[69] Zhang, Y. and Liu, S., *Design of conducting paths based on topology optimization*. Heat and Mass Transfer, 2008. **44**(10): p. 1217–1227.

[70] Yamasaki, S., et al. *A new structural optimization method based on the level set method for vibration problems and heat conduction problems*. in 12th AIAA/ISSMO Multidisciplinary Analysis and Optimization Conference. 2008.

[71] Iga, A., et al., *Topology optimization for thermal conductors considering design-dependent effects, including heat conduction and convection*. International Journal of Heat and Mass Transfer, 2009. **52**(11): p. 2721–2732.

[72] Anflor, C.T.M. and Marczak, R.J., *A boundary element approach for topology design in diffusive problems containing heat sources*. International Journal of Heat and Mass Transfer, 2009. **52**(19): p. 4604–4611.

[73] Dede, E.M. *Multiphysics topology optimization of heat transfer and fluid flow systems*. in *COMSOL Users Conference*. 2009. Boston.

[74] Kim, M.-G., Ha, S.-H. and Cho, S., *Level set-based topological shape optimization of nonlinear heat conduction problems using topological derivatives*. Mechanics Based Design of Structures and Machines, 2009. **37**(4): p. 550–582.

[75] Pingen, G. and Meyer, D., *Topology optimization for thermal transport*. in ASME 2009 Fluids Engineering Division Summer Meeting. 2009. ASME.

[76] Yoon, G.H., *Topological design of heat dissipating structure with forced convective heat transfer*. Journal of Mechanical Science and Technology, 2010. **24**(6): p. 1225–1233.

[77] Kim, M.-G., Kim, J.-H. and Cho, S., *Topology design optimization of heat conduction problems using adjoint sensitivity analysis method*. Computational Structural Engineering Institute of Korea, 2010. **23**(6): p. 683–691.

[78] Dede, E.M. *Multiphysics optimization, synthesis, and application of jet impingement target surfaces*. in 12th IEEE Intersociety Conference on Thermal and Thermomechanical Phenomena in Electronic Systems (ITherm). 2010. IEEE.

[79] Zhuang, C., Xiong, Z. and Ding, H., *Topology optimization of multi-material for the heat conduction problem based on the level set method*. Engineering Optimization, 2010. **42**(9): p. 811–831.

[80] Yamada, T., Izui, K. and Nishiwaki, S., *A level set-based topology optimization method for maximizing thermal diffusivity in problems including design-dependent effects*. Journal of Mechanical Design, 2011. **133**(3): p. 031011.

[81] Zhang, Y., Qiao, H., and Liu S., Design of the heat conduction structure based on the topology optimization in Developments in Heat Transfer. 2011, Croatia: INTECH Open Access Publisher. p. 523–536.

[82] Li, J.C., et al. *Heat conduction structural topology optimization based on RAMP*. in *Applied Mechanics and Materials*. 2011. Trans. Tech.

[83] Papoutsis-Kiachagias, E., et al., *Constrained topology optimization for laminar and turbulent flows, including heat transfer, in Evolutionary and Deterministic Methods for Design, Optimization and Control*. 2011, EUROGEN, Capua, Italy.

[84] Marck, G., et al., *Topology optimization using the SIMP method for multiobjective conductive problems*. Numerical Heat Transfer, Part B: Fundamentals, 2012. **61**(6): p. 439–470.

[85] Dede, E.M., *Optimization and design of a multipass branching microchannel heat sink for electronics cooling*. Journal of Electronic Packaging, 2012. **134**(4): p. 041001.

[86] Gregersen, M.M., Evgrafov, A. and Sorensen, M.P., *Finite volume based topology optimization of coupled fluid dynamic and thermal conduction systems*, in *Conference on Industrial Design Optimisation for Fluid Flow*. 2012, FLOWHEAD: Munich.

[87] Lee, K., *Topology optimization of convective cooling system designs*. 2012, Diss. The University of Michigan.

[88] Koga, A.A., et al., *Development of heat sink device by using topology optimization*. International Journal of Heat and Mass Transfer, 2013. **64**: p. 759–772.

[89] Burger, F.H., Dirker, J. and Meyer, J.P., *Three-dimensional conductive heat transfer topology optimisation in a cubic domain for the volume-to-surface problem*. International Journal of Heat and Mass Transfer, 2013. **67**: p. 214–224.

[90] Zhuang, C., Xiong, Z. and Ding, H., *Topology optimization of the transient heat conduction problem on a triangular mesh*. Numerical Heat Transfer, Part B: Fundamentals, 2013. **64**(3): p. 239–262.

[91] Dirker, J. and Meyer, J.P., *Topology optimization for an internal heat-conduction cooling scheme in a square domain for high heat flux applications*. Journal of Heat Transfer, 2013. **135**(11): p. 111010.

[92] Marck, G., Nemer, M. and Harion, J.-L., *Topology optimization of heat and mass transfer problems: laminar flow*. Numerical Heat Transfer, Part B: Fundamentals, 2013. **63**(6): p. 508–539.

[93] JING, G., et al., *Topology optimization for 2D heat conduction problems using boundary element method and level set method*. Transactions of JASCOME, 2013. **13**: p. 6.

[94] Matsumori, T., et al., *Topology optimization for fluid–thermal interaction problems under constant input power*. Structural and Multidisciplinary Optimization, 2013. **47**(4): p. 571–581.

[95] Kontoleontos, E., et al., *Adjoint-based constrained topology optimization for viscous flows, including heat transfer*. Engineering Optimization, 2013. **45**(8): p. 941–961.

[96] Zhuang, C. and Xiong, Z., *A global heat compliance measure based topology optimization for the transient heat conduction problem*. Numerical Heat Transfer, Part B: Fundamentals, 2014. **65**(5): p. 445–471.

[97] Cheng, C.-H. and Chen, Y.-F., *Topology optimization of heat conduction paths by a non-constrained volume-of-solid function method*. International Journal of Thermal Sciences, 2014. **78**: p. 16–25.

[98] Van Oevelen, T. and Baelmans, M., *Application of topology optimization in a conjugate heat transfer problem*. in 1st International Conference on Engineering and Applied Sciences Optimization. 2014. OPT-i 2014.

[99] Dede, E.M., Nomura, T. and Lee, J., *Thermal-composite design optimization for heat flux shielding, focusing, and reversal*. Structural and Multidisciplinary Optimization, 2014. **49**(1): p. 59–68.

[100] Alexandersen, J., et al., *Topology optimisation for natural convection problems*. International Journal for Numerical Methods in Fluids, 2014. **76**(10): p. 699–721.

[101] Lee, D., *Topology optimization of multi-material with heat conduction problem in Proceedings of the 5th European Conference of Computer Science (ECCS 2014)*. 2014, Switzerland: WSEAS press. p. 54–55

[102] Jing, G., et al., *Topological sensitivity of the objective function defined on morphing boundaries of two-dimensional heat conduction problems*. Boundary Elements and Other Mesh Reduction Methods, 2014. **57**: p. 3.

[103] Yaji, K., et al., *A topology optimization method for a coupled thermal–fluid problem using level set boundary expressions*. International Journal of Heat and Mass Transfer, 2015. **81**: p. 878–888.

[104] Zhuang, C. and Xiong, Z., *Temperature-constrained topology optimization of transient heat conduction problems*. Numerical Heat Transfer, Part B: Fundamentals, 2015. **68**(4): p. 366–385.

[105] Jing, G., et al., *Level set-based topology optimization for 2D heat conduction problems using BEM with objective function defined on design-dependent boundary with heat transfer boundary condition*. Engineering Analysis with Boundary Elements, 2015. **61**: p. 61–70.

[106] Cheng, C.-H. and Chen, Y.-F., *Topology optimization of conduction path in laminated metals composite materials*. International Journal of Thermal Sciences, 2015. **96**: p. 183–190.

[107] Dede, E.M., Joshi, S.N. and Zhou, F., *Topology Optimization, Additive Layer Manufacturing, and Experimental Testing of an Air-Cooled Heat Sink*. Journal of Mechanical Design, 2015. **137**(11): p. 111403.

[108] Alexandersen, J., *Topology optimisation of passive coolers for light-emitting diode lamps*. 11th World Congress of Structural and Multidisciplinary Optimisation, 2015.

[109] Jing, G., *A study on topology optimization for two-dimensional steady-state heat conduction problems using Boundary Element Method*. 2015, Nagoya University: Japan.

[110] Dede, E.M., Design optimization of a multi-device single-phase branching microchannel cold plate. Electronics Cooling Magazine. 2015.(3): p. 14–17.

7

Measurement of Transient Fluid Temperature in the Heat Exchangers

Magdalena Jaremkiewicz

Abstract

In the chapter, a method for measuring the transient temperature of the flowing fluid based on time temperature changes of the thermometer is described. In the presented method, the thermometer is considered as an inertial system of first and second order. To reduce the influence of random errors in the temperature measurement, the local polynomial approximation based on nine points is used. As a result, the first and second derivatives of a temperature, which indicate how the temperature of the thermometer varies over time, are determined very accurately. Next, the time constant is defined as a function of fluid velocity for sheathed thermocouples with different diameters. The applicability of the presented method is demonstrated on real data in the experiment. The air temperature is estimated from measurements carried out by the three thermocouples having different outer diameters when the air velocity varied in time. A comparison of the computed temperatures of air gives confidence to the accuracy of the presented method. The method presented in this chapter for measuring the transient temperature of the fluid can be used for the online monitoring of fluid temperature change with time.

Keywords: temperature measurement, transient conditions, first-order model, second-order model, time constant

1. Introduction

Under steady-state conditions, when the fluid temperature is constant, temperature measurement can be performed with the high degree of accuracy owing to the absence of damping and time lag. However, when fluid temperature varies rapidly, for example, during start-up, there is a significant difference between the actual and the measured temperature of the fluid. These differences occur because it takes time for heat to transfer through the heavy housing to the temperature sensor.

Most of the scientific publications concerning the measurement of temperature mainly discuss the problem of temperature measurement at steady state [1–9]. Only the step response of thermometers is studied to estimate the dynamic error of the temperature measurement. Few studies refer to the measurement of the transient fluid temperature, despite the high practical significance of the problem [10–13].

An example is the measurement of transient temperature steam or flue gases in power plants, which is very difficult. Measured temperature differs significantly from the real temperature of the fluid, which is caused by massive thermowells of the thermometers and their low heat transfer coefficients. Some thermometers may have a time constant of 3 min or more, which makes the implementation of a single temperature measurement requiring about 15 min [13].On the other side, some designs of thermometers need more than one time constant to describe the unsteady response of temperature sensor inserted into the thermowell. Measuring the fluid temperature in a controlled process may require the knowledge of two or three time constants, which describe the transient response of the thermometer [14].

The problem of dynamic errors in temperature measurements becomes particularly important in superheated steam temperature control systems, which use injection coolers (spray attemperators). Due to the large inertia of the thermometer, the measurements of transient fluid temperature can be inaccurate, which causes the automatic control system of the superheated steam temperature not to work properly.

A similar problem occurs in measuring the exhaust gas temperature as the time constants of the thermometer and the time delay are large.

2. Mathematical models of the thermometers

Generally, thermometers are considered as elements with lumped thermal capacity. It is assumed that the temperature of the thermometer is only a function of time and temperature differences inside it are neglected.

Based on these assumptions, the mathematical model of the thermometer is the differential equation describing the inertial system of the first order [14]:

$$\tau \frac{\mathrm{d}T(t)}{\mathrm{d}t} + T(t) = T_f(t) \tag{1}$$

where τ—time constant of the thermometer in the first-order model in s, $T(t)$—measured temperature in °C and $T_f(t)$—fluid temperature in °C.

The time constant is expressed by formula:

$$\tau = \frac{m_t c_t}{h_t A_t} \tag{2}$$

where m_t—thermocouple mass in kg, c_t—average-specific heat of the thermocouple in J/(kg·K), h_t—heat transfer coefficient on the outer surface of the thermocouple in W/(m^2·K) and A_t—outer surface area of the thermocouple in m^2.

The ordinary differential Eq. (1) was solved for the initial condition:

$$T(0) = T_0 = 0 \tag{3}$$

where T_0 signifies initial thermometer temperature in °C.

The initial problem, (Eqs. (1) and (3)), was solved using the Laplace transformation. The operator transmittance $G(s)$ then assumes the following form:

$$G(s) = \frac{T(s)}{T_f(s)} = \frac{1}{\tau s + 1} \tag{4}$$

where $T(s)$—Laplace transform of the thermometer temperature, $T_f(s)$—Laplace transform of the fluid temperature and s—complex variable.

For the step increase of the fluid temperature from $T_0 = 0$°C to the constant value T_s, the Laplace transform of the fluid temperature assumes the form $T_f(s) = T_s/s$ and the transmittance formula can be simplified to

$$\frac{T(s)}{T_s} = \frac{1}{s(\tau s + 1)} \tag{5}$$

After writing Eq. (5) in the form:

$$\frac{T(s)}{T_s} = \frac{1}{s} - \frac{1}{s + \frac{1}{\tau}} \tag{6}$$

it is easy to find the inverse Laplace transformation and determine the thermometer temperature as a function of time:

$$u(t) = \frac{T(t) - T_0}{T_s - T_0} = 1 - \exp\left(-\frac{t}{\tau}\right) \tag{7}$$

where $u(t)$ denotes unit-step response of the thermometer.

For structurally complex thermometers that measure the temperature of fluid under high pressure, the accuracy of the first-order model (1) is inadequate. The cross section of the temperature sensor placed in the massive housing is shown in **Figure 1**. This example will be analysed to show that the thermometers can be modelled as second-order inertial systems [14].

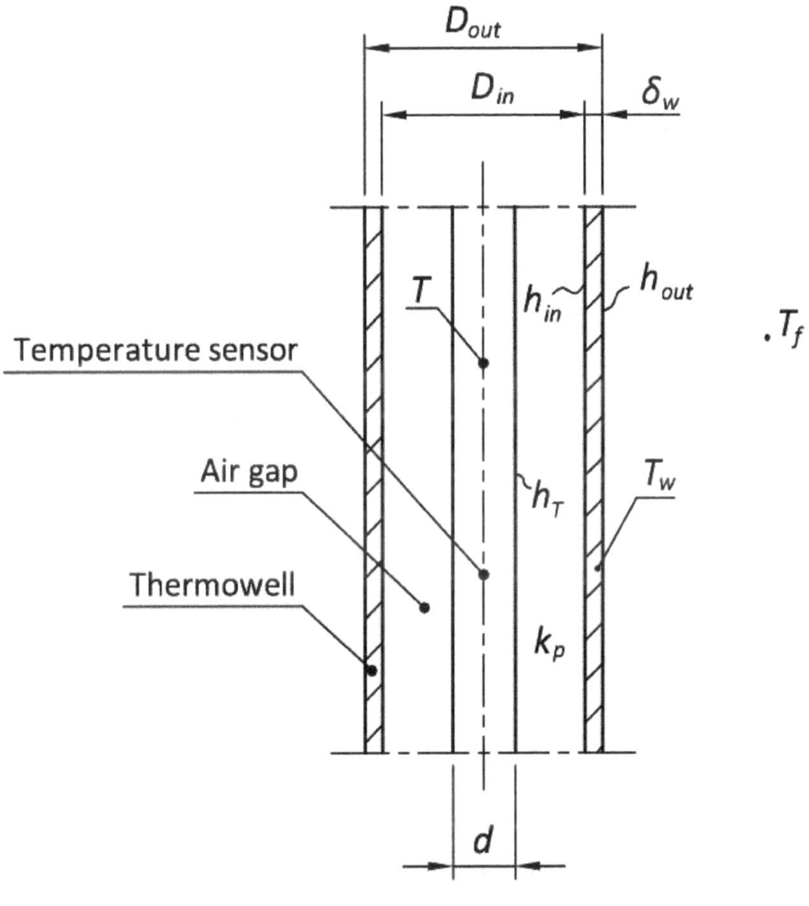

Figure 1. Part of the cross section of the temperature sensor together with the housing [14].

Air gap can occur between the outer thermowell and the temperature sensor (**Figure 1**). The discussion assumes that the thermal capacity $c\rho$ is neglected because of its small value.

Introducing the overall heat transfer coefficient u_{in} referenced to the inner surface of the housing

$$\frac{1}{u_{in}} = \frac{1}{h_{in}} + \frac{(1 + D_{in}/d)(D_{in} - d)}{4k_w} + \frac{D_{in}}{d} \cdot \frac{1}{h_T} \tag{8}$$

and accounting for the radiation heat transfer from the housing to the inner sensor, the heat balance equation for the sensor located within the housing assumes the form:

$$A\rho c\frac{dT}{dt} = P_{in}u_{in}\left(T_w - T\right) + C\left(T_w^4 - T^4\right) \tag{9}$$

where the symbol C denotes:

$$C = \frac{\pi d\sigma}{\frac{1}{\varepsilon_T} + \frac{d}{D_{in}}\left(\frac{1}{\varepsilon_w} - 1\right)} \tag{10}$$

In Eqs. (8) and (9), the symbols h_{in} and h_T represent heat transfer coefficient on the inner surface of the housing and the outer surface of the thermocouple in W/(m^{2}K), respectively, D_{in} inner diameter of the housing in m, d outer diameter of the thermocouple in m, k_o housing thermal conductivity in W/(m·K), A surface area of the thermocouple cross section in m^2, ρ average density of the thermocouple in kg/m^3, c average-specific heat of the thermocouple in J/(kg·K), P_{in} perimeter of the internal surface of the housing in m, T_w housing temperature in °C, $\sigma = 5.67 \times 10^{-8}$ W/(m^2·K^4) Stefan-Boltzmann constant and ε_w and ε_T emissivity of the inner housing and the thermocouple surface, respectively.

The convection and conduction of heat transfer between the fluid and the thermometer housing are characterised by the overall heat transfer coefficient u_{out} referenced to the outer housing surface:

$$\frac{1}{u_{out}} = \frac{1}{h_{out}} + \frac{1 + D_{out}/D_{in}}{2} \cdot \frac{\delta_w}{\lambda_w} \tag{11}$$

where h_{out} is the heat transfer coefficient on the outer surface of the housing in W/(m^2·K), D_{out} outer diameter of the housing in m, δ_w housing thickness in m and k_w housing thermal conductivity in W/(m·K).

The formulas (8) and (11) for the overall heat transfer coefficients were derived using the basic principles of heat transfer [2, 4, 8]. The heat transfer equation for the housing (thermowell) can be written in the following form:

$$A_w\rho_w c_w\frac{dT_w}{dt} = P_{out}u_{out}\left(T_f - T_w\right) - P_{in}u_{in}\left(T_w - T\right) - C\left(T_w^4 - T^4\right) \tag{12}$$

where A_w is the surface area of the housing cross section in m^2, ρ_w average density of the housing in kg/m^3, c_w average-specific heat of the housing, J/(kg·K) and P_{out} perimeter of the external surface of the housing in m.

Further analysis of the heat exchange between the housing of the thermometer and the temperature sensor omits the heat transfer by radiation [14]. This is possible when the gap between the thermowell and the temperature sensor is filled with a non-transparent substance or if one of the two emissivities ε_w and ε_T is near to zero.

Transforming Eq. (9), we get:

$$T_w = \frac{A\rho c}{P_{in}u_{in}}\frac{\mathrm{d}T}{\mathrm{d}t} + T \tag{13}$$

Substituting Eq. (13) into Eq. (12) yields:

$$\frac{(A_w\rho_w c_w)(A\rho c)}{(P_{in}u_{in})(P_{out}u_{out})}\frac{\mathrm{d}^2T}{\mathrm{d}t^2} + \frac{A_w\rho_w c_w}{P_{out}u_{out}}\left[1 + \frac{(P_{out}u_{out})(A\rho c)}{(P_{in}u_{in})(A_w\rho_w c_w)} + \frac{A\rho c}{A_w\rho_w c_w}\right]\frac{\mathrm{d}T}{\mathrm{d}t} + T = T_f \tag{14}$$

Introducing the following coefficients:

$$a_2 = \frac{(A_w\rho_w c_w)(A\rho c)}{(P_{in}u_{in})(P_{out}u_{out})}, \qquad a_1 = \frac{A_w\rho_w c_w}{P_{out}u_{out}}\left[1 + \frac{(P_{out}u_{out})(A\rho c)}{(P_{in}u_{in})(A_w\rho_w c_w)} + \frac{A\rho c}{A_w\rho_w c_w}\right] \tag{15}$$

the ordinary differential equation of the second order (14) can be written in the form:

$$a_2\frac{\mathrm{d}^2T}{\mathrm{d}t^2} + a_1\frac{\mathrm{d}T}{\mathrm{d}t} + T = T_f \tag{16}$$

The initial conditions are:

$$T(0) = T_0 = 0, \qquad \left.\frac{\mathrm{d}T(t)}{\mathrm{d}t}\right|_{t=0} = v_T = 0 \tag{17}$$

Equations (16) and (17) are solved using the Laplace transformation. The operator transmittance $G(s)$ is as follows:

$$G(s)=\frac{T(s)}{T_f(s)}=\frac{1}{(\tau_1 s+1)(\tau_2 s+1)} \tag{18}$$

The time constants τ_1 and τ_2 in Eq. (18) are:

$$\tau_{1,2}=\frac{2a_2}{a_1\pm\sqrt{a_1^2-4a_2}} \tag{19}$$

The differential Eq. (16) can be written in the following form:

$$\tau_1\tau_2\frac{\mathrm{d}^2T}{\mathrm{d}t^2}+(\tau_1+\tau_2)\frac{\mathrm{d}T}{\mathrm{d}t}+T=T_f \tag{20}$$

Equation (20) is solved for a step change in fluid temperature from T_0 = 0°C to the constant value T_f. Laplace transform of the constant fluid temperature T_f is $T_f(s) = T_s/s$, and the operator transmittance assumes the following form:

$$\frac{T(s)}{T_s}=\frac{1}{s(\tau_1 s+1)(\tau_2 s+1)} \tag{21}$$

Eq. (21) can be written in another form:

$$\frac{T(s)}{T_s}=\frac{1}{s}+\frac{\tau_1}{\tau_2-\tau_1}\cdot\frac{1}{\left(s+\frac{1}{\tau_1}\right)}-\frac{\tau_2}{\tau_2-\tau_1}\cdot\frac{1}{\left(s+\frac{1}{\tau_2}\right)} \tag{22}$$

Making inverse Laplace transformation of Eq. (22), the thermometer temperature as a function of time is obtained:

$$u(t)=\frac{T(t)-T_0}{T_f-T_0}=1+\frac{\tau_1}{\tau_2-\tau_1}\exp\left(-\frac{t}{\tau_1}\right)-\frac{\tau_2}{\tau_2-\tau_1}\exp\left(-\frac{t}{\tau_2}\right) \tag{23}$$

If we assume in Eq. (23) with $\tau_2 = 0$, then we obtain Eq. (7) with $\tau = \tau_1$.

A time delay in the time response of the first order inertial system does not appear. Measuring the temperature of the fluid at high pressure requires the use of thermometers with the massive housings. In such cases, there is a time delay between the temperature indicated by the sensor and the changing temperature of the fluid. The inertial system of second order is suitable to describe the response behaviour with a time delay [13].

3. Identification of time constants

The time constant τ in Eq. (7) or time constants τ_1 and τ_2 in Eq. (23) can be determined on the basis of experimental data. For this purpose, the method of least squares is proposed to use [13, 15]. Finding the minimum of the function S

$$S = \sum_{i=1}^{N} \left[u_m(t_i) - u(t_i) \right]^2 = \min \tag{24}$$

allows to determine values of the time constants. In Eq. (24), $u(t_i)$ denotes the approximating function given by Eq. (7), and N is the number of conducted measurements (t_i, $u_m(t_i)$). The sum of the squares of the differences of the measured values $u_m(t_i)$ and the fitted values $u(t_i)$ is minimised. When the time constant τ or time constants τ_1 and τ_2 are determined, their values can be substituted into Eq. (24) to calculate $S_{\min}$.

The uncertainties of the calculated time constant τ or time constants τ_1 and τ_2 are calculated using the mean square error [13, 16–18]:

$$S_N = \sqrt{\frac{S_{\min}}{N - m}} \tag{25}$$

where m denotes the amount of time constants (i.e. $m = 1$ for Eq. (7) and $m = 2$ for Eq. (7) and $m = 2$ for Eq. (23)). Based on the determined mean square error S_N, which is an approximation of the standard deviation, the uncertainties in the determined time constants can be calculated using the formulas given in the TableCurve 2D software [18].

4. Determination of transient fluid temperature

The fluid temperature can be determined on the basis of measured histories of the thermometer temperature $T(t)$ and known time constant τ, using Eq. (1), or time constants τ_1 and τ_2, using Eq. (20). Temperature of the thermometer $T(t)$ and its first- and second-order time derivatives can be smoothed using the formulas [3, 13]:

$$T(t)=\frac{1}{693}\cdot\left[-63f(t-4\cdot\Delta t)+42f(t-3\cdot\Delta t)\right.$$
$$+117f(t-2\cdot\Delta t)+162f(t-\Delta t)+177f(t)+162f(t+\Delta t)$$
$$\left.+117f(t+2\cdot\Delta t)+42f(t+3\cdot\Delta t)-63f(t+4\cdot\Delta t)\right] \tag{26}$$

$$T'(t)=\frac{\mathrm{d}T(t)}{\mathrm{d}t}=\frac{1}{1188\Delta t}\cdot\left[86f(t-4\cdot\Delta t)-142f(t-3\cdot\Delta t)\right.$$
$$-193f(t-2\cdot\Delta t)-126f(t-\Delta t)+126f(t+\Delta t)$$
$$\left.+193f(t+2\cdot\Delta t)+142f(t+3\cdot\Delta t)-86f(t+4\cdot\Delta t)\right] \tag{27}$$

$$T''(t)=\frac{\mathrm{d}^2T(t)}{\mathrm{d}t^2}=\frac{1}{462(\Delta t)^2}\cdot\left[28f(t-4\cdot\Delta t)+7f(t-3\cdot\Delta t)\right.$$
$$-8f(t-2\cdot\Delta t)-17f(t-\Delta t)-20f(t)-17f(t+\Delta t)$$
$$\left.-8f(t+2\cdot\Delta t)+7f(t+3\cdot\Delta t)+28f(t+4\cdot\Delta t)\right] \tag{28}$$

Symbol $f(t)$ in Eqs. (26)–(28) denotes the temperature measured by the thermometer, and Δt is a time step. Application of nine-point digital filter allows eliminating the influence of random errors of the measured temperature $T(t)$ on the determined temperature of the fluid $T_f(t)$. If the measured temperature is not disturbed by significant errors, the first and second derivatives of the temperature can be estimated using the central difference method [15]:

$$T'(t)=\frac{f(t+\Delta t)-f(t-\Delta t)}{2\Delta t} \tag{29}$$

$$T''(t)=\frac{f(t+\Delta t)-2f(t)+f(t-\Delta t)}{(\Delta t)^2} \tag{30}$$

5. Examples of application of the method for a step change in temperature of the fluid

The first example demonstrates the use of the described method of transient fluid temperature measurement in the case of thin sheathed thermocouple application. During the experiment, the thermocouple with outer diameter 1.5 mm at the ambient temperature was suddenly immersed into hot water at saturation temperature. The results are presented in **Figure 2**. The time step Δt of the measured temperature was 1.162 s. The measured history of the temperature was approximated by Eq. (7) in order to determine the time constant τ of the thermocouple. In the calculations, TableCurve 2D code [18] was used. The calculations of the time constant and the uncertainty at the confidence level of 95% gave the following results: $\tau = 1.54 \pm 0.09$ s.

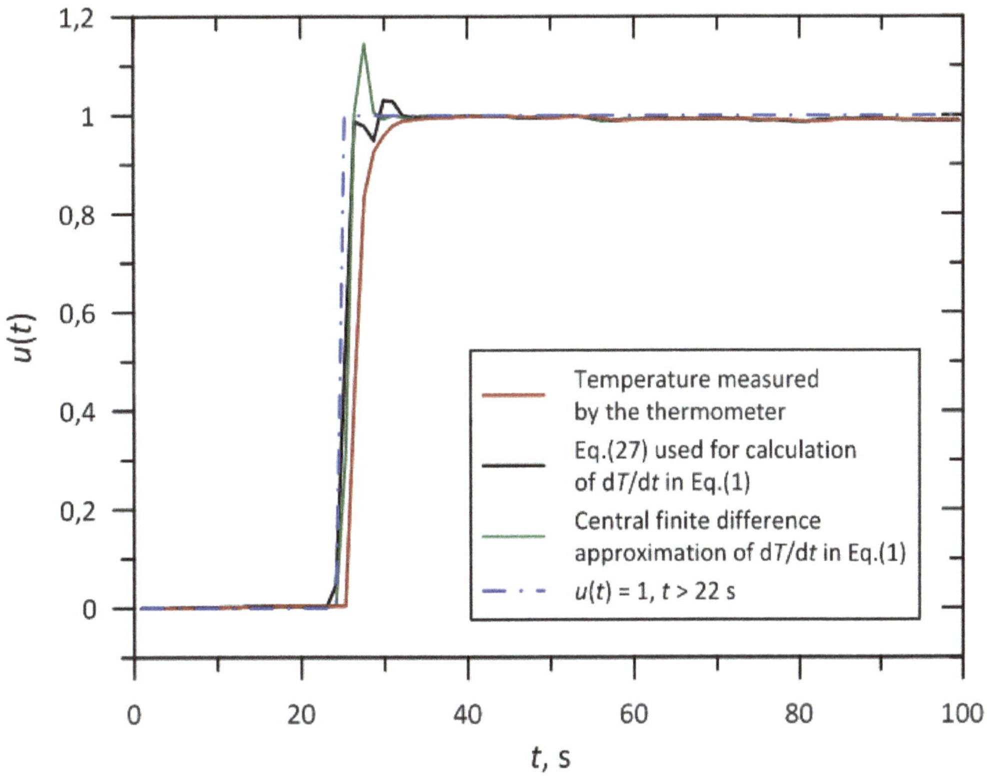

Figure 2. Fluid and thermometer temperature changes determined from the first-order equation (1) for the sheathed thermocouple with outer diameter 1.5 mm.

Next, the transient fluid temperature T_f was determined using Eq. (1) together with Eqs. (26) and (27). The first-order time derivative dT/dt in Eq. (1) was also calculated using the central difference quotient of Eq. (29). The results obtained show that the use of a first-order model for thin thermometers is sufficient (**Figure 2**). The results also indicate that the central difference approximation of the time derivative in Eq. (1) leads to less accurate results, since it is more sensitive to random errors in the experimental data.

Another example shows the application of the method for measuring transient fluid temperature with an industrial thermometer with massive thermowell and its complex construction (**Figure 3**). As in the previous example, the thermometer at the ambient temperature was suddenly immersed into hot water at a temperature of about 100°C. To compare the two methods of determining the unknown fluid temperature (using the first-and second-order model) for this thermometer, Eqs. (7) and (23) were used as the functions approximating the transient response of the thermometer. The following values with the 95% confidence uncertainty were obtained: $\tau = 14.07 \pm 0.39$ s, $\tau_1 = 3.0 \pm 0.165$ s and $\tau_2 = 10.90 \pm 0.2$ s.

Next, the fluid temperature changes were determined from Eq. (1) for the first-order model and from Eq. (20) for the second-order model (**Figure 4**).

The analysis of the results presented in **Figure 4** indicates that the second-order model delivers more accurate results than the first-order model.

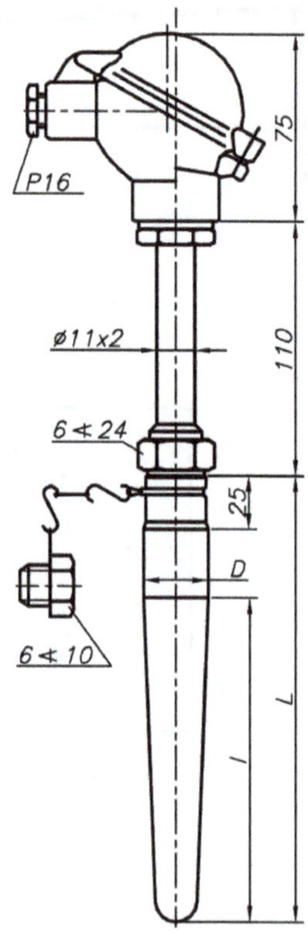

Figure 3. Industrial thermometer and its dimensions: $D = 18$ mm, $l = 65$ mm and $L = 140$ mm.

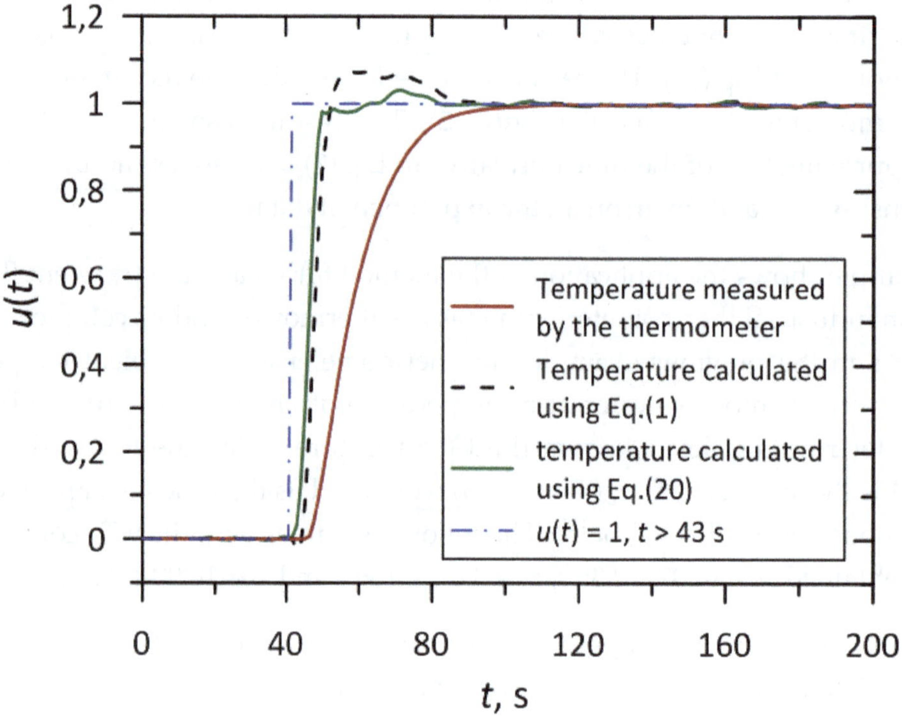

Figure 4. Fluid and industrial thermometer temperature changes determined from the first-order equation (1) and the second-order equation (20).

6. Experimental determination of time constant as a function of fluid velocity

The method presented in this chapter can also be used to determine the temperature of the flowing fluid. In this case, the time constant of the thermometer τ should be determined as a function of fluid velocity w. After specifying function $\tau(w)$, it should be substituted into Eq. (1).

Research has shown that for thin sheathed thermocouples with a diameter ranging from 0.5 to 6 mm, the equation of time constant as a function of fluid (air) velocity is as follows [15]:

$$\tau = \frac{1}{a + b\sqrt{w}} \tag{31}$$

where τ is the time constant of the thermometer in s, a constant in 1/s, b constant in $(\mathrm{m{\cdot}s})^{-1/2}$ and w fluid velocity in m/s.

As an example, the constants a and b were determined for K-type sheathed thermocouples with grounded hot junction with the outer diameter of 0.5, 1.0, 1.5 and 3.0 mm.

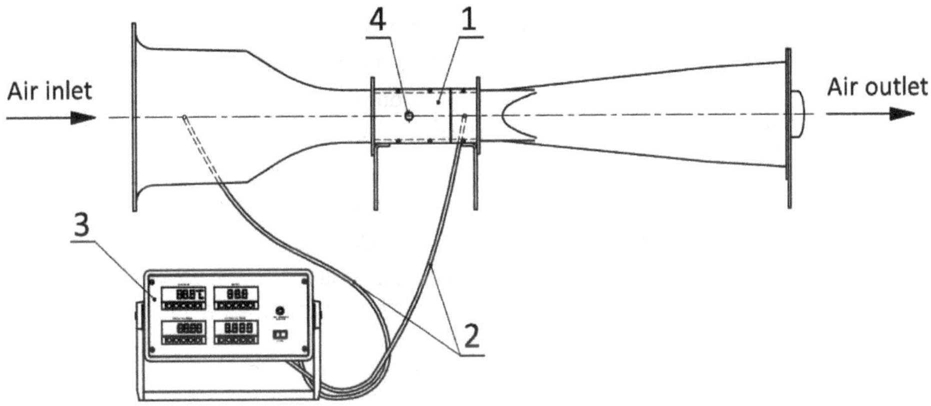

Figure 5. Wind tunnel used for determining thermocouple time constant—overall view [19]: (1) test chamber with an opening for the thermometer, (2) differential pressure measurement, (3) data acquisition system and (4) opening for the thermometer insertion.

The thermocouple time constant τ for various air velocities w was determined in an open benchtop wind tunnel (**Figure 5**). The WT4401-S benchtop wind tunnel is designed to give uniform flow rate over a 100 mm × 100 mm cross section [19].

The calculated time constants τ of the sheathed thermocouples with the outer diameter of 0.5, 1.0, 1.5 and 3.0 mm for different velocities of the air are shown in **Figure 6**. The experimental data collected for all thermocouples, as presented in **Figure 6**, were approximated by the least squares method. The best estimates for the constants a and b, with uncertainty at the 95% confidence level, for the tested thermocouples are [15]:

- $d_t = 0.5$ mm

  $a = 0.004337 \pm 0.000622$ 1/s and $b = 0.022239 \pm 0.001103$ $(\text{m·s})^{-1/2}$

- $d_t = 1.0$ mm

  $a = 0.020974 \pm 0.006372$ 1/s and $b = 0.103870 \pm 0.011240$ $(\text{m·s})^{-1/2}$

- $d_t = 1.5$ mm

  $a = 0.040425 \pm 0.003301$ 1/s and $b = 0.056850 \pm 0.004479$ $(\text{m·s})^{-1/2}$

- $d_t = 3.0$ mm

  $a = 0.128220 \pm 0.035716$ 1/s and $b = 0.220641 \pm 0.051122$ $(\text{m·s})^{-1/2}$

The time constant of the thermocouple τ strongly depends on the heat transfer coefficient h_t on the outer thermometer surface, which results from Eq. (2). The heat transfer coefficient is a function of Nusselt number, and this, in turn, is a function of fluid (air) velocity [20].

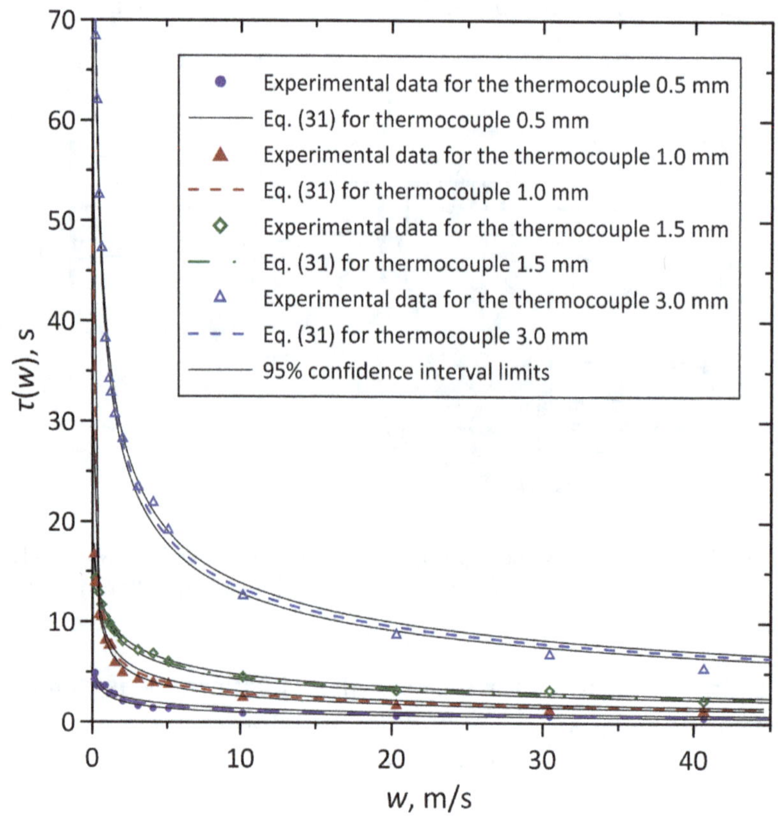

Figure 6. Time constants τ of sheathed thermocouple with outer diameters of 0.5, 1.0, 1.5 and 3.0 mm as a function of air velocity w with 95% confidence intervals.

When the velocity of the fluid and its temperature varies over time, time constant as a function of the velocity formulated by Eq. (31) can be used in Eq. (1) to determine the transient fluid temperature based on measurements made using a sheathed thermocouple.

To show how the described method can improve the temperature readings, the experimental measurements were presented [21]. The temperature of the flowing air in an open wind tunnel (**Figure 7**) was measured by K-type sheathed thermocouples with outer diameters of 0.5, 1.0 and 1.5 mm. During measurements, the temperature and velocity of air flowing through the tunnel were changed. The thermocouples were placed in the tunnel behind the heat exchanger and very close to each other (i.e. they measured the same temperature, but did not influence each other). The air velocity was measured by the vane anemometer FV A915 S220. Both the temperature and velocity data were collected using the Ahlborn ALMEMO 5990-0 data acquisition system.

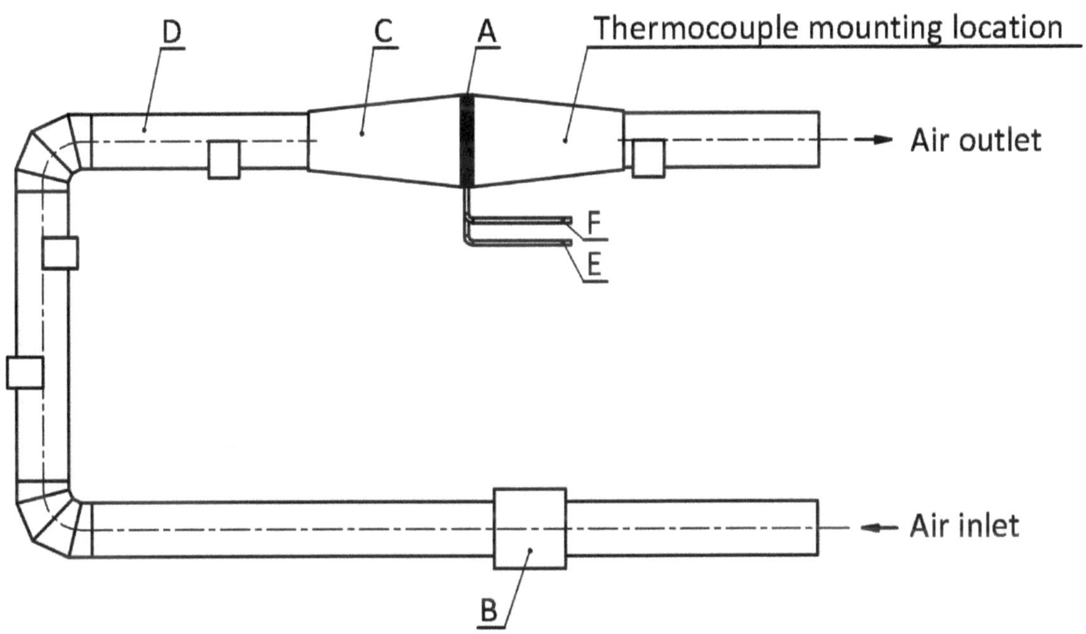

Figure 7. Diagram of an open wind tunnel [22]: (A) heat exchanger, (B) fan, (C) chamber, (D) air channel, (E) water outlet pipe and (F) hot water feeding pipe.

The comparison of the computed temperatures with the measured temperatures, when the time constants of the thermocouples are known, shows that the above method provides decent results (**Figures 8** and **9**).

The time histories of temperature obtained from calculations are very similar, especially for thermocouples with the outer diameter of 0.5 and 1.0 mm. In the small degree, the temperature calculated using the measurements with the thermocouple with the outer diameter of 1.5 mm deviates from them. This difference is due to the large time constant of the sheathed thermocouple with the outer diameter of 1.5 mm.

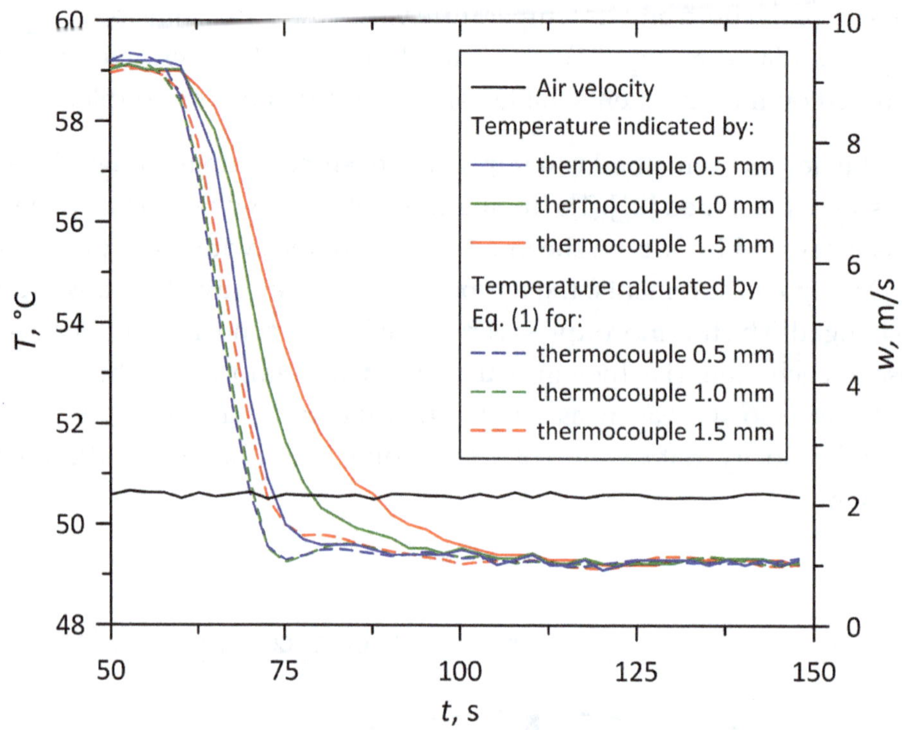

Figure 8. Temperature of the air measured by the thermocouples with outer diameters of 0.5, 1.0 and 1.5 mm and temperature calculated by Eq. (1) when the velocity of the air was constant.

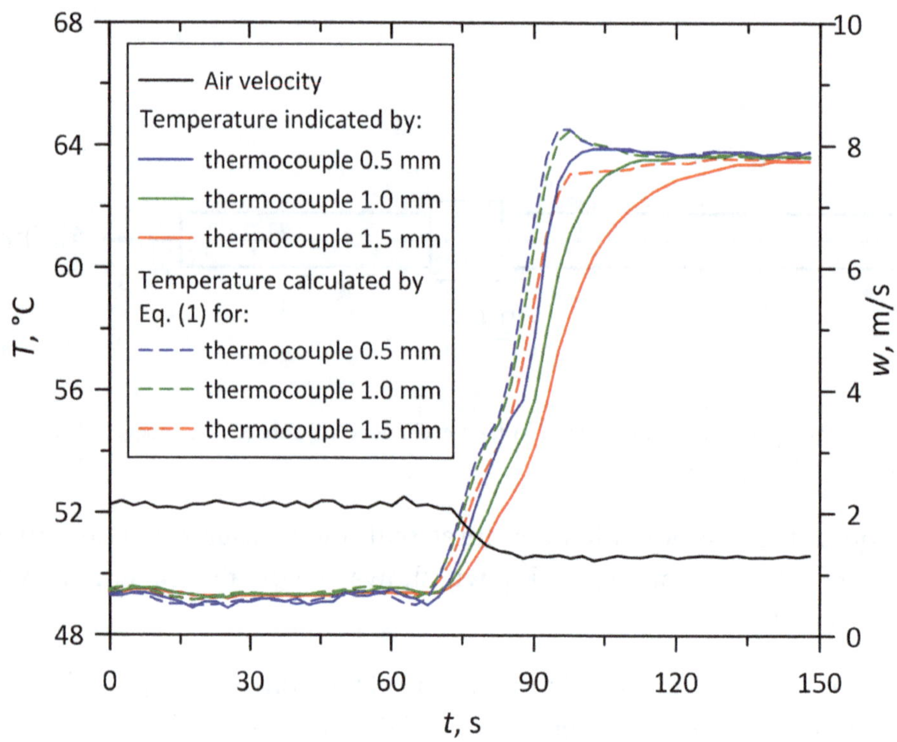

Figure 9. Temperature of the air measured by the thermocouples with outer diameters of 0.5, 1.0 and 1.5 mm and temperature calculated by Eq. (1) when the velocity of the air was changed.

7. Conclusions

Both methods of transient fluid temperature measurement presented in the chapter can be used online.

The first method, in which a mathematical model of the thermometer is first-order differential equation, is the most suitable for thermometers with very small time constants. In such cases, the delay of thermometer indication is small compared to changes in fluid temperature. In turn, the industrial thermometers, which are designed to measure the temperature of the fluid of high pressure, are characterised by a considerable delay of indications in reference to the actual changes of fluid temperature. For such thermometers, the second-order thermometer model, allowing for modelling the signal delay, is more adequate [14].

The method described in this chapter is the most suitable for measuring the transient temperature of gases, such as air or exhaust gases. This is due to the fact that the time constant depends on the heat transfer coefficient on the outer surface of the thermometer and in turn on the Reynolds and Prandtl numbers. For gases such as air or exhaust gas, Prandtl number varies slightly in a wide range of temperatures. However, during measurement of the transient steam temperature, the value of Prandtl number varies significantly depending on the pressure and temperature [23]. In this case, the inverse marching method described in [24] is more appropriate for correction of the dynamic error.

Author details

Magdalena Jaremkiewicz

Address all correspondence to: mjaremkiewicz@pk.edu.pl

Cracow University of Technology, Institute of Thermal Power Engineering, Cracow, Poland

References

[1] Nicholas JV, White DR. Traceable Temperatures. An Introduction to Temperature Measurement and Calibration. 2nd ed. New York: John Wiley & Sons; 2001.

[2] Michalski L, Eckersdorf K, Kucharski J, McGhee J. Temperature Measurement. 2nd ed. Chichester: John Wiley & Sons; 2001.

[3] Wiśniewski S. Temperature Measurement in Engines and Theramal Facilities. Warszawa: WNT; 1983.

[4] Taler J. Theory and Practice of Identification of Heat Transfer Processes. Wrocław: Zakład Narodowy imienia Ossolińskich; 1995.

[5] Kabza Z, Kostryko K, Zator S, Łobzowski A, Szkolnikowski W. Room Climate Control. Warszawa: Agenda Wydawnicza, Pomiary Automatyka Kontrola; 2005.

[6] Littler DJ et al. Instrumentation, Controls & Testing. Modern Power Station Practice. Oxford: Pergamon Press; 1991.

[7] Childs PRN. Practical Temperature Measurement. Oxford: Buterworth-Heinemann; 2001.

[8] Gerashchenko OA, Gordov AN, Lakh VI, Stadnyk BI, Yaryshev NA. Temperaturnye Izmereniya. Kiev: Naukova Dumka; 1984.

[9] Han JC, Dutta S, Ekkad SV.Gas Turbine Heat Transfer and Cooling Technology. 2nd ed. Boca Raton: CRC Press; 2012.

[10] Szekely V, Ress S, Poppe A, Torok S, Magyari D, Benedek Z, Torki K, Courtois B, Rencz M. New approaches in the transient thermal measurements. Microelectronics Journal. 2000;31(9):727–733.

[11] Crocker DS, Parang M. Unsteady temperature measurement in an enclosed thermo-convectively heated air. International Communications in Heat and Mass Transfer. 2001;28(8):1015–1024.

[12] Chau PC. Process Control: A First Course with MATLAB. Cambridge: Cambridge University Press; 2002.

[13] Taler J, Jaremkiewicz M, Taler D, Sobota T. Fluid temperature measurement under transient conditions. Archives of Thermodynamics. 2009;30(3):75–88.

[14] Jaremkiewicz M, Taler D, Sobota T. Measuring transient temperature of the medium in power engineering machines and installations. Applied Thermal Engineering. 2009;29(19):3374–3379.

[15] Jaremkiewicz M. Reduction of dynamic error in measurements of transient fluid temperature. Archives of Thermodynamics. 2011;32(4):55–66.

[16] ASME. Policy on reporting uncertainties in experimental measurements and results. Journal of Heat Transfer. 2000;122:411–413.

[17] Moffat RJ. Describing the uncertainties in experimental results. Experimental Thermal and Fluid Science. 1988;1:3–17.

[18] TableCurve 2D v. 5.1. Automated Curve Fitting & Equation Discovery. San Jose, California: Systat Software Inc.; 2002.

[19] WT4401-S & WT4401-D Benchtop Wind Tunnels. Stamford: Omega; DOI: https://www.omega.com/manuals/manualpdf/M1776.pdf–Accessed July 15, 2016

[20] Sanitjai S, Goldstein RJ. Forced convection heat transfer from a circular cylinder in crossflow to air and liquids. International Journal of Heat and Mass Transfer. 2004;47:4795–4805.

[21] Jaremkiewicz M. Measurement of transient fluid temperature. In: Richard B. Hetnarski, editor. Encyclopedia of Thermal Stresses. Dordrecht: Springer; 2014. p. 2932–2938.

[22] Taler D. The Dynamics of Tubular Heat Exchangers. Cracow: Uczelniane Wydawnictwa Naukowo-Dydaktyczne AGH; 2009.

[23] Meyer CA, McClintock RB, Silvestri GJ, Spencer RC. ASME Steam Tables. New York: The American Society of Mechanical Engineers; 1993.

[24] Jaremkiewicz M, Taler D, Sobota T. Measurement of transient fluid temperature. International Journal of Thermal Sciences. 2015;87:241–250.

8

Direct-Contact Heat Exchanger

Hua Wang, Qingtai Xiao and Jianxin Xu

Abstract

Direct-contact heat transfer involves the exchange of heat between two immiscible fluids by bringing them into contact at different temperatures. There are two basic bubbling regimes in direct-contact heat exchanger: homogeneous and heterogeneous. Industrially, however, the homogeneous bubbling regime is less likely to prevail, owing to the high gas flow rates employed. The mixture homogeneity and the non-homogeneity of the mixture can be characterized by the Betti numbers and the mixing time can be estimated relying on image analysis and statistics in a direct-contact heat exchanger. To accurately investigate the space-time features of the mixing process in a direct contact heat exchanger, the uniformity coefficient method based on discrepancy theory for assessing the mixing time of bubbles behind the viewing windows is effective. Hence, the complexity of the bubble swarm patterns can be reduced and their mechanisms clarified, and the heat transfer performance in a direct-contact heat exchanger can be elucidated.

Keywords: direct-contact heat transfer, flow pattern, Betti numbers, discrepancy, mixing uniformity

1. Introduction

1.1. Direct-contact heat exchanger

Direct transfer involves two immiscible fluids under different temperatures in contact for heat exchange [1]. Compared with the traditional direct-contact heat exchanger, heat transfer means has more advantages due to a more simple design, low temperature driving force and higher heat transfer efficiency [2, 3]. Direct-contact heat exchangers (DCHEs) make use of gas-liquid phase change heat exchanger within the working fluid. That is to say, DCHEs put to use heat transfer between two kinds of fluid in the absence of a partition. A direct contact heat exchanger can be used for seawater desalination, heat recovery, ocean thermal energy conversion, thermal energy storage systems, etc.[4, 5]. In addition, DCHEs have been applied to give a good solution in harnessing the solar energy [6] and provide a better understanding of ice formation, growth and detachment from the droplets producing ice slurry [7].

1.2. Mixing efficiency assessment

Mixing plays a fundamental role in many industrial applications, such as chemical engineering, metallurgical process, printing process, medical and bio-medical industries, and has a decisive impact on the overall performance of reaction processes. The purpose of mixing is to obtain a homogeneous mixture; however, many researchers have pointed out that the local mixing and the flow pattern has significant effects on the properties of the final products [8]. There is an increased want for measuring and comparing mixing performance. An efficient evaluation of mixing effects is required in those various fields, but as a result of its intricacy, theoretical methods are very limited. Monitoring or measuring the mixing appropriately is of much concern from the practical point of view and for the confirmation of theoretical models as much [9]. The existence of a second phase that makes the continuous phase flow and mixing process more complicated, especially for a direct contact with the boiling heat transfer process. The boiling heat transfer process, in which mixing efficiency assessment is common, is one of the most efficient kinds of heat transfer processes widely used in numerous engineering systems. Hence, the work of characterizing the homogeneous bubbling regimes in a DCHE is one of the most useful and instructive topics in DCHE.

1.3. Bubbling regimes

There are two basic types of bubbling regimes in DCHE: homogeneous and heterogeneous. In the homogeneous bubbling regime, there are few diversifications in the size of the bubbles, and breakage and coalescence phenomena are inappreciable [10–12]. Industrially, nevertheless, the homogeneous bubbling regime is not likely to prevail, thanks to the high gas flow rates used. This is good for the heterogeneous bubbling regime, characterized by a widespread of bubble sizes and crucial frequencies of breakage and coalescence [13]. For an air-water system, Ribeiro and Lage [13] employed transient experimental measuring of the temperature of the liquid, bubbling height, evaporation rate, gas volume fraction and bubble size distributions in a direct-contact evaporator for four surface gas velocities including operation in both homogeneous and heterogeneous bubbling regimes. Ribeiro et al. [14] also analysed the photographs of homogeneous and heterogeneous bubbling regimes using different liquids in a DCHE handling with a perforated-plate sparger. Le Coënt et al. [15] studied the compounding of two staves and a viscous liquid in a classical reactor. He found that there was an alleged "pseudo-homogeneous" state before it was mixed completely homogeneously. In reality, a pseudo-homogeneity was achieved much more quickly (<40 s), but subsequent images revealed that polymers still remained in the reactor. The time of the pseudo-homogeneous state begins is called the pseudo-homogeneous time. In our DCHE, we found that there was a comparatively stable state in the completely heterogeneous bubbling regimes also. Consequently, we defined this completely stable state as pseudo-homogeneous. Peyghambarzadeh et al. [16] found that bubble growth was a considerably complicated process, and detecting distinguishable bubbles was scarcely possible at high heat fluxes, while in this experiment, we have captured the rough sketch of bubbles.

1.4. Image analysis

A literature survey showed that image analysis has been used in transparent laboratory vessels to circumvent the drawback of subjectivity of measurement interpretation. Fortunately, the image processing technology has been widely used for feature extraction in medical and chemical industries. Thus, just that technology of image intensification, these bubble images can be computed with the following methods. Bubble growth is severely a function of flow of heat and liquid flow rate [16]. If the flow rate is lower, larger bubbles are observed at constant heat fluxes. This may be due to the fact that the growth of bubbles weakens with the time which is necessary at the velocity of flow is higher. Hence, the bubbles are smaller than those observed at higher high velocity. Similarly, according to the results of Ref. [16], the effect of heat flux is more meaningful than that of flow velocity. Many small bubbles are created on the heat transfer surface, inventing high turbulence flow at high heat fluxes. Consequently, heterogeneous and pseudo-homogeneous bubbling regimes are necessary and worth learning in a DCHE. At the meantime, it is one of the most challenging tasks of direct-contact heat transfer. The current commonly used method is to do with image processing techniques to acquire the features of bubbling regimes.

1.5. Betti numbers

In 1995, Hyde et al. [17] recommended the topological invariant features the topology penetration structure complexity, in the number of micro-structure processing is one of the two material phases. From the perspective of theory, Betti numbers are the number of handles a special case of a topological invariants in a micro-structure [18]. Algebraic topology provides measurable information on complex objects, and Betti numbers are rough measures of this information. Gameiro et al. [18] came up with a method using the Betti numbers to describe the geometry of the fine-grained and snake-like micro-structures created in the process of spinodal decomposition. The zeroth Betti numbers β_0 figure the number of connected components (pieces) in the space Ω. More accurately, if $\beta_0 = k$, then Ω has exactly k components. The first Betti numbers β_1 state a measure of the number of tunnel structure. In a two-dimensional field, tunnels are decreased to loops. It is a remarkable fact that the size and the shape of the component and the loops and do not affect the number of Betti numbers. Friedrich [19] proposed the same chemical group which used the structure descriptors to distinguish un-related chemical group of chondrite and the application of Betti numbers to research the difference in rock chondrite meteorites. A multiphase mixture usually shows a macroscopic homogeneity consisting scattered fine pieces. With an increase in dispersity of the pieces, the homogeneity of the mixture was increased. In the cracks of the pieces are the blowholes, and a polymerization of blowholes gets an aggregation. The more frequently the pores appear, the more likely agglomeration is to occur at the surface. Moreover, increase in the number of pores showed that the heterogeneity of the mixture was mixed more evenly.

In our previous work, using the Betti numbers for gas-liquid-solid three-phase mixing effects of molten salt system based on the reaction of CH_4 + ZnO were characterized. Nitrogen was used to imitate the gas phase (CH_4) and mainly mixing effect in the sink. The zeroth Betti numbers were used to measure the number of pieces in the patterns, bring about beneficial parameter to describe the mixture homogeneity, which was the number of masters in the micro-structure occupied by one of the two phases. The first Betti numbers were introduced to describe the mixing heterogeneity of mixture. Because we only quantified the solid-liquid mixed flow pattern, the mixture of nitrogen bubble will disappear after image binarization.

1.6. Heat transfer performance

It must be pointed out that Gulawani et al. [20] studied and founded that the turbulent flow pattern in a gas-liquid interface heat transfer coefficient and the immersed surface has a significant impact. Under Gulawani et al. [20, 21] inspiration and guidance, our work is mainly described the flow pattern characteristics of bubbles under the effect of heat transfer in the DCHE. Dahikar et al. [22] and Tayler et al. [23] used the Betti numbers to represent the heterogeneous and pseudo-homogeneous of bubbles. In addition, the relationship between the Betti numbers and the heat transfer coefficient has been obtained in a DCHE.

1.7. Mixing time.

Both mixing speed and phase transition time in the direct contact boiling heat transfer process are fast. An accurate mixing time is critical to appropriately evaluate computational fluid dynamics models and then enhances equipment understanding and develops scale down models for process characterization and design space definition during late stage process development. In the past few years, many researchers have studied the mixing time and many methods were proposed to measure mixing time. But at present, there is no generally accepted method of measuring mixing time, mainly because of each method is not universal, that is each method has its own limitations, such as conductivity [24], pH [25], the dual indicator system method [26], tracer concentration [27–30], electrical resistance tomography [9], coloration decolouration methods [31], the box counting with erosions method [15] and Betti numbers with image analysis [32]. The limitation of each method has been described in details [31]. In all of the above-mentioned technologies, the Betti numbers are one of the most worthy methods to measure the mixing time and get further information of the mixing process. The Betti numbers can be effectively quantitative mixing time, the development process of mixing and degree of homogeneity. It has been used to characterize the evolution of the bubble group in direct contact with the boiling heat transfer process. But, we found that the Betti number method to be used for mixing time and the different evaluation indexes for mixing time have a similar trend, such as the slope p [15], pH, tracer concentration ct, the percentage of mixed pixels, $M(\%)$ [31], and the standard deviation (σG) [26]. These indicators change at the beginning of the mixing and quickly tend to be stable after fluctuations. The mixture of non-uniformity caused these fluctuations. This is the inevitable process of pseudo-homogeneous critical point to determine the influence accurate estimates of mixing time, which has often been overlooked. A literature survey shows that it is mainly used for determining the critical point of mixing time, including the mean value of Betti number (mean method) [32], slope p (slope method) [31], and standard deviation (SD) and the selected threshold [33]. Accurate estimation about the mixing time of work was published less than others, especially the critical point determination impact. The idea of a three-sigma method is inspired and motivated by statistical process control (SPC). According to Woodall [34], SPC can commonly be divided into two phases. The data of phase I are clean gathered under stable operating conditions, whereas the major of phase II is to detect any changes. The 3σ principle is that if the sample data come from a normal distribution $N(\mu, \sigma^2)$, most of the data (99.73%) will lie within the range [μ-3σ, μ+3σ]. It is imposed to detect outliers in the quality control of samples. If the result is normal, the process of the product specification will lie within the scope μ±3σ of the standard value. Otherwise, the production process is considered to be abnormal. Homoplastically, a changed three-sigma edit test has been successfully used in distributed self-fault diagnosis algorithm for large-scale wireless sensor networks [35]. Our research confirmed that the critical point of the response time ignored may result in significant error in mixing time estimation.

1.8. L_2-star discrepancy

The mixing process in DCHE has been studied by many experiments. Similarly, at present, there is no generally accepted way to measure mixing homogeneity, mainly because each method has its own deficiencies, such as thermal method, conduct metric method, pH method, decolourization methods, Schlieren method, Betti numbers method [36], etc. In all of the above-mentioned technologies, the Betti numbers have been used to characterize the evolution of the heterogeneous and pseudo-homogeneous bubbling regimes. But, with the Betti numbers for characterization of mixing uniformity have a space-time limitation; it may lead to significant errors in the evaluation of mixing uniformity. The key question is how to measure the random bubble swarm of minimum difference of space-time consistency bubble swarm of domain. Fang and Wang putted forward the concept of UD (uniform design) that dispersed experimental points uniformly scattered on the domain. One should choose a set of given all possible designs with amount of minimum difference of laboratories under the design of all possible factors and experimental runs. The above is the basic idea of UD [37]. UD has been widely used since 1980 [38]. Inspired and motivated by Fang [37–39], our main research objects are the study of characteristics of time-space features and analyse the mixing process of numerical simulation and experimental analysis. Uniform design theory and image analysis have been applied to quantitative uniformity of time and space in a DCHE.

1.9. Modified L_2-star discrepancy

Recently, we were vitalized and motivated by Xu et al. [39], by the literature that introduces the relatively not complicated and accurately uniformity coefficient (UC) technology, which is based on image processing technology and the theory of uniform design to determine the mixing time and uniformity in a DCHE. The space-time characteristics can be quantified by means of the uniformity coefficient method, which based on L_2-star discrepancy (UC-LD) and provides a method of direct measurement about the macro-mixing evolution. With the same Betti number just is aimed to separate the local and global uniform [40]. Whereas, the nature of the UC such as rotation invariance has not been explored and it has a lot dependence of calculating the initial conditions of UC, namely UC-LD. Clearly, the L_2-star discrepancy is much easier to calculate numerically according to Heinrich [41]. Unfortunately, the L_2-star discrepancy shows some shortcomings, as pointed out by Hickernell [42]. For instance, it is influenced by all the sub-dimensions of the projection uniformity. In order to overcome these shortcomings, Hickernell studied uniformity of some new measure methods, which are also associated with the L_2 norm, namely, the centred discrepancy (CD) and the wrap-around discrepancy (WD). The centred discrepancy (CD) and the wrap-around discrepancy (WD) satisfy a Koksma-Hlawka type inequality according to Xu et al. [39]. According to the theory of UC-LD, if the image shows a low CD or WD that can be called homogeneous mixing of a set or bubble swarm pattern, and the use of UC-CD and UC-WD provides at least likely to get a good approximation of mixing bubble swarm of spatial distribution. Additionally, UC-CD and UC-WD exhibit some advantages including rotation invariance, reflection invariance and projection uniformity [39].

1.10. Chapter structure

The chapter is organized as follows. In the next section, experiments and methodology are presented; the results and discussion are presented subsequently; the conclusion is briefly summarized in this section finally. Then the acknowledgements and references are presented in the end.

2. Measurement of mixing uniformity in DCHE

2.1. Experimental

2.1.1. Apparatus

The schematic of the experiment employed in the present research is sketched in **Figure 1** [32]. There are two circulation loops in the test device for this experiment. The first loop, which consists of the DCHE (1), electric heater (2), heat transfer fluid (HTF), pump (7) and connecting inlet and outlet pipelines, is a continuous-phase circulation loop for fluid flow, and the other, which consists of the DCHE (1), centrifugal pump (4), plate condenser (5), centrifugal pump (6) and connecting inlet and outlet pipelines, is a dispersed-phase circulation loop for working medium flow. The temperature control device, gear oil pump (3), regulates the initial temperature difference arising from heat exchange. The frequency control cabinet, gas mass flow-meter (8), regulates the rates of flow of the HTF and working medium. The patterns were imaged by a high-speed shutter video camera, which was placed at the second viewing window. In the bubble evaporation process, we could observe the most active stage of the bubbling regime. HTF and the refrigerant R-245fa (1, 1, 1, 3, 3 pentafluoropropane) were used as the continuous phase and the dispersed phase in all runs, respectively.

2.1.2. Experimental design

The settings of the experimental plan affecting the heat transfer capacity of the tested DCHE are determined through the orthogonal array (OA) experimental design method.

As listed in **Table 1**, design parameters with four factors and three levels were selected to investigate the influence of heat transfer capacity. The $L_9(3^4)$ orthogonal array table was chosen for designing the experiment. The interaction between the design parameters was neglected in the present study. H is the height of HTF in the DCHE, ΔT is the initial heat transfer temperature difference, U_g is the refrigerant flow rate, and U_0 is the flow rate of the HTF.

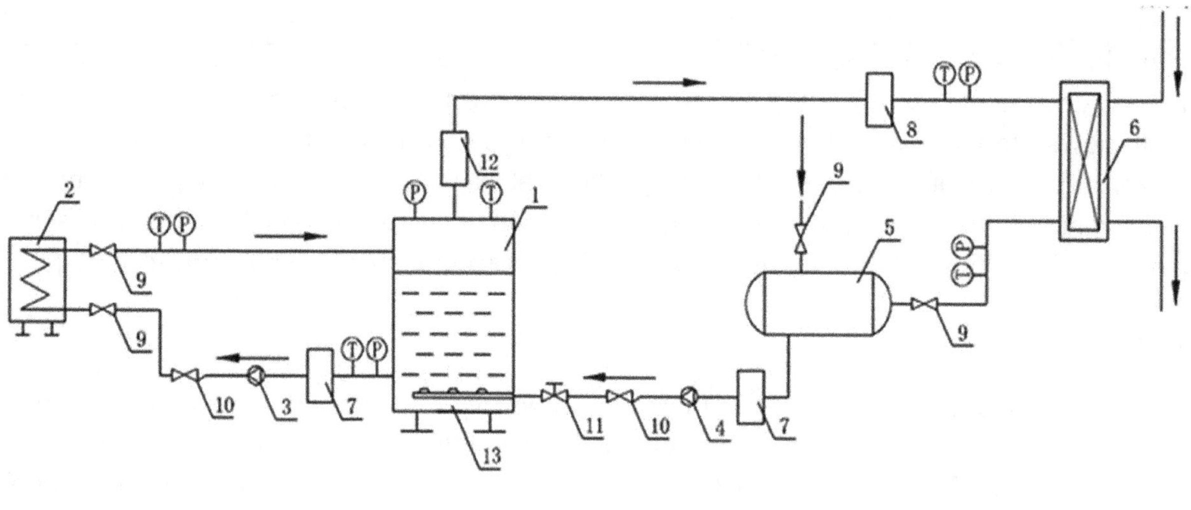

Figure 1. Experimental equipment for direct-contact heat transfer.

Symbol	Design parameter	Unit	Level		
A	H	mm	460	530	600
B	ΔT	K	80	100	120
C	U_g	m^3/s	1×10^{-4}	2×10^{-4}	3×10^{-4}
D	U_0	kg/s	0	0.15	0.3

Table 1. Design parameters and levels.

As shown in **Table 2**, the numbers E_1–E_9 denote different experimental levels according to the orthogonal array table.

2.2. Pattern acquisition and processing

A high-speech video camera was employed to obtain the patterns, and the brand used was PRAKTICA from Germany, with resolution 4 million pixels with no LED light. The images,

Experimental	E_1	E_2	E_3	E_4	E_5	E_6	E_7	E_8	E_9
H(mm)	460	460	460	530	530	530	600	600	600
ΔT(K)	80	100	120	80	100	120	80	100	120
U_g (× $10^4 m^3/s$)	1	2	3	2	3	1	3	1	2
U_0 (kg/s)	0	0.15	0.3	0.3	0	0.15	0.15	0.3	0

Table 2. Design experiments according to four factors and three levels orthogonal table.

which were blurred in photographing, can be improved using some image processing techniques. It takes 8 minutes to shoot in each occasion of the orthogonal experiment. Because of difficulties in storing and calculating these images, we choose equal interval sampling from 6000 images, in total, 12,000 images are collected.

Figure 2 is randomly obtained in the present image-processing process. In order to suppress the background of the original image, eliminate noise and enhance the image, gray-scale transformation, top-hat transform is used here. The binaruzation operation was used to calculate the Betti numbers. With a dilation process, an erosion process named as an opening was executed. This process, aiming at removing tiny or isolate points at the finer locations, and smoothing the boundaries of larger points, could not change the size of the image significantly. In contrast, with a dilation erosion process, a dilation process named as an opening was executed. This operation, aiming at filling up tiny pores within the points, connecting nearby points, and smoothing the borders, could not alter the size of the image significantly. The opening is used here to remove small holes representing sile bubbles or small bubble swarms of the binarization images. Since the behavioural characteristics of bubble swarms could not be accurately portrayed by binarized images with noise, an opening operation must be executed to eliminate image noise by the appropriate thresholds selected.

Thus, the white area indicates the bubble swarm, and the black area refers to the continuous phase. As the experimental conditions, the captured image is relatively fuzzy; however, its quality can be improved by using the digital image processing techniques. The resultant image that could be used for the following analysis was identifiable.

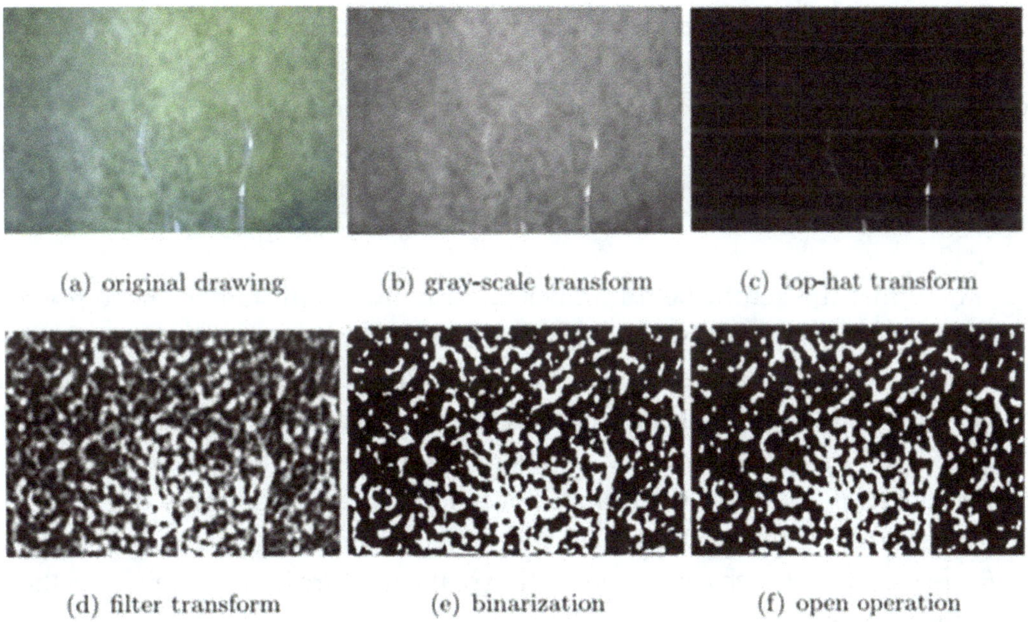

Figure 2. Treatment for one piece of bubble swarm patterns.

2.3. Methodology

2.3.1. Performance evaluation of the DCHE model

Owing to the complexity of the DCHE multiphase structure, heat exchange performance has often been expressed in terms of the volumetric heat transfer coefficient, h_V, which is given by [43]:

$$h_V = \frac{Q}{V \times LMTD} \tag{1}$$

where V is the volume of the continuous phase in the DCHE, and Q is the rate of heat transfer from the continuous phase to the dispersed phase, given by:

$$Q = m(h_{do} - h_{di}) \tag{2}$$

where m is mass flow-rate of the dispersed phase steam, and h is the enthalpy of the dispersed phase. The *LMTD* in Eq. (1) is the logarithmic mean temperature difference, which is defined as:

$$LMTD = \frac{(T_{ci} - T_{do}) - (T_{co} - T_{di})}{\ln \frac{(T_{ci} - T_{do})}{(T_{co} - T_{di})}} \tag{3}$$

where T is temperature. In all the equations, the subscript c refers to the continuous phase, d refers to the dispersed phase, i refers to the inlet, and o refers to the outlet.

2.3.2. Computational homology (Betti numbers)

Box-counting with erosions method, which was developed by Le Coënt et al. [15], can be applied to quantify the mixture homogeneity; however, it is not available for quantifying the mixture non-homogeneity. As shown in the experiment, some agglomerates still exist in the vessel after stirring for quite a long time. With computational homology, an original analysis method aiming at getting the quantification of the mixture homogeneity and non-homogeneity was proposed.

As we all know that the zeroth Betti number and the first Betti number have the following information [18, 44]: β_0 equals the number of connected components that make up the space, and β_1 provides a measure of the number of tunnels in the structure. In a two-dimensional domain, tunnels are reduced to loops. Since an image is three-dimensional, it has three Betti numbers: β_0, β_1, and β_2. β_2 measures the number of completely enclosed cavities, such as the interior of a sphere. β_0 indicates the number of pieces, and β_1 represents the number of the holes. In other words, the mixing effect will vary with the number of pieces in the glass vessel. So β_0 and β_1 are used to get such a characterization of the mixture homogeneity and the mixture non-homogeneity, respectively.

β_0, β_1 and their averages $\bar{\beta}_0$, $\bar{\beta}_1$ of the binary images of the patterns can be obtained at different submerged lengths of the lance and flow rates of the gas. Also, we may obtain the value of time T (time unit: seconds) at which β_0 of the black/white image is equal to $\bar{\beta}_0$. The time T can be employed to obtain the minimum mixing time.

Set

$$\chi^{+}:=\{t|\beta_0(t)>\bar{\beta}_0, t>T\} \quad \chi^{-}:=\{t|\beta_0(t)<\bar{\beta}_0, t>T\} \quad A:=\frac{1}{2}\left[\frac{1}{m}\sum_{t\in\chi^{+}}\beta_0(t)-\frac{1}{n}\sum_{t\in\chi^{-}}\beta_0(t)\right] \quad (4)$$

where $\beta_0(t)$ denotes the zeroth Betti number of the binary image of the pattern, which is captured at the time t, and m, n are the numbers of elements in χ^{+}, χ^{-}. A is used to estimate the deviation amplitude of β_0(t) from their average $\bar{\beta}_0$.

In two-dimensional cases, β_0 is the number of connected components, such as black regions. The number of these holes, which is completely enclosed by cubes/pixels, is measured by β_1, and β_1 represents the number of the holes in the domain. One can easily count these white regions. As shown in **Figure 2**, β_0 represents the number of continuous phases, whereas β_1 represents the number of bubble swarms.

The calculation of Betti number is difficult, and the methods are only in their early stages [44]. The free software package CHomP was used to calculate Betti numbers [44, 45]. We could compute β_0 and β_1 of the open operation images of the patterns at different experimental levels using the CHomP software package [45].

Subsequently, we obtained the value of time t (seconds) that can be used to estimate the pseudo-homogeneous time with β_1 representing the average of β_1 of the open operation image after the pseudo-homogeneous process. The bubble sizes were found to be almost the same by inspecting many test images. With the pseudo-homogeneous time, the entire visible area was covered by the bubbles [36]. Just as we all known, combining the evolution of Betti numbers, we can distinguish whether the distribution is uniform or not. In the beginning, the Betti number increases and then rapidly stabilizes after fluctuations.

As **Figure 3** shows, a conversion operation of open operation images was performed. The results showed that a black-and-white conversion directly leads to a switch between the corresponding objects of the zeroth and first Betti numbers [46]. To illustrate, β_0 and β_1 represent the number of the continuous phase in Figure 3a, and the opposite in Figure 3b, respectively. Since it is the white pores that most directly reflect the flow patterns of the bubble swarms, β_1 is still used to characterize the number of bubble swarms.

2.3.3. *Three-sigma method*

Let X be a normally $N(\mu, \sigma^2)$ distributed random variable. For any $k > 0$, $P\{|X- \mu|< k\sigma\} = 2\Phi\{k\}-1$, where $\Phi\{\cdot\}$ is the distribution function of the standard normal law; whence, in particular, for $k = 3$ it follows that $P\{\mu-3\sigma < X < \mu+3\sigma\} = 0.9973$. The latter equation means that X can differ from its expectation by a quantity exceeding 3σ on the average in not more than 3 times in a thousand trials [47]. This circumstance is sometimes used by an experimenter in certain problems, by assuming that $\{|X- \mu|> 3\sigma\}$ is practically impossible, and consequently, $\{|X-\mu|< 3\sigma\}$ is practically certain. The probability of exceeding the range of "$\mu\pm3\sigma$" occurring twice is 7.29×10^6. Indeed the experimental time series of Betti numbers approximate normal distribution, as shown in **Figure 4** [36].

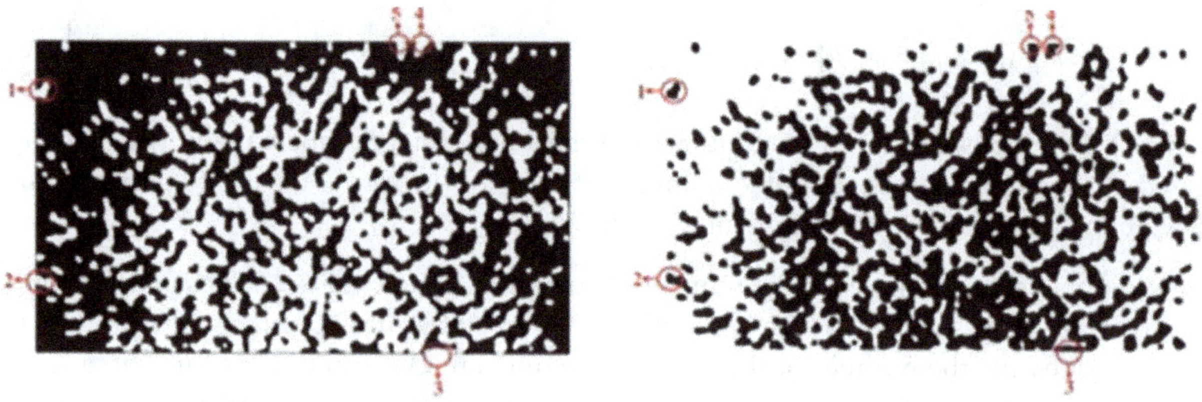

(a) Original drawing (b) Convert selection

Figure 3. Influence of the boundary on the Betti numbers.

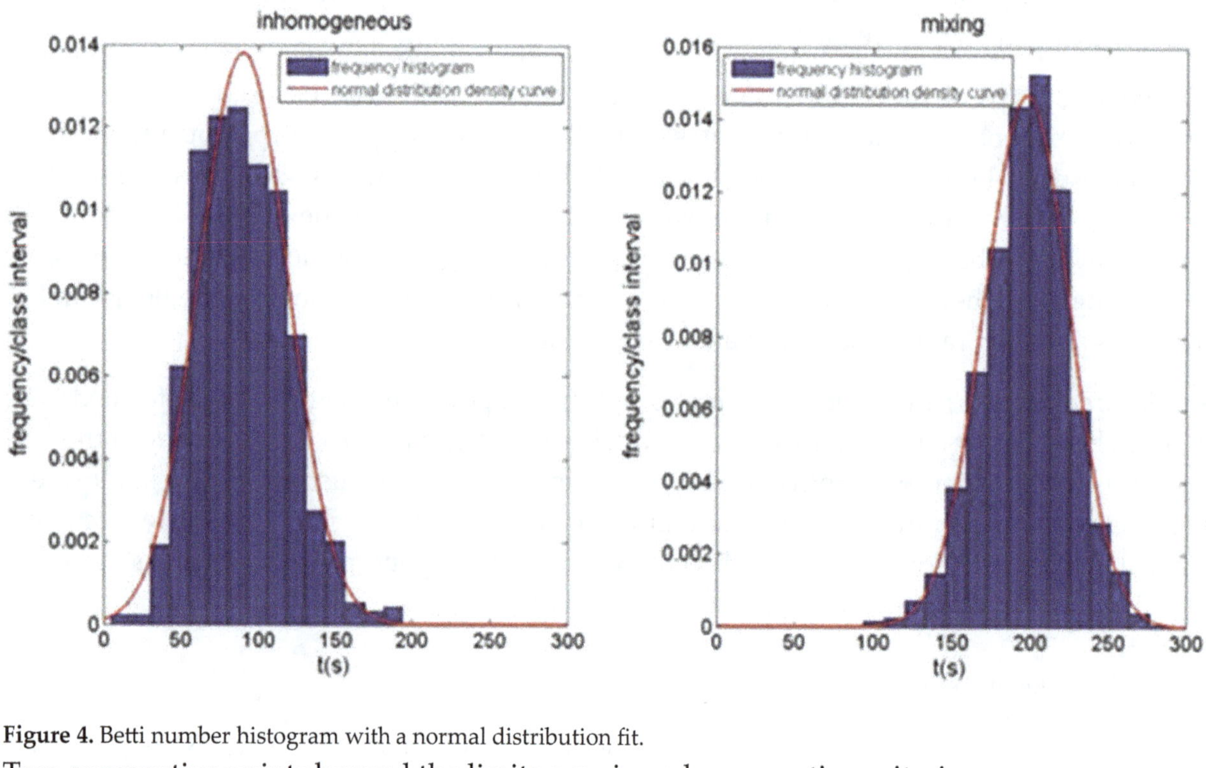

Figure 4. Betti number histogram with a normal distribution fit.

Two consecutive points beyond the limits are viewed as exception criteria.

Step 1: Giving a time point $t_0 > t$, and t is the mixing time, a homogeneous trend was presented during the evolution of Betti number time series of bubble swarm.

Step 2: Calculating the mean t and standard deviation σ of Betti numbers time series after the time t_0.

Step 3: Determining whether an event exceeds the range of "$\mu \pm 3\sigma$" occurring twice, as t_0 in reverse order. If so, then the moment is defined as mixing time, t. In order to quantify the macro-mixing efficiency using the Betti numbers, the data of the Betti numbers satisfying approximately normal distribution are collected from mixing homogeneity process, then let μ represents the estimated mean and σ represents the estimated standard deviation, and the mixing time is the time when the critical point exceeds the range of μ-3σ in reverse order twice.

The technique by itself is not limited to transparent tanks. It can be used in conjunction with electrical resistance tomography (ERT), position emission tomography (PET) and magnetic resonance imaging (MRI) [36].

2.3.4. *Measures of uniformity*

A popular figure of merit is the star discrepancy [48] and its generalization the L_p-star discrepancy. Let $F_u(\boldsymbol{x})=x_1x_2\ldots x_s$ be the uniform distribution function on C^s, where $\boldsymbol{x} = (x_1, x_2, \ldots, x_s)$. Let $F_p(\boldsymbol{x})$ be the empirical distribution function of $P = \{x_1, x_2, \ldots, x_n\}$:

$$F_p(x) = \frac{1}{n}\sum_{i=1}^{n} 1_{[x_i,\infty)}(\mathbf{x}) \tag{5}$$

where $\mathbf{1}_A(\boldsymbol{x})$ is the indicator function. Then the L_p-star discrepancy can be defined as the L_p-norm of difference between uniform and empirical distribution function, and then the L_p discrepancy can be defined as:

$$D_p(P) = \left[\int_{C^s} \left| F_u(\mathbf{x}) - F_p(\mathbf{x}) \right|^p dx\right]^{\frac{1}{p}} \tag{6}$$

By taking $p = \infty$, L_∞ discrepancy, which defined as the maximum deviation between these two distributions, is called the star discrepancy [48]. It is probably the most commonly used and can be expressed in another way as follows:

$$D_p^*(P) = \sup_{x \in C^s} \left| F_u(\mathbf{x}) - F_p(\mathbf{x}) \right| \tag{7}$$

With the discrepancy criterion in mind, we next discuss how to construct a uniformity coefficient. $\boldsymbol{x}=(x_1, x_2, \ldots, x_s) \in C^s$, $[0, \boldsymbol{x}] = [0, x_1]\times[0, x_2]\times\ldots\times[0, x_s]$ is the rectangle determined by the origin O and x decided on C^s. $Vol([0, \boldsymbol{x}])$ denotes the volume of the rectangular solid $[0, \boldsymbol{x}]$, where $Vol([0, \boldsymbol{x}])=x_1x_2\ldots x_s=F_u(x)$. Let $|\,|$ be the gained number of points in a group. The function of $|P\cap [0, \boldsymbol{x}]/n|$ represents an empirical distribution, as shown below:

$$F_p(x) = \frac{1}{n}\sum_{i=1}^{n} 1_{[x_i,\infty)}(x) = \frac{|P\cap[\mathbf{0},\mathbf{x}]|}{n} \tag{8}$$

Definition 2.1. The local discrepancy function is

$$dis\ c^*(x) = F_u(x) - F_p(x) = Vol([\mathbf{0},\mathbf{x}]) - \frac{|P\cap[\mathbf{0},\mathbf{x}]|}{n} \tag{9}$$

The difference between theory and empirical distribution can be used to measure the local discrepancy function with a rectangle $[0, \boldsymbol{x}]$. It can be expressed in another way as follows:

$$dis\, c_i(s,t) = \frac{vol_i(s,t)}{vol_i(\cdot,t)} - \frac{hol_i(s,t)}{hol_i(\cdot,t)} \tag{10}$$

where i denotes the four corners of an image, $i = 1, 2, 3, 4$, vol denotes the volume of the rectangular solid, hol is the number of bubbles.

Definition 2.2. The mean absolute discrepancy is often defined as follows:

$$MAD(s,t) = \frac{1}{4}\sum_{i=1}^{4}\left| dis\, c_i(s,t) \right| \tag{11}$$

In **Figure 5**, the influence of iteration steps on the measurement is not pronounced. The MAD (mean absolute discrepancy) is conducted by the four corners of an image.

Definition 2.3. Uniformity coefficient (UC) at time t is often defined by

$$UC(t) = 1 - Median\{MAD(s,t)\} \tag{12}$$

$$UC_1(t) = 1 - \sqrt{\frac{1}{S}\sum_{s=1}^{S} MAD(s,t)} \tag{13}$$

In every case, the degree of mixing uniform could be detected successfully by the uniformity coefficient method (**Figure 6**). After certain processing, the value range of UC is usually [0, 1]. We also denote that the measurement is not pronouncedly affected by the iterative steps.

In **Figure 7**, when the pixels sizes are reduced from 16:9 to 4:3, the influence of homogenization curve by the uniformity coefficient method is not reduced [40]. However, the trend of homogenization curve by Betti numbers method becomes unclear.

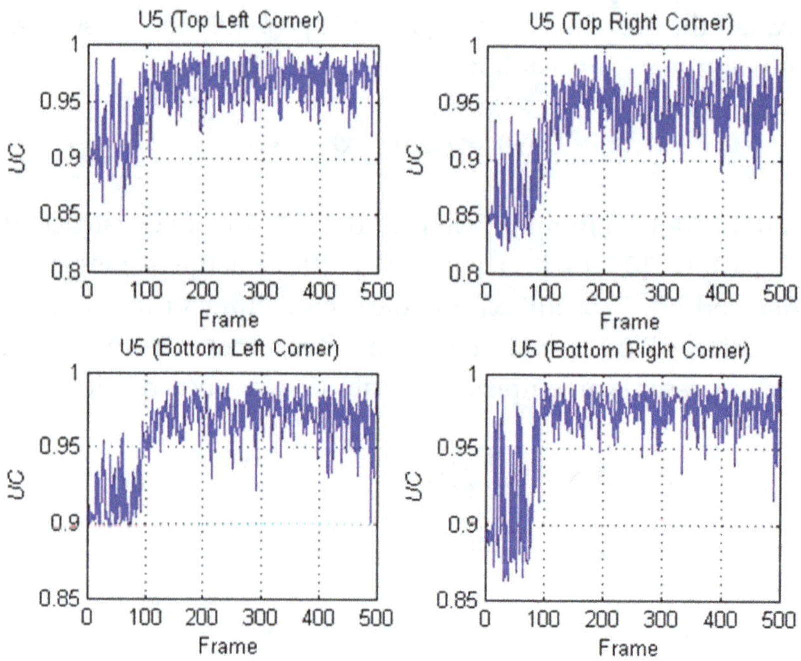

Figure 5. Effect of initial positions on uniformity coefficient.

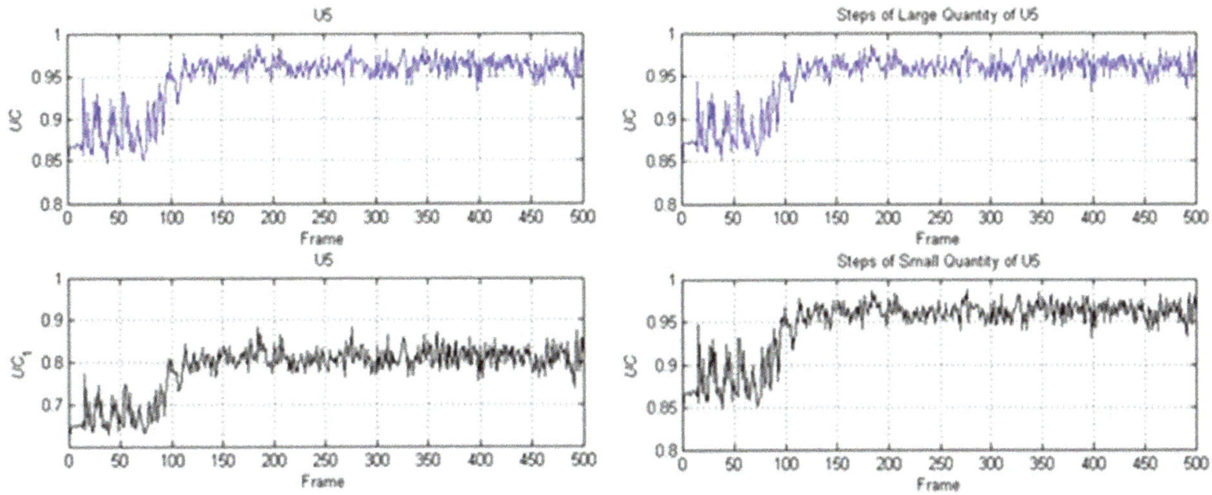

Figure 6. Effect of expressions (left) and iteration steps (right) on uniformity coefficient.

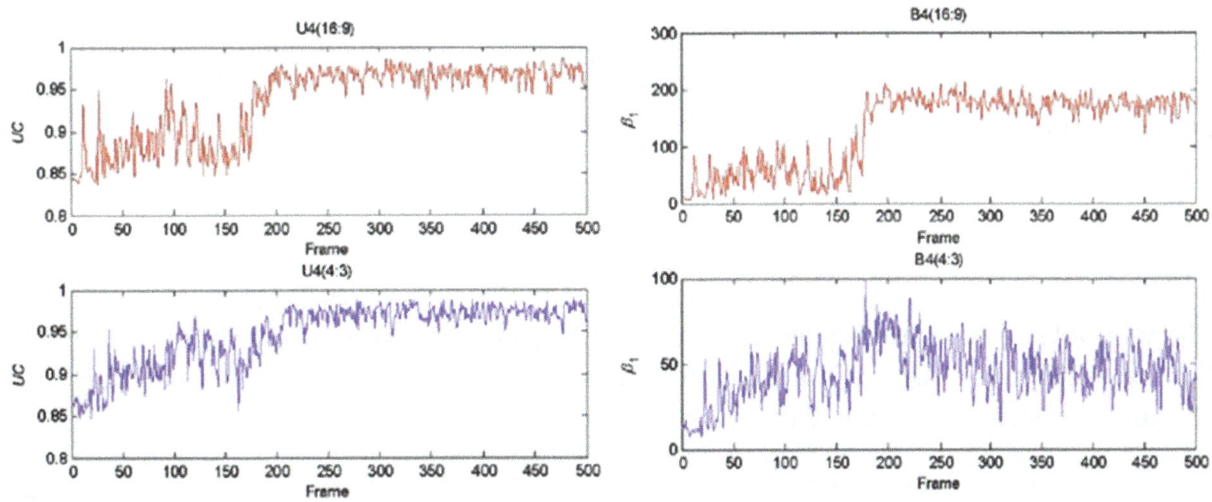

Figure 7. Effect of different pixels (16:9 and 4:3) on uniformity coefficient.

The evolution of the UC and Betti numbers of binary images at different image sizes was clearly shown in **Figure 7**.

Quasi-Monte Carlo method is the most commonly used measure of uniformity in the literature, especially when $p = \infty$ and 2. When $p = 2$, Warnock [49] gave an analytic and simple formula for calculating L_2-star discrepancy as follows:

$$D_2^*(P) = \left\{ \left(\frac{1}{3}\right)^s - \frac{1}{n}\sum_{i=1}^{n}\prod_{j=1}^{s}\frac{1-x_{ij}^2}{2} + \frac{1}{n^2}\sum_{i,l=1}^{n}\prod_{j=1}^{s}\left[1-\max\left(x_{ij}, x_{lj}\right)\right] \right\}^{\frac{1}{2}} \tag{14}$$

where $x_k=(x_{k1}, x_{k2}, \ldots, x_{ks})$. Unfortunately, the L_2-star discrepancy exhibits some limitations, as pointed out by Heinrich and Hickernell [41, 42]. To overcome these disadvantages, other discrepancies were proposed. From the definition of discrepancy, its formula is calculated as follows if objective function takes the uniform distribution function on X:

$$D(P,K) = \left\{ \int_{X^2} K(x,y)\mathrm{d}F_u(x)\mathrm{d}F_u(y) - \frac{2}{n}\sum_{i=1}^{n}\int_X K(x,y)\mathrm{d}F_u(y) + \frac{1}{n^2}\sum_{i,k=1}^{n} K(x_i, x_k) \right\}^{\frac{1}{2}} \tag{15}$$

According to Fang et al. [50], the reproducing kernel functions are taken, respectively, as follows,

$$K^c(z,t) = 2^{-s}\prod_{j=1}^{s}\left(2 + \left|z_j - \frac{1}{2}\right| + \left|t_j - \frac{1}{2}\right| - \left|z_j - t_j\right|\right) \tag{16}$$

$$K^w(z,t) = \prod_{j=1}^{s}\left(\frac{3}{2} - \left|z_j - t_j\right| + \left|z_j - t_j\right|^2\right) \tag{17}$$

hence, the analytical expressions for centred discrepancy and wrap-around discrepancy are as follows:

$$\begin{aligned}\mathrm{CD}(t) = &\left\{ \left(\frac{13}{12}\right)^s - \frac{2}{n}\sum_{i=1}^{n}\prod_{j=1}^{s}\left(1 + \frac{1}{2}\left|x_{ij} - 0.5\right| - \frac{1}{2}\left|x_{ij} - 0.5\right|^2\right)\right. \\ &\left. + \frac{1}{n^2}\sum_{i=1}^{n}\sum_{k=1}^{n}\prod_{j=1}^{s}\left(1 + \frac{1}{2}\left|x_{ij} - 0.5\right| + \frac{1}{2}\left|x_{kj} - 0.5\right| - \frac{1}{2}\left|x_{ij} - x_{kj}\right|\right)\right\}^{\frac{1}{2}}\end{aligned} \tag{18}$$

$$\mathrm{WD}(t) = \left\{ -\left(\frac{4}{3}\right)^s + \frac{1}{n}\left(\frac{3}{2}\right)^s + \frac{2}{n^2}\sum_{i=1}^{n-1}\sum_{k=i+1}^{n}\prod_{j=1}^{s}\left(\frac{3}{2} - \left|x_{ij} - x_{kj}\right| + \left|x_{ij} - x_{kj}\right|^2\right)\right\}^{\frac{1}{2}} \tag{19}$$

UC-CD and UC-WD related to time t are defined and denoted by

$$\mathrm{UC-CD}(t) = 1 - \mathrm{CD}(t), \mathrm{UC-WD}(t) = 1 - \mathrm{WD}(t) \tag{20}$$

where CD(t) and WD(t) denote the modified discrepancy.

Many bubble patterns are related to time t and one piece of patterns corresponds to one t. For the calculation of UC–CD and UC–WD, using the x-axis and y-axis values of a Cartesian coordinate system to determine the position of bubble swarm is necessary.

Step 1: Transform image matrix to 0–1 matrix.

Step 2: Search the coordinates of one bubble located in top-left and bottom-right corners.

Step 3: Calculate the mean values of rows and columns of the two above coordinates.

In our work, $M = 1280$ and $N = 720$, then readjusting every piece of bubble patterns to unified pixel size of $M \times N$. The image pixel matrix is transformed into coordinate within [0,1]

$$x = \frac{x_0 - 1}{M - 1}, \quad y = \frac{N - y_0}{N - 1} \tag{21}$$

and where x_0 and y_0 denote the column and row of one image matrix, x and y denote the coordinates of horizontal axis and vertical axis in Cartesian coordinate system. Furthermore, in this work, $s = 2$, so (x, y) of one experimental point is equal to (x_{i1}, x_{i2}) of Eqs. (7) and (8).

The origin of coordinates lies in the bottom-left (BL) corner of one piece of pattern. Certainly, other three groups of transformation ways are used to make origin of coordinates locate in bottom-right (BR), top-right (TR) and top-left (TL) coiners' of one piece of patterns, respectively. Detailed formulas as follows,

$$x = \frac{M - x_0}{M - 1} \quad y = \frac{N - y_0}{N - 1} \quad x = \frac{M - x_0}{M - 1} \quad y = \frac{y_0 - 1}{N - 1} \quad x = \frac{x_0 - 1}{M - 1} \quad x = \frac{y_0 - 1}{N - 1}. \tag{22}$$

More interesting, these transform methods are different but corresponding to the coordinates rotate operation for the rectangular plane coordinate system. Hence, we will talk about the rotational invariance and neglect the different transform methods in next section.

2.4. Mixing quantification by Betti numbers

2.4.1. Multiphase mixing quantification

Now this new method is used to study the influence of the flow rate and the submerged length on the degree of the mixing homogeneity and non-homogeneity of solid and liquid. The acquisition system was shown in **Figure 8**. The patterns were gained at the speed of 30 frames per second by a camera taking 10,000 images in each experiment.

Figure 9 shows that an initial image was subtracted from each image.

2.4.2. Characterization of heat transfer process

Figure 10 shows the evolution of β_1 in the open operation images produced in representative experimental cases. β_1 for each pattern is shown as a solid line, and the horizontal dotted line corresponds to the average of β_1 [32]. The vertical dotted line corresponds to the pseudo-homogeneous time t.

Experiments indicate that, in **Figure 11**, volumetric heat transfer coefficient shows good correlation with average Betti number and pseudo-average-time value [32]. An interesting tendency is found in the better cases of L_6 and L_2, in which the larger first Betti numbers averages and

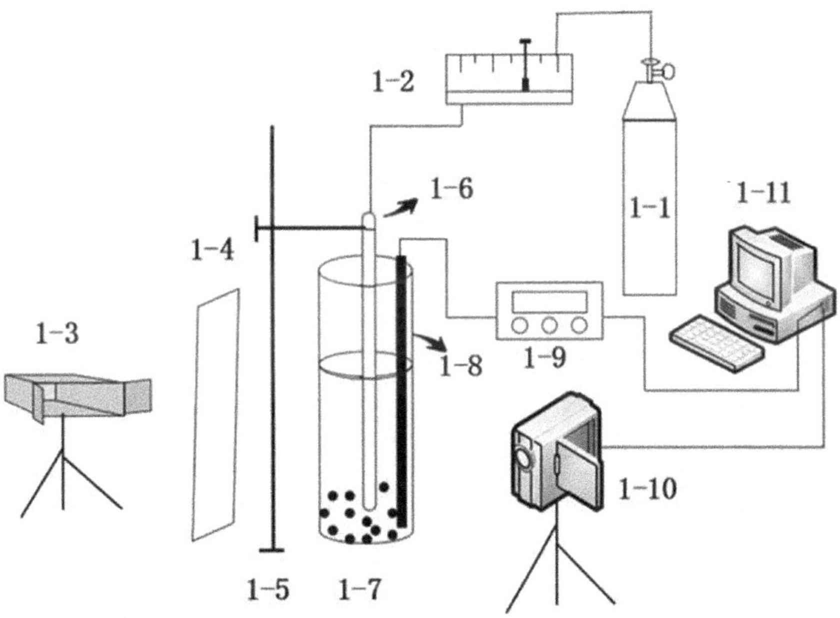

1-1: N_2、1-2: rotermeter、1-3: light source、1-4: white board、1-5: iron support stand、1-6: spray pipe、1-7: reactor、1-8: probe、1-9: conductivity meter(DDSJ-308A)、1-10: camera、1-11: computer

Figure 8. Scheme of experimental equipment.

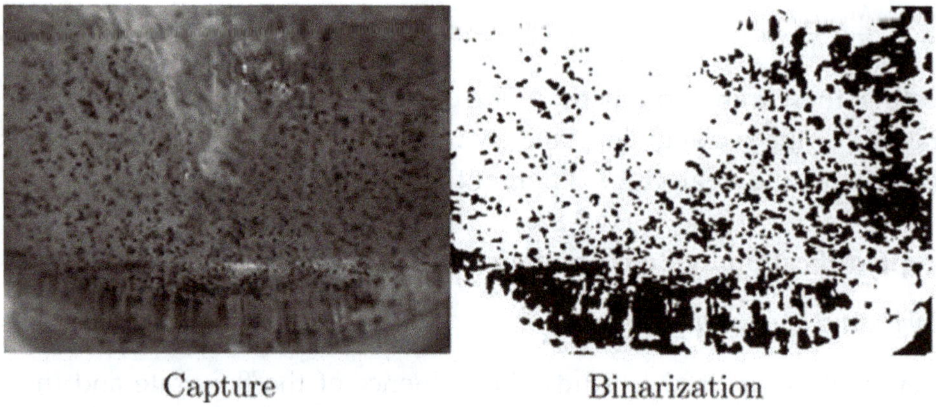

Figure 9. Binarization for one piece of images.

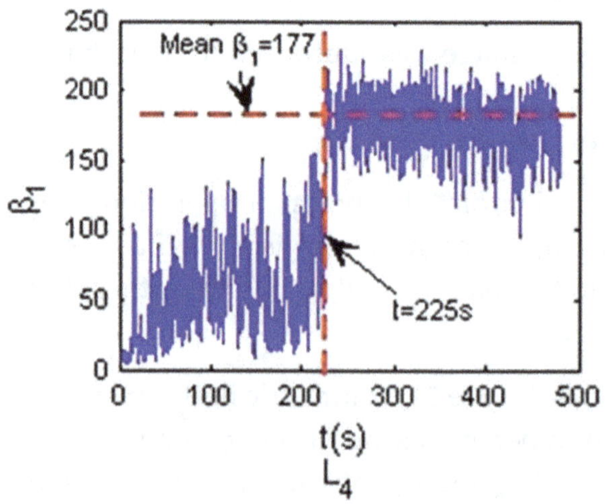

Figure 10. The evolution of Betti numbers at E_4.

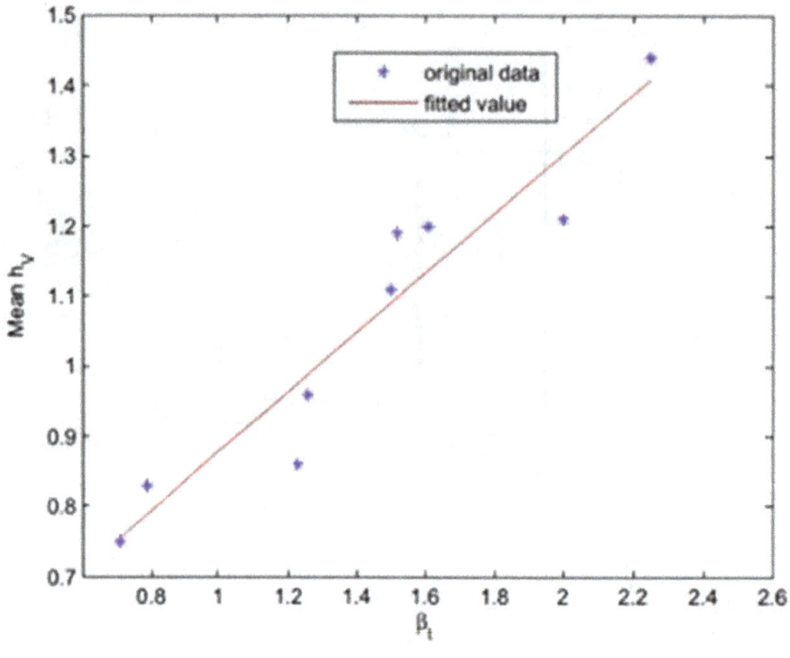

Figure 11. Fitting of β_t and $h^- V$.

shorter pseudo-homogeneous times correspond to a higher volumetric heat transfer coefficient h_V, while L_4 and L_7 are the worse cases. As shown in **Table 3**, t and $\bar{h}_V$ are nearly opposite in all cases except L_5, L_8 and L_9. Therefore, the parameters t and $\bar{\beta}_1$, which can be used to characterize a bubbling flow pattern, are both related to $\bar{h}_V$.

Parameter	E_1	E_2	E_3	E_4	E_5	E_6	E_7	E_8	E_9
T	159	93	120	226	126	92	264	105	145
$\bar{\beta}_1$	197	186	208	177	197	189	187	176	196
β_t	1.26	2	1.23	0.79	1.61	2.25	0.71	1.50	1.52
$\bar{h}_V$	0.96	1.21	0.86	0.83	1.20	1.44	0.75	1.11	1.19

Table 3. The data of the t, β^- 1, β_t, and h^- V for the entire orthogonal array table.

Let $\beta_t = \bar{\beta}_1 \cdot t^{-1}$, the tendency of β_t is consistent with that of $\bar{h}_V$. According to our analysis, a linear relation between β_t and $\bar{h}_V$ seems to be the outcome. Let $\bar{h}_V = a \times \beta_t + b$. The least-squares fitting method was used to obtain the parameters a and b between β_t and $\bar{h}_V$. In this work, $a = 0.4241$, $b = 0.4547$. The linear relationship is illustrated in **Figure 11**. The correlation coefficient is 0.95. In the end, we constructed a model on the parameters β_t and $\bar{h}_V$, which points out the relationship between the flow pattern of a bubble swarm and the heat transfer performance of a DCHE.

2.4.3. Accurate estimation of mixing time

(1) Mixing time estimations by different methods

Based on the above, β_t has been defined by the Betti number average as well as the mixing time, which is synergistic with $\bar{h}_V$. Correlation degree and correlation coefficient are used to investigate about the bubble swam patterns and heat transfer performance for the mixing time evaluation effectiveness. According to reference [36], the computing results show that the synergy by our 3σ method between β_t and $\bar{h}_V$ is much better than the other methods. In **Figure 12**, the plots show the evolution and determination of mixing time measured by different methods [36].

It is found that the correlation coefficient between $t_{3\sigma}$ and mixing time estimated by these methods is −0.2304 (mean method), 0.9494 (slope method), 0.9265 (SD method) and 0.9731 (3σ method).

In **Table 4**, μ is the mean of Betti number time series after the time t_0. t_{mean} is the mixing time by mean method. t_{slope} is the mixing time by slope method. t_{sd} is the mixing time by SD method. $t_{3\sigma}$ is the mixing time defined by 3σ method. $T_{3\sigma}$ is the inhomogeneous time by 3σ method. δ_t is the difference between $t_{3\sigma}$ and $T_{3\sigma}$. β_{mean} is calculated by mean method, and the others are defined similarly [40]. It can be quantified by time intervals δ_t between inhomogeneous time and mixing time. From the view point of the time interval, the transitional state appears the following forms: sudden change case ($\delta_t = 0$); interval case ($\delta_t > 0$); overlapping case ($\delta_t < 0$).

(2) Simulation experiments

By real data analysis of the Bitti number data, we have compared the proposed method with mean method, slope method and SD method. In order to assess the effectiveness of the new

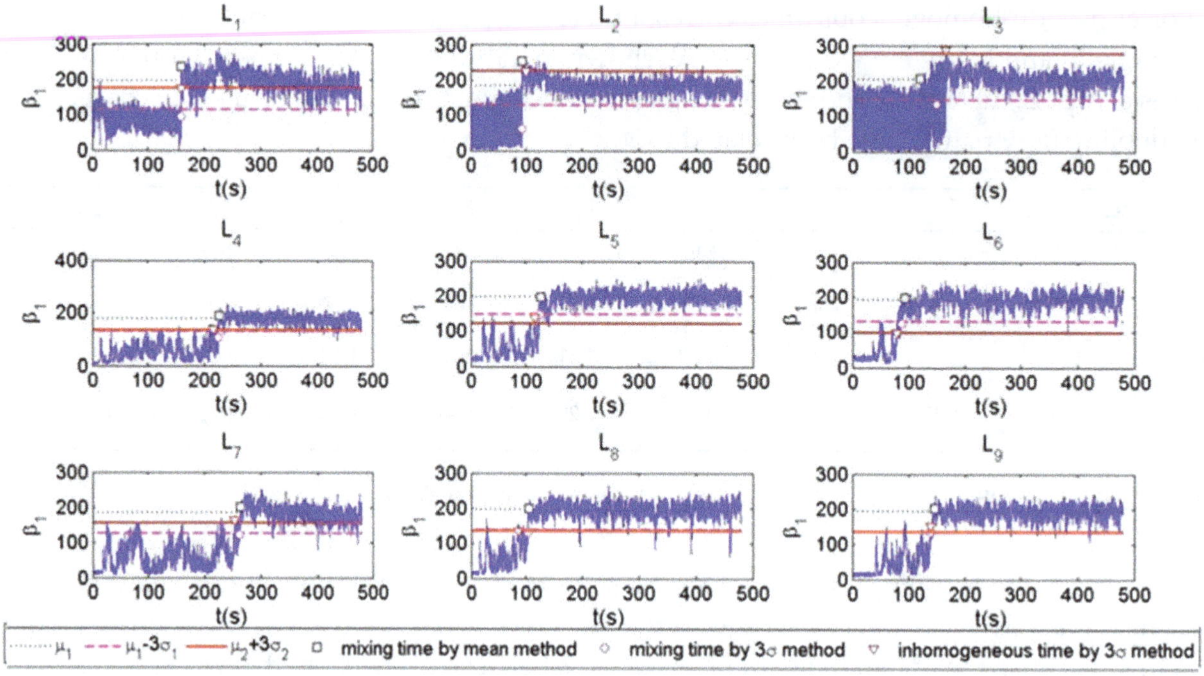

Figure 12. Evolution and determination of mixing time measured by two methods.

method and provide more evidences of good performance of this method, the mean absolute error (MAE) and the mean square error (MSE) are often used.

$$MAE = \frac{1}{n}\sum_{i=1}^{n}\left|t_i - \hat{t}_i\right| \quad MSE = \frac{1}{n}\sum_{i=1}^{n}\left(t_i - \hat{t}_i\right)^2 \tag{23}$$

where t_i is the real mixing time, and $\hat{t}_i$ is the estimate of t_i.

Index	E_1	E_2	E_3	E_4	E_5	E_6	E_7	E_8	E_9
M	197	185	206	177	197	190	186	197	192
t_{mean}	159	93	120	226	126	92	264	105	145
t_{slope}	153	80	112	216	123	83	249	97	137
t_{sd}	135	71	158	177	134	101	232	99	133
$t_{3\sigma}$	159	93	150	225	124	87	262	103	138
$T_{3\sigma}$	159	99	165	215	115	79	252	85	139
δ_t	0	-6	-15	10	9	8	10	18	-1
β_{mean}	1.24	1.99	1.72	0.78	1.56	2.07	0.70	1.88	1.32
β_{slope}	1.29	2.31	1.84	0.82	1.60	2.29	0.75	2.03	1.40
β_{sd}	1.46	2.61	1.30	1.00	1.47	1.88	0.80	1.99	1.44
$\beta_{3\sigma}$	1.24	1.99	1.37	0.79	1.59	2.18	0.71	1.91	1.39

Table 4. Computing results of mixing performance by four methods.

From **Table 5**, we can see that proposed method has a distinct advantage [36]. **Figure 13** shows an example of 1000 simulation results.

Index	Mean method	Slope method	SD method	3σ method
MAE	6.22	6.66	8.07	3.38
MSE	41.78	44.85	101.85	12.11

Table 5. Comparison of computer simulation results by 1000 times.

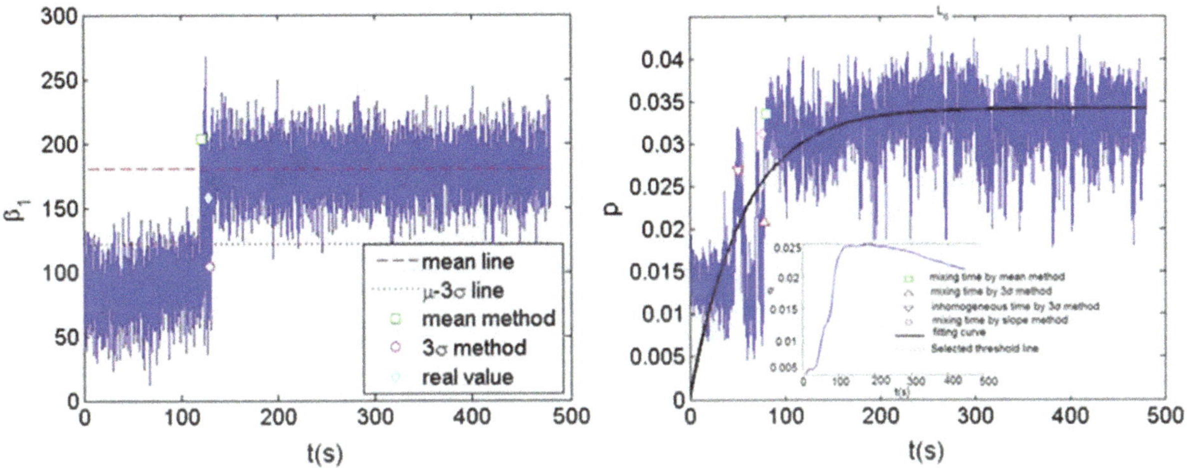

Figure 13. One of these simulation results by 1000 times.

2.5. Measuring bubbles uniformity by discrepancy

2.5.1. Quantifying mixing efficiency using L_2-star discrepancy

(1) Quantification of mixing efficiency

The variation of the uniformity coefficient with frames can be an effective method to determine the critical mixing time and mixing uniform.

In **Figure 14**, quantitative comparisons of the homogenization curve and mixing time predicted by the uniformity coefficient method are conducted with reported experimental data and other predictions by the Betti numbers method.

The comparisons show that good agreements of the mixing time obtained by Betti numbers method and uniformity coefficient method have also been achieved as given in **Table 6** [40].

(2) Recognition of local and global uniformity

Numerical simulations. Generated small sets were used with the same Betti number randomly to assess the performance of the UC implementation for approximating the discrepancy of a given set of points. By checking a large number of experimental images, we found the sizes

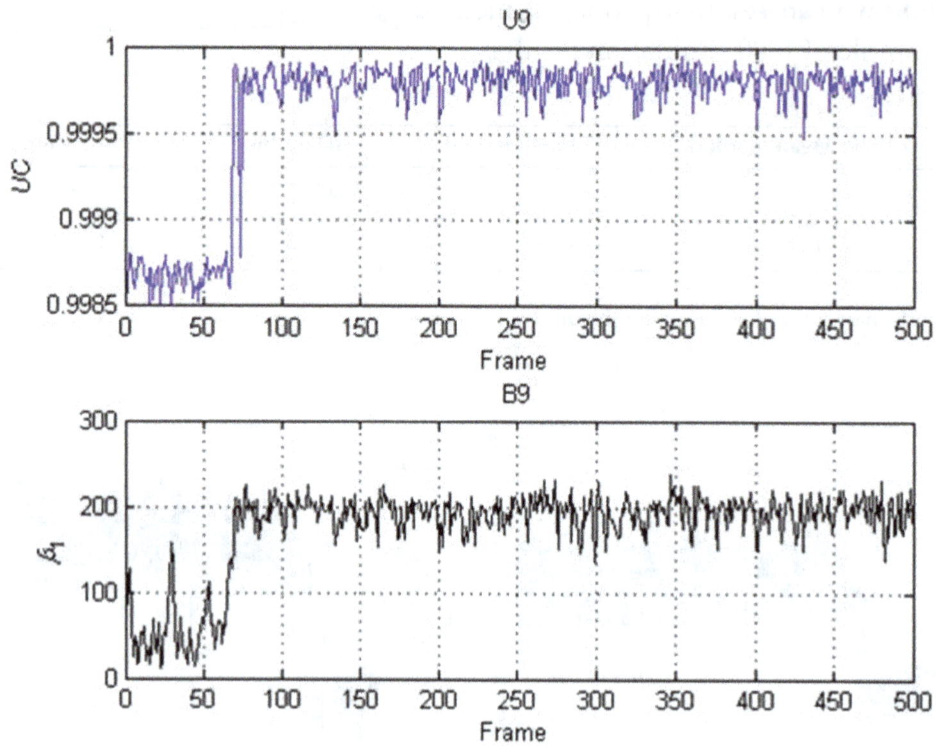

Figure 14. Evolution of mixing efficiency by the uniformity coefficient method (denoted as U9) and Betti numbers method (denoted as B9).

Parameter	E_1	E_2	E_3	E_4	E_5	E_6	E_7	E_8	E_9
t	159	93	120	226	126	92	264	105	145
t^*	151	97	172	224	120	84	259	118	127

Table 6. Computing results of mixing time by the Betti numbers method (t) and uniformity coefficient method (t^*) at the whole orthogonal arrays table.

of the bubbles are nearly the same. Two hundred and thirty-four bubbles have the same small blank area of the radius of the circle. MATLAB software randomly selects the centre. Although 234 bubbles have the same size, they are spread in different places.

These plots of **Figure 15** are got by simulation with 720 lines and 1280 rows. Among them, the simulation 1 (**Figure 15a**) is the corresponding local uniform description, simulation 2 (**Figure 15b**) corresponds to the portrayal of global uniform. The lattice points generate the example of lattice uniform used in the demonstration of the algorithm. Under the guidance of the lattice points method, the simulation 3 (**Figure 15c**) performs for a lattice points set, which has 234 bubbles. It is can be seen that the lattice uniform is most accurate uniformly. One can take set of points or objects, which are generated by these experiments and simulations to check the algorithm.

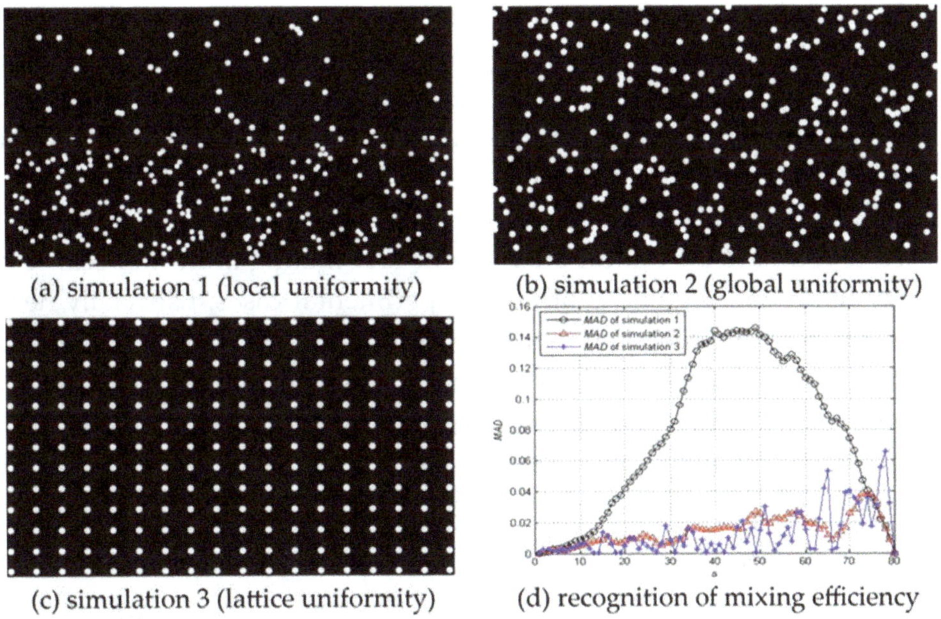

Figure 15. Numerical simulations and MAD evolutions of different mixing efficiency.

Experimental examples. The plots in **Figure 16** are obtained by experiments E_3 and E_6. The experiments are response to the worse and better cases. **Figure 16a–c** shows the difference of mixing uniformity in case 1 of E_6 with $\beta_1 = 160$, whereas **Figure 16e**, **d**, and **f** shows the difference of mixing uniformity in case 2 of E_3 with $\beta_1 = 194$. Comparison results show that the different experimental cases with the same Betti numbers can be identified by the MAD evolutions. The results in **Figure 16** indicate that the uniformity coefficient method might be enough to obtain a good estimation and quantification of multiphase mixing effects.

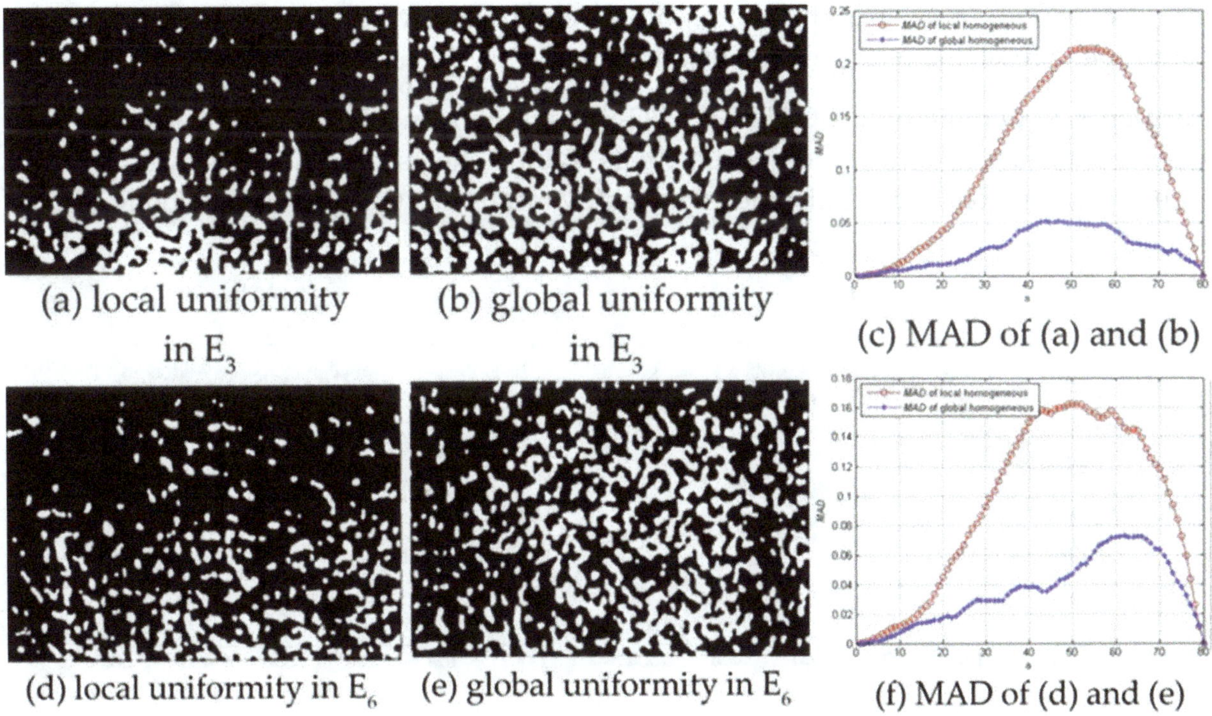

Figure 16. MAD evolutions of different experimental cases with the same Betti numbers.

2.5.2. Modified L_p-star discrepancy for measuring mixing uniformity

(1) Video-frequency image sequence of experimental cases

In **Figure 17**, quantitative comparisons of the homogenization curve utilizing uniformity coefficient with modified discrepancy methods are conducted with reported experimental data and the other method is UC-LD. The variation of the UC-LD and UC-WD versus i can be effectively used to estimate the critical mixing time and mixing uniform because of the similar evolutionary trend and regularity of most experimental cases. Especially here there is a clearly and distinctively different at the early phase of mixing process of experimental case E_2 (see **Figure 17a**), and obvious differences in numerical performance have a deep significant role to assess the mixing efficiency. For another experimental case E_7, UC-CD and UC-WD successfully measured the mixing process from no uniformity to uniformity clearly, just like UC-LD (see **Figure 17b**).

(2) Verification of properties

Suppose coordinates of bubble swarms in **Figure 2** can be denoted as follows,

$$X_p = \begin{pmatrix} x_{11} x_{21} \\ x_{21} x_{22} \\ \vdots \ \vdots \\ x_{n1} x_{n2} \end{pmatrix} \tag{24}$$

where n is the number of bubble swarms or experimental points noted earlier. With one important note as mentioned above, the first rank elements $x_{11}, x_{21}, \ldots, x_{n1}$ of X_P correspond to the x values identified in Eq. (21) the second rank elements $x_{12}, x_{22}, \ldots, x_{n2}$ of X_P correspond to the y values identified in Eq. (21). UC-CD and UC-WD are formulated as the two methods of measures of uniformity. In addition, they are dimensionless, which vary from 0 to 1. And it is hard for it to reach the certain endpoint values.

Invariance to permutation. UC-CD and UC-WD are permanent for disrupted order of the experimental factors or the experimental points. **Table 7** shows verification data about invariance to permutation. Bubbles disordered in **Table 7** means that the operator randomly shuffles

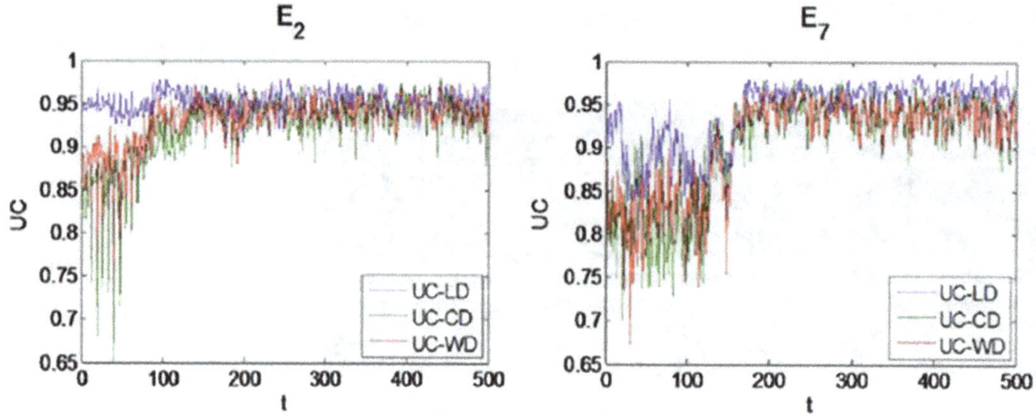

Figure 17. Uniformity coefficients of E_2 and E_7.

rows in Eq. (24). Coordinates disordered in **Table 7** means that the operator randomly shuffles columns in Eq. (24). It is quite clear found that the two parameters identically equal to 0.9751 and 0.9723 individually even though the disorder of rows or ranks happened.

Modified UC	In proper order	Disordered Bubble	Coordinates	Both
UC-CD	0.9289	0.9289	0.9289	0.9289
UC-WD	0.9311	0.9311	0.9311	0.9311

Table 7. The data of verification of invariance to permutation.

Invariance under reflection. In theory, UC-CD(P) and UC-WD(P) are invariant if x_{i1} and x_{i2} are replaced individually by 1-x_{i1} and 1-x_{i2}, $1\le i\le n$. The data of verification of invariance under reflection are depicted in **Table 8**. More in detail, x = 1/2 in **Table 8** means that the operator rotates the origin of coordinates system from bottom-left to bottom-right in a piece of patterns, y = 1/2 in **Table 8** means that the operator rotates the origin of coordinates system from bottom-left to top-left in a piece of patterns, x = 1/2 and y = 1/2 (both) in **Table 8** means that the manipulator spins the origin of coordinates system from bottom-left to top-left. It is obviously found that the two parameters individually equal to 0.9751 and 0.9723 even though the origin of coordinates system is rotated.

Projection uniformity. The projection uniformity over all sub-dimensions can be considered, and UC-CD(P) and UC-WD(P) are also invariant to it in theory. The data of verification of projection uniformity are depicted in **Table 9**. Projected to x-axis in **Table 9** refers to set x_{i2} = 0 in Eq. (24), projected to y-axis refers to x_{i1} = 0 and projected to origin refers to x_{i1} = 0 and x_{i2} = 0. It is quite clear that the projection uniformity over all sub-dimensions can be obtained. All this to say the consideration is not insignificant for high-dimensional cases.

Modified UC	No reflected	Reflected		
		x = 1/2	y = 1/2	Both
UC-CD	0.9289	0.9289	0.9289	0.9289
UC-WD	0.9311	0.9311	0.9311	0.9311

Table 8. The data of verification of invariance under reflection.

Modified UC	No projected	y = 0	x = 0	Projected to origin
UC-CD	0.9289	0.3946	0.3936	0.0554
UC-WD	0.9311	0.5263	0.5253	0.3128

Table 9. The data of verification of projection uniformity.

(3) Time complexity. The time complexity of different methods is shown in **Table 10**. Through experimental comparison, we may draw the conclusion that UC-CD and UC-WD can replace

UC-LD and Betti numbers to some extent. Determination of the position of bubble swarms spends too much time, which leads to make the upper time complicated. But, other progressive technology can change this disadvantage.

Single image	Betti numbers	UC-LD	UC-CD	UC-WD
Reckoning	$\beta_1 = 174$	0.9632	0.9289	0.9311
Running time (s)	3.07	2.61	18.33	18.23

Table 10. The data of time of different methods.

(4) Numerical simulations and experimental examples. In order to assess the performance of UC-CD and UC-WD implementation for approximating the discrepancy of a given set of points, the three sets in **Figure 14** were used. **Table 11** shows that UC-LD of the three simulated images are affected by initial position, but UC-CD and UC-WD not. Comparing the modified UC of **Figure 14b** and c, the absolute difference |0.9751–0.9657| = 0.0094 is less than |0.9623–0.9464| = 0.0159 since **Figure 14b** and **c** seems to have the same degree of mixing uniformity. So it is concluded that UC-CD may outperform UC-WD and perform more sensitive for practical engineering application in some sense. The data in **Table 11** also show that the difference of mixing uniformity coefficients including UC-LD, UC-CD and UC-WD with the same Betti numbers in **Figure 16a**, b, d and e. Meanwhile, it is noticed that different initial positions are response to different UC-LDs, which bring unreasonable and bias measurement of uniformity in practice. In other words, UC-LD may result in multiple values, but UC-CD and UC-WD do not have this problem. Moreover, the absolute difference of UC-CDs is larger than that of UC-WDs. The comparison result shows that UC-CD performs more sensitive than UC-WD in identifying the different patterns with the same Betti numbers. Those are the major of our presented work in this part.

2.6. Conclusions

1. Because a new technique based on algebraic topology was introduced for quantifying the efficiency of multiphase mixing, the mixture homogeneity and the non-homogeneity

Plots		UC-LD BL	BR	TR	Figure 15	UC-CD	UC-WD	TL (a)
0.9326	0.9033	0.9381	0.9189	0.8052	0.8493	(b)	0.9854	0.9866
0.9854	0.9955	0.9751	0.9623	(c)	0.9925	0.9932	0.9930	0.9925
0.9657	0.9464	Figure 16	(a)	0.8699	0.9077	0.9094	0.9033	0.9102
0.9170	(b)		0.9826	0.9523	0.9823	0.9869	0.9474	0.9436
(d)	0.8686		0.9507	0.8645	0.9056	0.8793	0.9121	(e)
0.9676	0.9794		0.9504	0.9825	0.9429	0.9306		

Table 11. The data of numerical simulations and experimental examples.

of the mixture can be characterized by the Betti numbers for binary images of the patterns. The zeroth Betti numbers β_0 are used to estimate the numbers of pieces in the patterns, leading to a useful parameter to characterize the mixture homogeneity. The first Betti numbers β_1 are introduced to characterize the non-homogeneity of the mixture. This novel method may be applied for studying a variety of multiphase mixing problems in which multiphase components or tracers are visually distinguishable.

2. In a DCHE, Betti number can estimate the number of bubbles assembling in flow patterns and to get the pseudo-homogeneous time. Experimental analysis constructs a simple linear model representing a bubble swarm and the heat transfer performance of a DCHE. In addition, the Betti number average and the pseudo-homogeneous time t define a new index. A better fitting curve about and the volumetric heat transfer coefficient average is received and its correlation coefficient is 0.95. A paradigm is established on the basis of this novel method for the study of flow patterns and heat transfer performance. And the paradigm offers an optional route to study the relationship of flow patterns and heat transfer in other heat transfer processes.

3. A novel method relying on image analysis and statistics was developed to estimate the mixing time accurately in a DCHE. The three sigma method researches the critical point determination of the pseudo-homogeneous process, which satisfies approximately normal distribution and surpasses the range of occurring twice. The mean value method, slope method and standard deviation method make quantitative comparisons of the mixing time. In addition, time intervals between in-homogeneous time and mixing time quantify the quasi-steady state. Neglecting critical point could make substantial errors in mixing time estimation, which is proved.

4. A straightforward method, uniformity coefficient (UC) method based on L_2-star discrepancy (UC-LD), presented for assessing the uniformity and mixing time of bubbles behind the viewing windows in a DCHE is effective. An imaging technique processed in the MATLAB software tracks the evolution of bubbles movement. The local discrepancy of a set of bubbles seems to be helpful to judge the difference between theory and empirical distribution. The UC links to a discrepancy, leading to a useful parameter which expresses the mixture homogeneity and mixing time. A comparison was made between the mixing time and uniformity obtained by UC method and the data obtained by Betti numbers method. Discussing the simulation and experiments conducted between local and global uniform (with the same Betti numbers) and examples are given for illustration. UC method calculates the space-time features of the mixing process successfully. The UC curves can study and compare mixing efficiency of different systems with the novel method, which can generate accurate mixing information and has a well reliability.

5. The properties of UC have been explored and there was a great influence of calculating the initial position on the original UC, namely UC-LD. The UC-LD method applies to the modified uniformity coefficient based on modified L_2-star discrepancies. A Koksma-Hlawka-type inequality is applicable to uniformity coefficient based on wrap-around discrepancy (UC-WD) as well as uniformity coefficient based on centred discrepancy(UC-CD) theoretically. In addition, they show some advantages includ-

ing rotation invariance (reflection invariance), permutation invariance and the ability to measure projection uniformity. Analysing real experimental cases and simulating to evaluate the performance of the novel method. The experimental results show that UC-CD presents more sensitive performance than UC-WD so the UC-CD is more appropriate for industry.

In summary, we believe that on the basis of a large amount of previously published works, the complexity of the bubble swarm patterns can be reduced and their mechanisms clarified, and the heat transfer performance in a DCHE can be elucidated.

Acknowledgement

This work is supported by the National Natural Science Foundation of China (51666006, 51406071) and Scientific and Technological Leading Talent Projects in Yunnan Province (2015HA019).

Author details

Hua Wang*, Qingtai Xiao and Jianxin Xu

*Address all correspondence to: wanghua65@163.com

Kunming University of Science and Technology, Kunming, PR China

References

[1] S. Kar, X.D. Chen, M.I. Nelson, Direct-contact heat transfer coefficient for condensing vapour bubble in stagnant liquid pool, Chem. Eng. Res. Des. 85 (3) (2007) 320–328.

[2] C.P. Ribeiro Jr., P.L.C. Lage, Population balance modeling of bubble size distributions in a direct-contact evaporator using a sparger model, Chem. Eng. Sci. 59 (12) (2004) 2363–2377.

[3] T. Lemenand, C. Durandal, D. Della Valle, H. Peerhossaini, Turbulent directcontact heat transfer between two immiscible fluids, Int. J. Therm. Sci. 49 (10) (2010) 1886–1898.

[4] T. Nomura, M. Tsubota, T. Oya, N. Okinaka, T. Akiyama, Heat storage in direct contact heat exchanger with phase change material, Appl. Therm. Eng. 50 (1) (2013) 26–34.

[5] H.B. Mahood, Direct-contact heat transfer of a single volatile liquid drop evaporation in an immiscible liquid, Desalination 222 (1–3) (2008) 656–665.

[6] Y.J. Hyun, J.H. Hyun, W.G. Chun, Y.H. Kang, An experimental investigation into the operation of a direct contact heat exchanger for solar exploitation, Int. Commun. Heat Mass Transfer 32 (3–4) (2005) 425–434.

[7] M.N.A. Hawlader, M.A. Wahed, Analyses of ice slurry formation using direct contact heat transfer, Appl. Energy 86 (7–8) (2009) 1170–1178.

[8] Nguyen D, Rasmuson A, Niklasson Bjorn I, Thalberg K. CFD simulation of transient particle mixing in a high shear mixer. Powder Technol 258 (2014) 324–330.

[9] S. Kim, A.N. Nkaya, T. Dyakowski, Measurement of mixing of two miscible liquids in a stirred vessel with electrical resistance tomography, Int. Commun. Heat Mass Transfer 33 (9) (2006) 1088–1095.

[10] T. Maruyama, S. Yoshida, T. Mizushina, The flow regime transition in a bubble column, J. Chem. Eng. Jpn. 14 (1981) 352–357.

[11] H.T. Bi, J.R. Grace, Regime transitions: analogy between gas-liquid cocurrent upward flow and gas-solid upward transport, Int. J. Multiphase Flow 22 (2000) 1–19.

[12] M.C. Ruzicka, J. Zahradnik, J. Drahos, N.H. Thomas, Homogeneous heterogeneous regime transition in bubble columns, Chem. Eng. Sci. 56 (15) (2001) 4609–4626.

[13] C.P. Ribeiro Jr., P.L.C. Lage, Direct-contact evaporation in the homogeneous and heterogeneous bubbling regimes. Part I: experimental analysis, Int. J. Heat Mass Transfer 47 (17–18) (2004) 3825–3840.

[14] C.P. Ribeiro Jr., C.P. Borges, P.L.C. Lage, Sparger effects during the concentration of synthetic fruit juices by direct-contact evaporation, J. Food Eng. 79 (3) (2007) 979–988.

[15] A.-L. Le Coënt, A. Rivoire, S. Briancon, J. Lieto, An original image-processing technique for obtaining the mixing time: The box-counting with erosions method, Powder Technol. 152 (1–3) (2005) 62–71.

[16] S.M. Peyghambarzadeh, M.M. Sarafraz, N. Vaeli, E. Ameri, A. Vatani, M. Jamialahmadi, Forced convective and subcooled flow boiling heat transfer to pure water and n-heptane in an annular heat exchanger, Ann. Nucl. Energy 53 (0) (2013) 401–410.

[17] J.M. Hyde, M.K. Miller, M.G. Hetherington, A. Cerezo, G.D.W. Smith, C.M. Elliott, Spinodal decomposition in Fe-Cr alloys: experimental study at the atomic level and comparison with computer models III. Development of morphology, Acta Metall. Mater. 43 (9) (1995) 3415–3426.

[18] M. Gameiro, K. Mischaikow, T. Wanner, Evolution of pattern complexity in the Cahn-Hilliard theory of phase separation, Acta Mater. 53 (2005) 693–704.

[19] J.M. Friedrich, Quantitative methods for three-dimensional comparison and petrographic description of chondrites, Comput. Geosci. 34 (12) (2008) 1926–1935.

[20] S.S. Gulawani, S.K. Dahikar, C.S. Mathpati, J.B. Joshi, M.S. Shah, C.S. RamaPrasad, D.S. Shukla, Analysis of flow pattern and heat transfer in direct contact condensation, Chem. Eng. Sci. 64 (8) (2009) 1719–1738.

[21] S.S. Gulawani, J.B. Joshi, M.S. Shah, C.S. RamaPrasad, D.S. Shukla, CFD analysis of flow pattern and heat transfer in direct contact steam condensation, Chem. Eng. Sci. 61 (16) (2006) 5204–5220.

[22] S.K. Dahikar, M.J. Sathe, J.B. Joshi, Investigation of flow and temperature patterns in direct contact condensation using PIV, PLIF and CFD, Chem. Eng. Sci. 65 (16) (2010) 4606–4620.

[23] A.B. Tayler, D.J. Holland, A.J. Sederman, L.F. Gladden, Exploring the origins of turbulence in multiphase flow using compressed sensing MRI, Phys. Rev. Lett. 108 (26) (2012) 1–5. 264504.

[24] Q. Zhang, Y. Yong, Z.S. Mao, C. Yang, C. Zhao, Experimental determination and numerical simulation of mixing time in a gas–liquid stirred tank, Chem. Eng. Sci. 64 (12) (2009) 2926–2933.

[25] A. Zhang, V.L. Tsang, R. Korke-Kshirsagar, T. Ryll, Effects of pH probe lag on bioreactor mixing time estimation, Process Biochem. 49 (6) (2014) 913–916.

[26] Y. Zhao, X. Li, J. Cheng, C. Yang, Z.S. Mao, Experimental study on liquid–liquid macromixing in a stirred tank, Chem. Eng. Sci. 50 (10) (2011) 5952–5958.

[27] D.M. Fries, P.R.V. Rohr, Liquid mixing in gas–liquid two-phase flow by meandering microchannels, Chem. Eng. Sci. 64 (6) (2009) 1326–1335.

[28] D. Nguyen, A. Rasmuson, I.N. Björn, K. Thalberg, Cfd simulation of transient particle mixing in a high shear mixer, Powder Technol. 258 (258) (2014) 324–330.

[29] S.L. Yeoh, G. Papadakis, M. Yianneskis, Determination of mixing time and degree of homogeneity in stirred vessels with large eddy simulation, Chem. Eng. Sci. 60 (s 8–9) (2005) 2293–2302.

[30] J. Behin, N. Farhadian, Residence time distribution measurements in a two dimensional rectangular airlift reactor by digital image processing, Exp. Thermal Fluid Sci. 51 (11) (2013) 244–250.

[31] F. Cabaret, S. Bonnot, L. Fradette, P.A. Tanguy, Mixing time analysis using colorimetric methods and image processing, Ind. Eng. Chem. Res. 46 (14) (2007) 5032–5042.

[32] J. Huang, J. Xu, X. Sang, H. Wang, W. Hua, Quantifying the synergy of bubble swarm patterns and heat transfer performance using computational homology, Int. J. Heat Mass Transf. 75 (4) (2014) 497–503.

[33] G. Rodriguez, T. Anderlei, M. Micheletti, M. Yianneskis, A. Ducci, On the measurement and scaling of mixing time in orbitally shaken bioreactors, Biochem. Eng. J. 82 (3) (2014) 10–21.

[34] W.H. Woodall, Controversies and contradictions in statistical process control, J. Qual. Technol. 1 (4) (2000) 341–350.

[35] M. Panda, P.M. Khilar, Distributed self fault diagnosis algorithm for large scale wireless sensor networks using modified three sigma edit test, Ad Hoc Netw. 25 (2015) 170–184.

[36] Xu J, Xiao Q, Fei Y, et al. Accurate estimation of mixing time in a direct contact boiling heat transfer process using statistical methods, Int. Commun. Heat Mass Transf. 75 (2016) 162–168.

[37] Fang KT, Wang Y. Number theoretic methods in statistics. Chapman & Hall CRC Press London, UK: 1994.

[38] Fang KT. Uniform design application of number theory to experimental design. Acta Math Appl Sin 3 (1980) 363e72.

[39] Xu J, Xiao Q, Chen Y, et al. A modified L_2-star discrepancy method for measuring mixing uniformity in a direct contact heat exchanger,Int. J. Heat Mass Transf. 97 (2016) 70–76.

[40] Y. Fei, Q.-T. Xiao, J.-X. Xu, J.-X. Pan, S.-B. Wang, H. Wang, J.-W. Huang, A novel approach for measuring bubbles uniformity and mixing efficiency in a direct contact heat exchanger, Energy 93 (2) (2015) 2313–2320.

[41] S. Heinrich, Efficient algorithms for computing the -discrepancy, Math. Comput. Am. Math. Soc. 65 (216) (1996) 1621–1633.

[42] F. Hickernell, A generalized discrepancy and quadrature error bound, Math. Comput. Am. Math. Soc. 67 (221) (1998) 299–322.

[43] S. Thongwik, N. Vorayos, T. Kiatsiriroat, A. Nuntaphan, Thermal analysis of slurry ice production system using direct contact heat transfer of carbon dioxide and water mixture, Int. Commun. Heat Mass Transfer 35 (6) (2008) 756–761.

[44] T. Kaczynski, K. Mischaikow, M. Mrozek, Computational Homology, SpringerVerlag, New York, 2004.

[45] CHomP. http://chomp.rutgers.edu/software.

[46] N.W. Smirnow, I.W. Dunin-Barkowski, Mathematische statistik in der technik, Veb Deutscher Verlag Der Wissenschaften, Berlin, 1969.

[47] J. Xu, H. Wang, H. Fang, Multiphase mixing quantification by computational homology and imaging analysis, Appl. Math. Modell. 35 (5) (2011) 2160–2171.

[48] Niederreiter H. Random number generation and quasi-Monte Carlo methods. Philadelphia: SIAM; 1992. MR 93h 62135.

[49] T.T. Warnock, Computational investigations of low-discrepancy pointsets, in: S.K. Zaremba (Ed.), Applications of Number Theory to Numerical Analysis, Academic Press, New York, 1972, pp. 319–343.

[50] K.-T. Fang, R. Li, A. Sudjianto, Design and Modeling for Computer Experiments, CRC Press, Boca Raton, 2005.

9

Basic Aspects of Gas Turbine Heat Transfer

Shailendra Naik

Abstract

The use of gas turbines for power generation and electricity production in both single cycle and combined cycle plant operation is extensive and will continue to globally grow into the future. Due to its high power density and ability to convert gaseous and liquid fuel into mechanical work with very high thermodynamic efficiencies, significant efforts continue today to further increase both the power output and thermodynamic efficiencies of the gas turbine. In particular, the aerothermal design of gas turbine components has progressed at a rapid pace in the last decade with all gas turbine manufacturers, in order to achieve higher thermodynamic efficiencies. This has been achieved by using higher turbine inlet temperatures and pressures, advanced turbine aerodynamics and efficient cooling systems of turbine airofoils, and advanced high temperature alloys, metallic coatings, and ceramic thermal barrier coatings. In this chapter, issues related to the thermal design of gas turbine blades are highlighted and several heat transfer technologies are examined, such as convective cooling, impingement cooling, film cooling, and application of thermal barrier coatings. Typical methods for validating the thermal designs of gas turbine airofoils are also outlined.

Keywords: gas turbines, airofoils, efficiency, heat transfer, aerodynamics, film cooling, convective cooling, impingement, turbulator, pins, thermal barrier coating, compressible flows

1. Introduction

The aerothermal design of advanced gas turbines has progressed significantly in the last decade, primarily due to the requirement of increased turbine efficiencies and power. Performance increases are driven not only for reducing the consumption of fuel and the subsequent cost benefits, but also to reduce the emissions of CO_2, which is a primary component for the increased global warming. Over the last decade, major gas turbine performance enhancements have been achieved by the use of higher turbine inlet temperatures and pressures, design of advanced turbine aerodynamics, through reductions in turbine cooling and leakage air, and via the introduction of new high temperature alloys, metallic antioxidation coatings, and thermal barrier coatings. In today's energy market, there is wide range of gas turbines ranging from 1 to 500 MW and can operate with low and high calorific fuels. **Figure 1** shows the GT26 heavy duty gas turbine [1, 2].

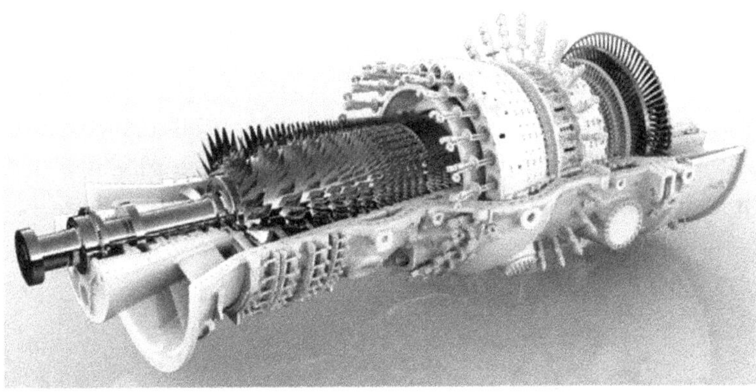

Figure 1. Heavy duty gas turbine, GT26.

The operation of a gas turbine, which essentially consists of four major components; compressor, combustor, turbine, and the exhaust diffuser, is governed by the Brayton thermodynamic cycle. For simple power generation applications, a generator is normally coupled to the gas turbine, whereby the mechanical work generated by the turbine is converted to useful electrical energy. In today's energy market, most gas turbine-based power plants are operated in combined cycle operation mode. **Figure 2** shows a typical component layout of a combined cycle plant, whereby the gas turbine plant is coupled to a steam turbine plant via the heat recovery steam generator. Thermodynamically, the gas turbine operates in a Brayton cycle, whereas the steam turbine operates in a Rankine cycle. Due to this combination, **Figure 2** highlights that combined cycle efficiencies are significantly higher than that of a gas turbine in simple operation. There are many variations of the gas turbine combined cycle plant and the interdependency of the component efficiencies and plant operating conditions. An extensive overview of industrial gas turbine combined cycle plant is given in Ref. [1].

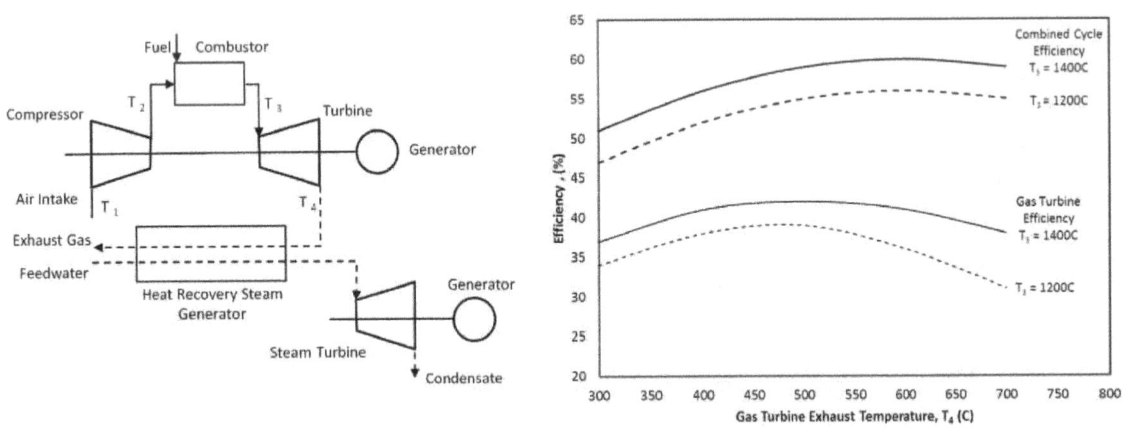

Figure 2. Basic combined cycle plant arrangement, adapted from Ref. [1].

The historical progress in increased combined cycle plant efficiencies was recently reviewed in Ref. [3] and is highlighted in **Figure 3**, which predicts a continuous growth in cycle efficiencies approaching 65% over the next decades. **Figure 3** also shows that a major part of the current growth in combined cycle efficiencies is attributed to improvements in gas turbine thermodynamic efficiencies, particularly with the H and J class gas turbines. A key driver for the latter has been the increased turbine inlet temperatures, and as shown in **Figure 4**, this has also resulted in the development of high-grade alloy, coatings, and very efficient airofoil cooling designs which can maintain the blade metal temperatures and structural integrity for long continuous operating periods. In this chapter, issues related to the thermal design of gas turbine blades are examined and the various cooling technologies are outlined. In addition, typical methods for validating the thermal designs of gas turbine airofoils are also outlined.

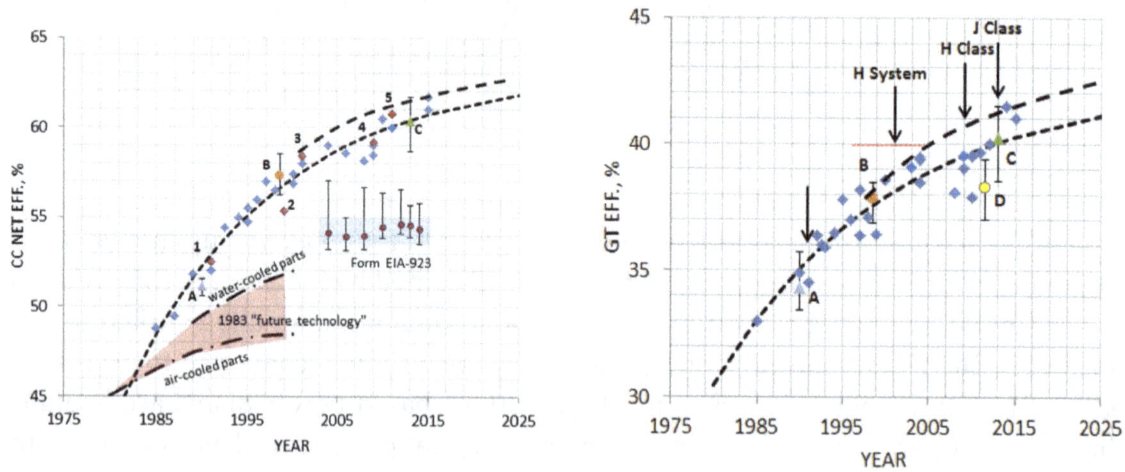

Figure 3. Performance evolution of combined cycle and single cycle gas turbine [3].

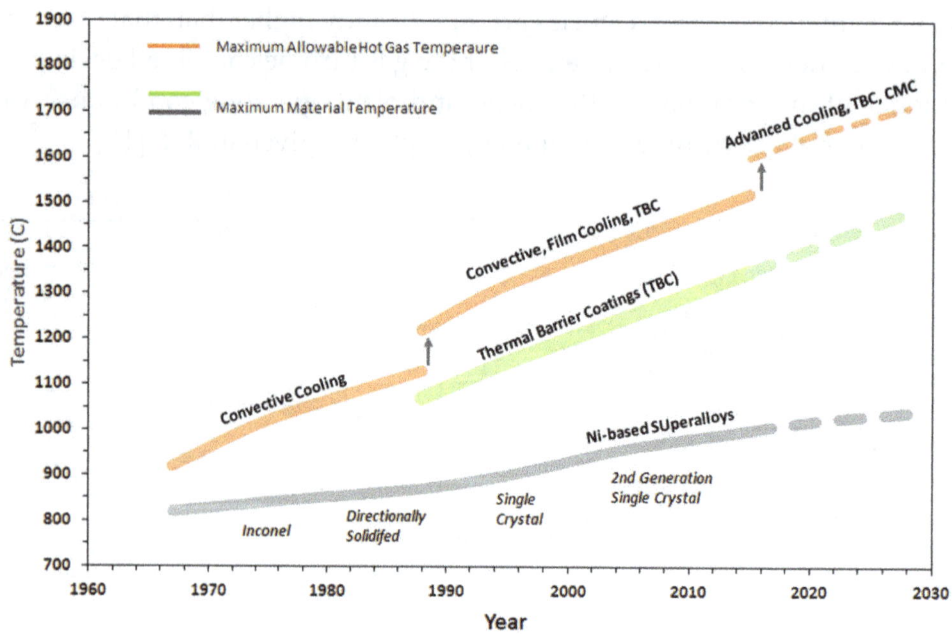

Figure 4. Evolution of gas turbine hot gas temperatures, materials, and cooling technology, adapted from Ref. [7].

2. Design consideration of cooled turbine blades

In the design of air cooled gas turbine blades, there are several different factors related to the integration of a turbine blade thermal design into the overall gas turbine. Some of the key factors which influence the overall design of the turbine blade include [4–6, 51],

- Overall gas turbine performance (power output and efficiency) and airofoil component lifetime requirements.
- Variation of ambient conditions, start-up load gradients, and shut-down conditions.
- Turbine aerodynamics, external heat loads to airofoils and turbine inlet temperatures.
- Hot gas temperature, pressure, and velocity profiles from the combustor chamber, and the expansion characteristics of the hot gas within the turbine.
- Choice of coolant from the compressor bleeds and the supply conditions over the entire operating envelope of the gas turbine.
- Geometrical clearances and gaps.
- Blade material and its properties at elevated temperatures.
- Manufacturing capability of the blade internal cooling core, machining of film cooling holes, application of thermal barrier coatings, and overall manufacturing costs.
- Maintenance methods and reconditioning of the turbine blades.

In **Figure 5**, some of the above parameters are highlighted, such as the impact of the coolant extracted from the different stages of the compressors. The front stage of the turbine will normally use the coolant extracted with the highest pressures, while the middle and rear turbine stages progressively use coolant extracted with lower pressures and temperatures. For the rear stage airofoils, the cooling systems are normally low pressure drops systems and do not have features such as film cooling and impingement cooling. The front stage airofoils however do have cooling systems with film cooling and impingement, as they are generally fed with the high pressure compressor end air. In addition to the impact of the air flow system, another major interface parameter for designing the airofoil cooling system is the combustor hot gas temperature and its distribution [8]. **Figure 5** shows a schematic of the hot gas distribution at the turbine inlet, which is generally nonuniform, and dependant on the upstream combustor and burner design. As **Figure 5** shows, typically there is a radial distribution of the hot gas temperature, which is commonly referred to as the profile factor or the radial temperature distortion factor (RTDF). In addition to this, there is also a circumferential temperature distribution which is referred to as the pattern factor or the outer temperature distortion factor (OTDF). In the thermal design of gas turbine airofoils, blade tips, and endwalls, these radial and circumferential temperature distributions are always considered in the design process, and are normally based on in-situ engine measurements and high fidelity CFD predictions.

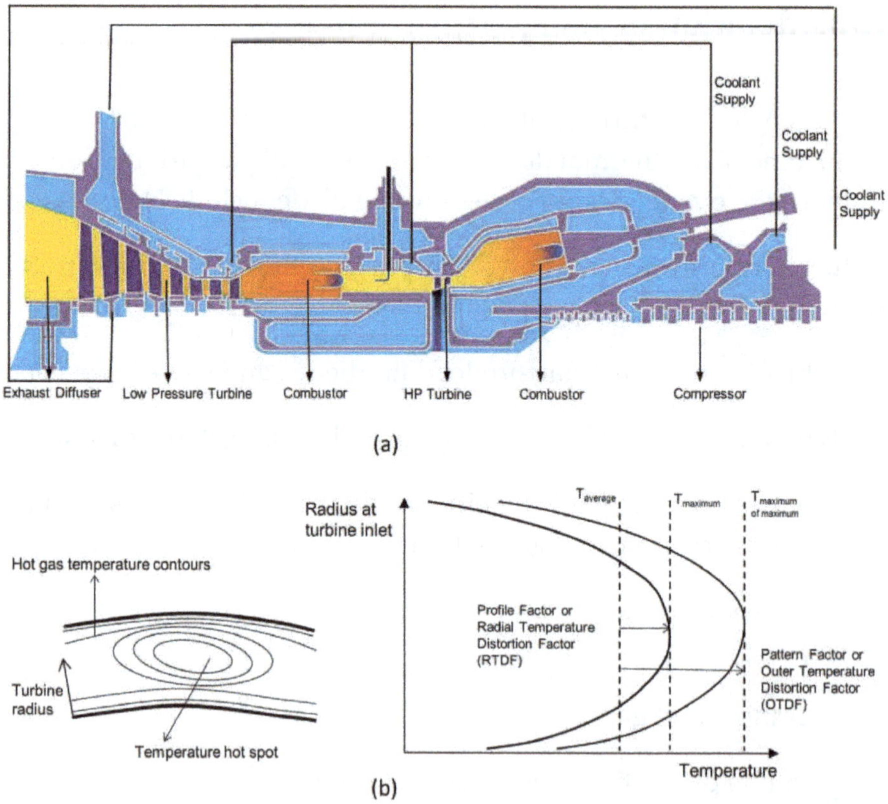

Figure 5. Major design factors influencing the gas turbine overall aerothermal design, (a) coolant supply system, (b) combustor hot gas temperature profiles.

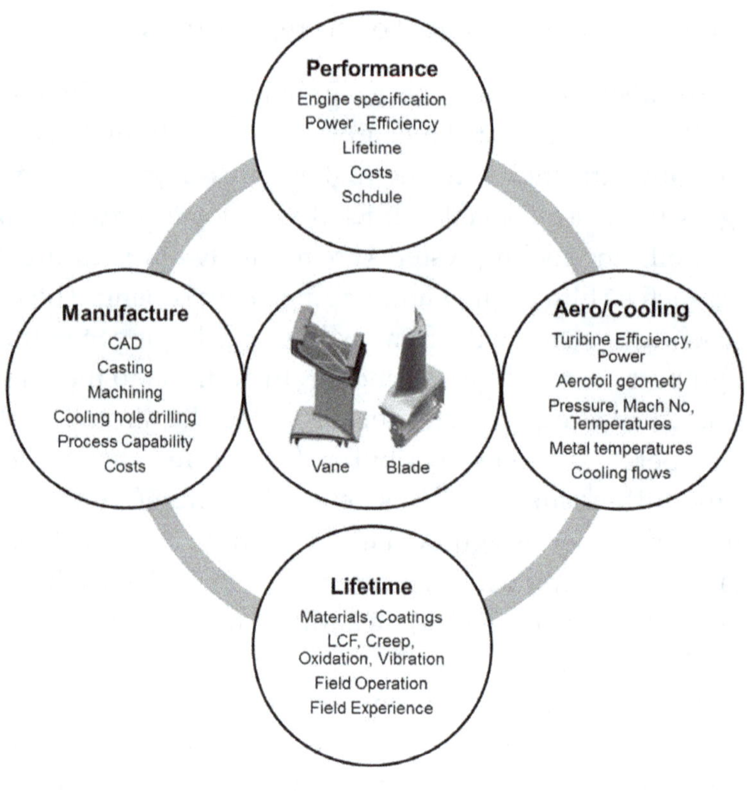

Figure 6. Major design interfaces for overall airofoil designs.

At the airofoil component level design, **Figure 6** shows an overview of several other interface considerations which needs to be accounted for in the overall optimization of the airofoil thermal design. The major design drivers for an optimized airofoil design include engine performance targets, aerothermal targets, component lifetime and mechanical integrity targets, and the manufacturing and cost constraints. Within these global requirements, **Figure 6** also highlights that are also many subtargets, such as manufacturing capability and field experience.

3. Turbine blade thermal analysis

3.1. Global thermal assessments

Due to the large number of operating and geometrical parameters that influence the heat transfer mechanism in gas turbine blades, simplified zero-dimensional relationships and design charts are often utilized. This allows for assessing the impact of various operational and geometrical parameters on a given blade cooling system. For such zero-dimensional analysis, it is important to have the detailed 2D and 3D thermal analysis results of the specific turbine blade or vane, and which has effectively been proven for meeting the design performance and lifetime in field gas turbines. This is commonly referred to as the reference blade from which new designs and concepts can be developed with a sufficient degree of confidence.

The thermal analysis is based on a simplified conjugate heat transfer analysis of flow in a cooling passage of a turbine blade as shown in **Figure 7** and assumes that the airofoil; (a) metal temperature is the average surface temperature at the airofoil midspan, (b) is exposed to the maximum hot gas temperature profile at the blade inlet, and (c) the coolant enters at the blade root and exits at the blade trailing edge. Then by performing a simple energy balance, it can be observed that,

Heat transferred from the hot gas to the airofoil = heat gained by the airofoil = heat gained by the coolant in the airofoil.

$$Q = h_g S_g L (T_f - T_m) = h_c S_c L (T_m - T_c) = m_c C_{pc} (T_{co} - T_{ci}) \tag{1}$$

where Q is the total heat transferred to the airofoil, h_g and h_c are respectively the hot gas and coolant heat transfer coefficients, L is the airofoil height, S_g and S_c are respectively the total airofoil perimeter on the gas and coolant sides, T_f is the average film cooling temperature, T_m is the average airofoil metal temperature, and T_{ci} and T_{co} are the coolant inlet and out temperatures. T_c is the average of the coolant inlet and outlet temperatures. For film-cooled airofoils, a film cooling effectiveness is additionally defined, which essentially modifies the driving hot gas temperature, T_g, with a film temperature, which is defined by;

$$\text{Film Cooling Effectiveness,} \quad \eta_f = \frac{T_g - T_f}{T_g - T_{co}} \tag{2}$$

After rearranging the above equations, the following relationships can be derived;

$$\text{Cooling Effectiveness,} \quad \varepsilon = \frac{T_g - T_m}{T_g - T_{ci}} \tag{3}$$

$$\text{Mass flow function,} \quad m^* = \frac{m_c \; C_{pc}}{h_g \; S_g \; L} \tag{4}$$

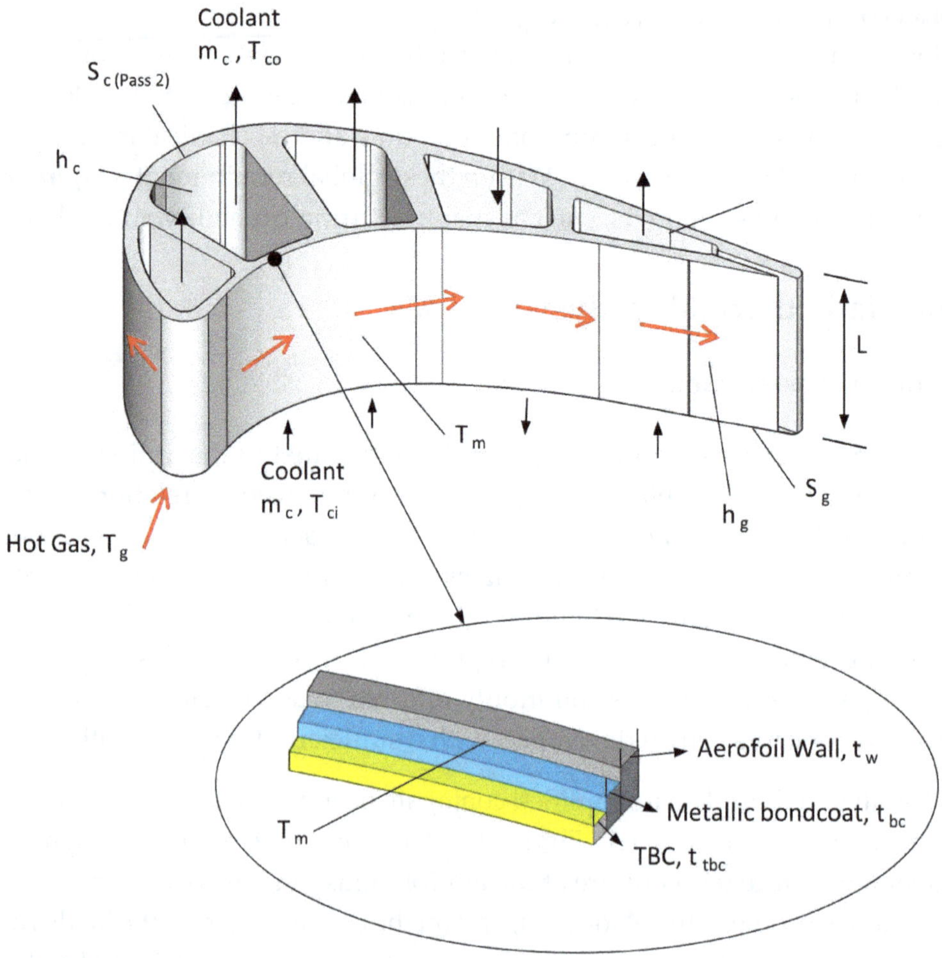

Figure 7. Thermal design parameters of a gas turbine airofoil.

Cooling Efficiency, $$\eta = \frac{T_{co} - T_{ci}}{T_m - T_{ci}} \tag{5}$$

By further combining for the effectiveness, massflow function and efficiency, the following practical engineering formulations can be derived.

Overall effectiveness, $$\varepsilon = \frac{m^* \eta}{1 + m^* \eta} \tag{6}$$

Overall efficiency, $$\eta = 1 - \exp\left[-\frac{A}{m^*}\right] \quad \text{Where } A = \frac{h_c \, . \, S_c}{h_g \, . \, S_g} \tag{7}$$

To represent thermal barrier coatings (TBC) and the airofoil wall thickness, the hot gas and coolant heat transfer coefficients in the above equations are replaced by effective heat transfer coefficients, i.e.,

$$h_{g,eff} = \frac{h_g}{1 + Bi_{tbc}}$$ (8), where the TBC Biot number, $Bi_{tbc} = \frac{h_g t_{tbc}}{k_{tbc}}$

$$h_{c,eff} = \frac{h_c}{1 + Bi_w}$$ (9), where wall Biot number, $Bi_w = \frac{h_c t_w}{k_w}$.

Where, t_{tbc} and k_{tbc} are the thermal barrier coating thickness and thermal conductivity, respectively. Similarly, t_w and k_w are the metal wall thickness and thermal conductivity. From the above relationship, it can be observed that for the extreme cases, when $\varepsilon = 0$, $T_m = T_{hg}$, the airofoil metal temperature equal the gas temperature, and when $\varepsilon = 1$, $T_m = T_c$ and the airofoil metal temperature equals the coolant temperature. For most gas turbine blades ranging from the rear to the front turbine stages, the effectiveness values are respectively in the range of 0.1–0.7.

Figure 8 shows the relationship between the cooling effectiveness, mass flow function and the cooling efficiency, which is shown in an alternative form to that normally highlighted in Refs. [4–6]. Here, the cooling efficiency and the massflow function parameters are plotted on the horizontal and vertical axis respectively, which makes it easier to compare the cooling efficiencies of different turbines. From **Figure 8**, it is clear the front stages of the gas turbine, which are exposed to the highest hot gas temperatures, will generally have the highest cooling effectiveness and efficiency levels as their cooling designs will include film cooling, thermal barrier coatings, impingement cooling, turbulator convective cooling, and advanced alloys. The rear stages which are generally exposed to the lowest hot gas temperatures are generally convectively cooled, consume the least amount of cooling air, and are represented by the lowest effectiveness and efficiency values.

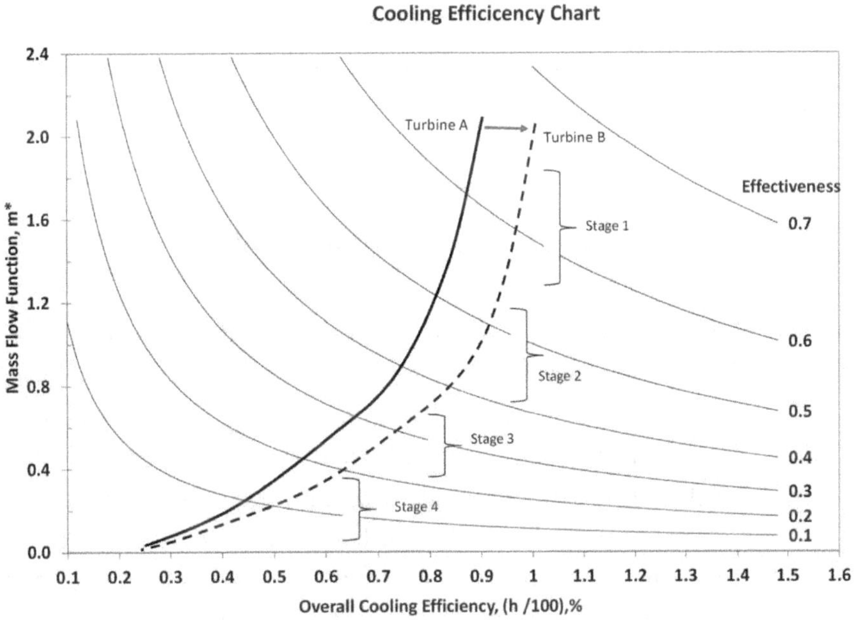

Figure 8. Heat transfer performance chart for gas turbine blades.

3.2. Detailed aerothermal designs

During the detailed design phases, the design of cooled turbine airofoils is normally done using design systems which incorporate the effect of all three-dimensional geometrical and aerothermal effects. There is extensive use of computational fluid dynamics, as part of the overall turbine design process and the thermal analyses are based on conjugate heat transfer-based model. **Figure 9(a)** shows a typical example of a gas turbine blade conjugate heat transfer model [9], where both the internal coolant flows in the internal cooling passages and the external heat

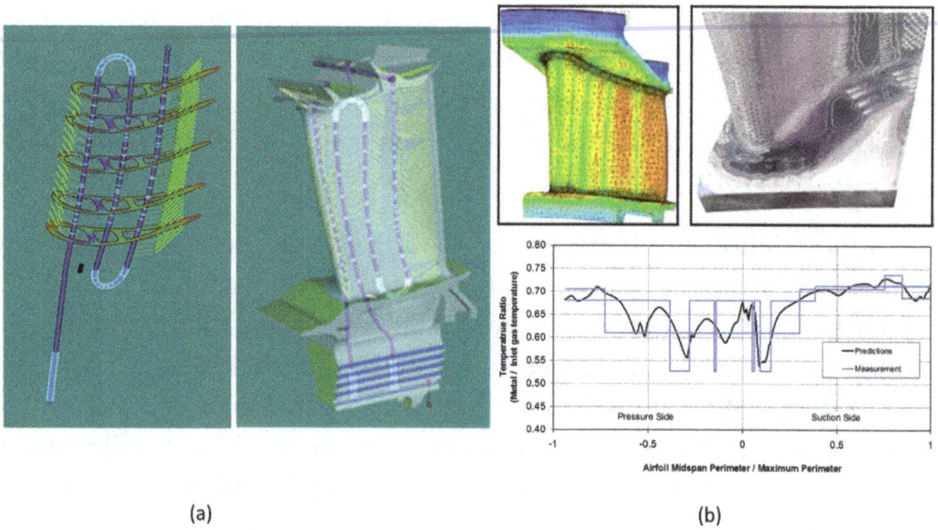

Figure 9. Detailed airofoil aerothermal design using (a) conjugate thermal modelling [9], and (b) 3D thermal modelling and comparisons with measured engine data [2, 10].

loads on the airofoil hot gas surfaces are directly simulated. **Figure 9(b)** shows a typical example of the predicted metal temperature on a turbine vane based on a conjugate heat transfer model and compared to measured metal temperatures from a test engine [2, 10].

4. External heat transfer of cooled turbine airofoils

The aerodynamics of the gas path flows through the static turbine vanes and rotating blades consist of a range of flow phenomena and flow structures such as accelerating sonic and transonic flows, unsteady flows, separated flows, secondary flows, overtip leakage flows, and interacting flows between the main gas path flows and coolant and leakage flows. To enhance the turbine aerodynamic efficiency and manage the external heat loads, significant research efforts have been made over the past decade to minimize the energy losses which are associated with the latter flow phenomenon. Similarly, there has been a significant research effort [4–6, 11, 12], in understanding and minimizing the external heat transfer on the turbine airofoils and endwalls, which is essentially defined by the gas path aerodynamics, thermodynamics, turbine geometrical annulus, and the geometrical profiles of the airofoils.

4.1. Airofoil external heat loads

The aerodynamic development of the boundary layer on the turbine static and rotating airofoils is highly nonuniform, and it largely determines the absolute levels of the external heat transfer coefficient to which it will be exposed. Other factors that significantly influence the airofoil heat transfer include the mainstream turbulence, profile curvature, streamwise pressure gradients, surface roughness, upstream wakes, and film cooling. **Figure 10** shows the Mach number measured on a turbine vane and blade [13], and highlights; (a) the strong Mach

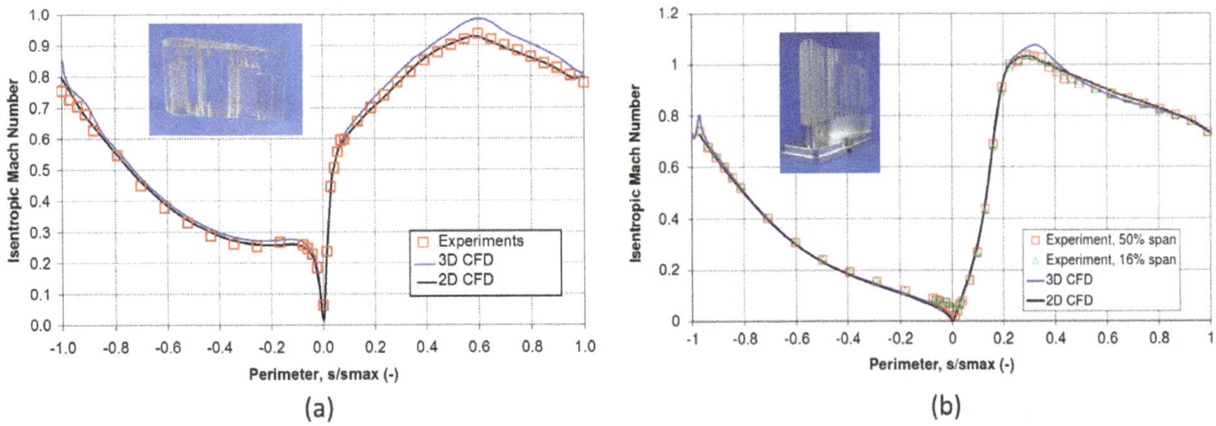

Figure 10. Aerodynamic measurements and predictions on a 1st stage (a) vane and (b) blade [12].

number variations near the leading edge stagnation point, (b) accelerating flow on the pressure and suction sides immediately downstream of the leading edge, (c) region of transitional boundary layer, (d) regions of accelerating turbulent flows on the pressure side, and (e) regions of peak Mach numbers on the suction side followed by decelerating flows towards the trailing edge. It is this variation in the profile Mach number which largely determines the vane and blade external heat transfer coefficients.

The detailed distribution of the heat transfer coefficients on the turbine vane and blade of a high pressure turbine was measured by Tallman et al. [12] for a range of operating conditions. **Figure 11** shows the distribution of the measured and predicted Stanton numbers, (St = Nu/ Re.Pr) at 50% airofoil span and at Re/L = 3.1 e6, and clearly highlights the differences in the heat transfer distribution between the vane and the blade. This is largely due to the different profile shapes, leading edge diameters, Mach number distributions, and the overall pressure ratio across the vane and blade. **Figure 11** highlights the heat transfer distribution associated with the various aerodynamic flow regimes on the airofoil. For the vane, the heat transfer coefficient increases from the leading edge to the suction side, reaches a peak value, and then decelerates towards the trailing edge. On the pressure side, it reduces from the leading edge and after transition, continuously accelerates up to the trailing edge. For the blade, the peak heat transfer coefficient value is at the leading edge, which then decreases gradually on the

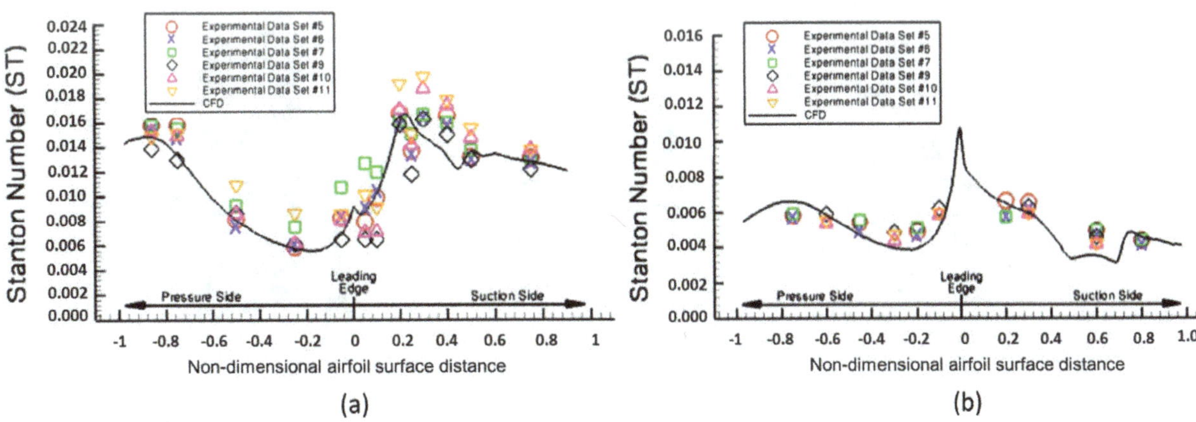

Figure 11. External heat transfer measurements at 50% span, (a) 1st stage vane, and (b) 1st stage blade. [12].

suction side until the trailing edge. However, on the pressure side, the heat transfer coefficient reduces rapidly from the leading edge, and then there is a transition to higher values until the trailing edge. These typical trends in the nonuniformity of the heat transfer coefficient are generally observed on most turbine vanes and blades. However, in addition to these generalized airofoil heat transfer distributions, actual industrial gas turbines blades are also affected by several other parameters, such as; inlet pressure and temperature profiles, airofoil shape and curvature, position of film cooling holes, thermal barrier coating roughness, transient wakes from upstream vanes, and blade passage turbulence intensity levels.

4.2. Endwall external heat loads

At the vane and blade endwall or platform, the aerodynamic flows are highly three-dimensional, transonic, and consist of areas where the hot gas flow strongly interacts with cooler rim purge and leakage flows. **Figure 12(a)** highlights the salient features of the hot gas path flow interactions on the platforms, which are largely pressure driven flows generated by the crosspassage pressure differences on the pressure and suction side of neighbouring airofoils. Over the last decade, there has been a significant experimental and numerical research effort to understand the behavior and impact of these high speed endwall flows on the platform heat transfer [4–6, 11, 16, 17]. Some of the key factors which define the platform heat transfer include the inlet profile of hot gas temperature, pressure and turbulence intensities, film cooling, platform contouring, and impact of leakage and rim purge flows.

Figure 12(b) shows the heat transfer and film cooling distributions on a first stage vane [15]. The heat transfer coefficient distribution shows that the suction side shoulder and the pressure side trailing edge regions experience the highest heat transfer coefficients, which also correspond to the areas with the highest Mach numbers. For the vane platform film cooling effectiveness without the upstream purge flows, **Figure 12(c)** shows that the measured and predicted film cooling effectiveness compares quite well, and the films remain attached to the passage wall and are very effective in cooling the platform. Due to the three-dimensional nature of the endwall flows, **Figure 12** highlights that the magnitude and directions of the local velocity, temperature, and pressure distributions play a dominant role in the heat transfer distributions on airofoil platforms.

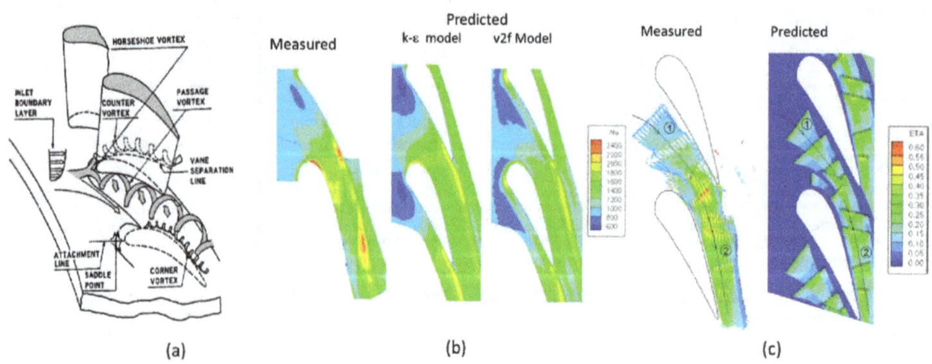

Figure 12. Endwall flow and heat transfer, (a) flow structures [14], (b) heat transfer coefficients [15], and (c) film cooling effectiveness [15].

4.3. Blade tip and endwall external heat loads

The blade tip and its neighbouring endwall regions are one of the most complex aerodynamic and heat transfer areas of the gas turbine. **Figure 13** shows some typical blade tips designs ranging from flat tips, squealers and shrouded tips, and its impact on the turbine efficiency. For flat tip and squealer tip designs, this region is dominated by the pressure driven overtip leakage flow from the airofoil pressure side to the suction side. This flow then travels through the narrow gap between the rotating blade and the static casing endwall, and subsequently interacts with the main cross passage flows to form a high speed vortex on the tip suction side. For the shrouded blade, the gas flow is from the leading to the trailing edge. The hot gas flows then interact within the rotating shroud fins with the shroud cooling air. Due to the complex flow structure and high heat loads at the blade tips, an accurate knowledge of the local aerodynamics and heat transfer is important for ensuring that the mechanical integrity of the blade tips are ensured for long operating periods, especially at higher gas turbine operating temperatures.

Figure 14 shows the sensitivity of the key parameters which influence the metal temperature of a typical squealer blade tip design. The main parameters influencing the tip metal temperatures are the hot gas temperatures and the cavity mixed temperatures. Other parameters such as the wall thickness and the heat transfer coefficients also play a major role in determining the tip metal temperatures. The heat transfer distribution on the blade tip and the endwall is highly nonuniform and driven largely by the local Mach number distributions and the tip geometry [4–6, 11, 53].

Figure 15 shows the flow distributions for two squealer tip designs and highlights the complex flow structure within the tip crown and the flow interactions between the tip leakage, main hot gas flows, and the coolant within the blade passage [18]. For these two blade tip designs, **Figure 16** also shows the measured and predicted heat transfer coefficients on the

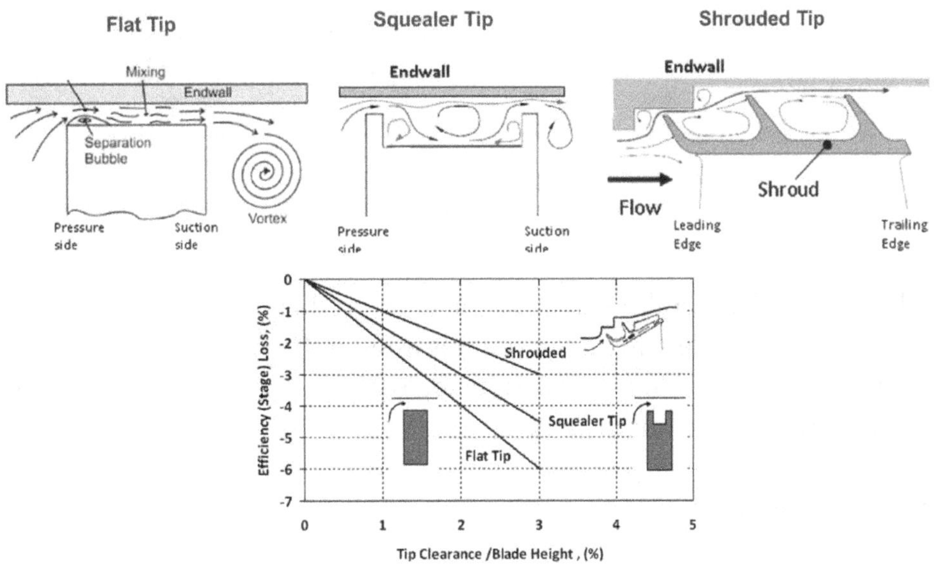

Figure 13. Typical blade tip designs and performance characteristics.

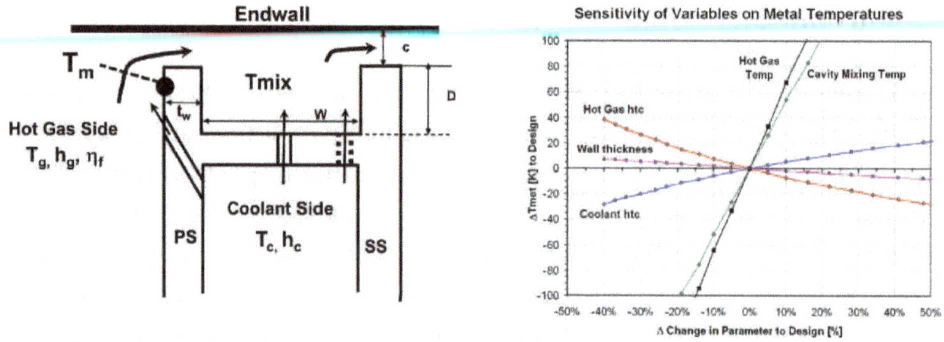

Figure 14. Sensitivity of operating conditions on blade tip heat transfer.

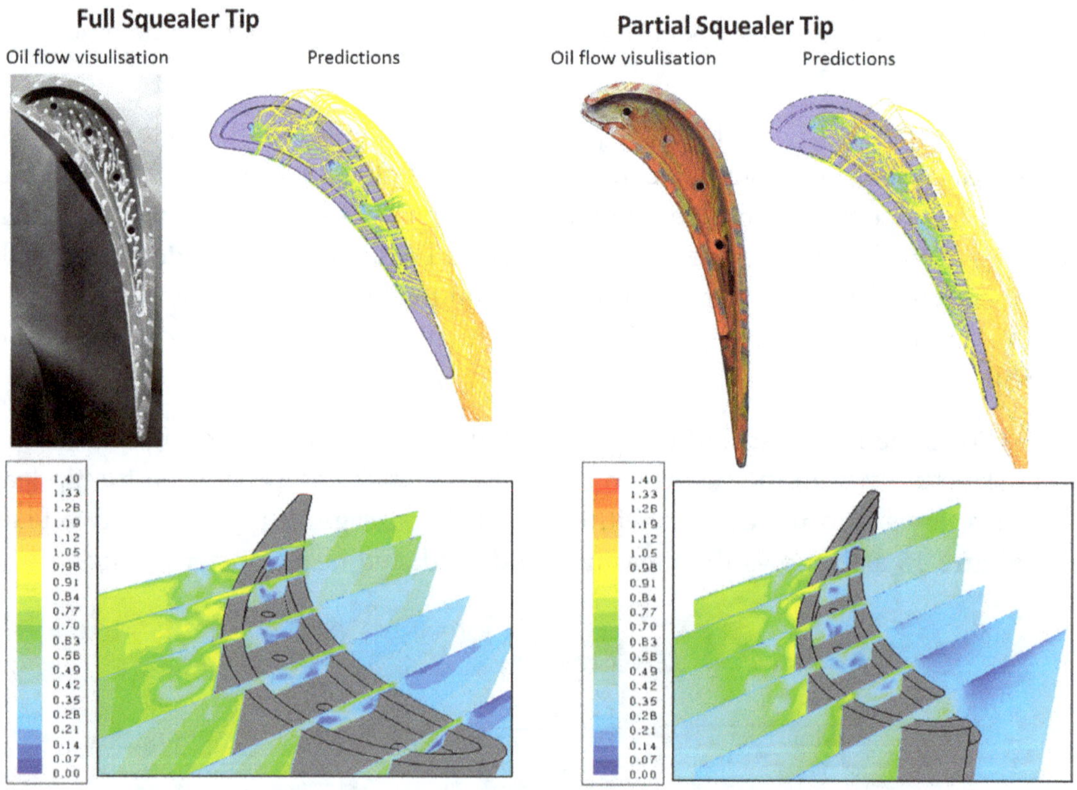

Figure 15. Flow structure and Mach number distributions for full and partial squealer blade tips [18].

blade with film cooling [18]. Both measurements and predictions show that on the blade tip, very high values exist in the leading edge regions and on the suction side rims. However, on the neighbouring endwall, the high heat transfer regions are largely location on the blade pressure side and towards the trailing edge.

4.4. Thermal barrier coatings

The use of high temperature thermal barrier coatings (TBC) for reducing the incident heat flux on both static and rotating gas turbines blades is extensive in gas turbines, particularly

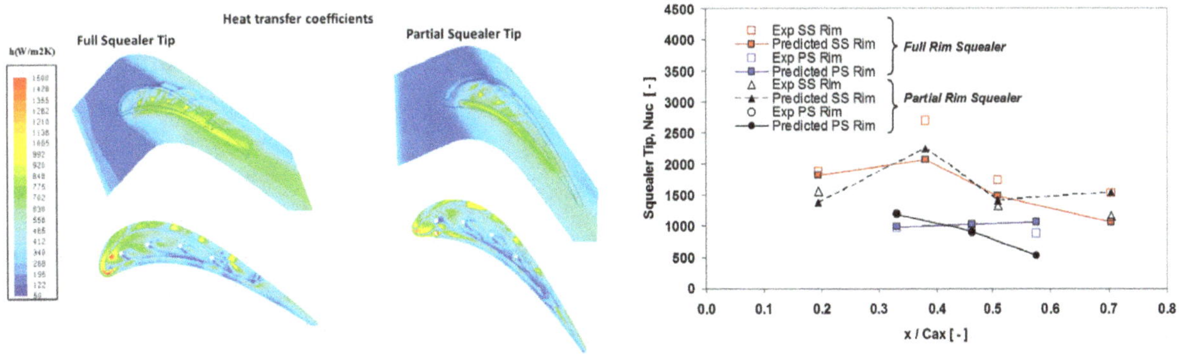

Figure 16. Heat transfer coefficient distributions on blade tip and endwall for full and partial squealer blade tips [18].

in the first and second turbine stages. There are essentially two main types of TBC, which are in widespread use in the gas turbine industry, namely air plasma sprayed (APS) and electron beam physical vapour deposition (EBPVD) [7]. For heavy duty gas turbines, the APS TBC is widely used with thickness which can range from 100 to up to 600 μm. The thermal impact of thermal barrier coatings on the turbine blade thermomechanical integrity is significant, and they therefore play an important role as a thermal protection system for gas turbine components. As highlighted previously, in the thermal analysis of turbine blades, the thermal barrier coating is generally represented as a thermal resistance to the incident heat flux, by modifying the hot gas transfer coefficient via the thermal barrier coating Biot number. **Figure 17** shows that by increasing the thickness of the thermal barrier coating and reducing its thermal conductivity, the effective hot gas heat transfer coefficient can be significantly reduced. This results in a direct reduction of the incident heat flux on the turbine blade.

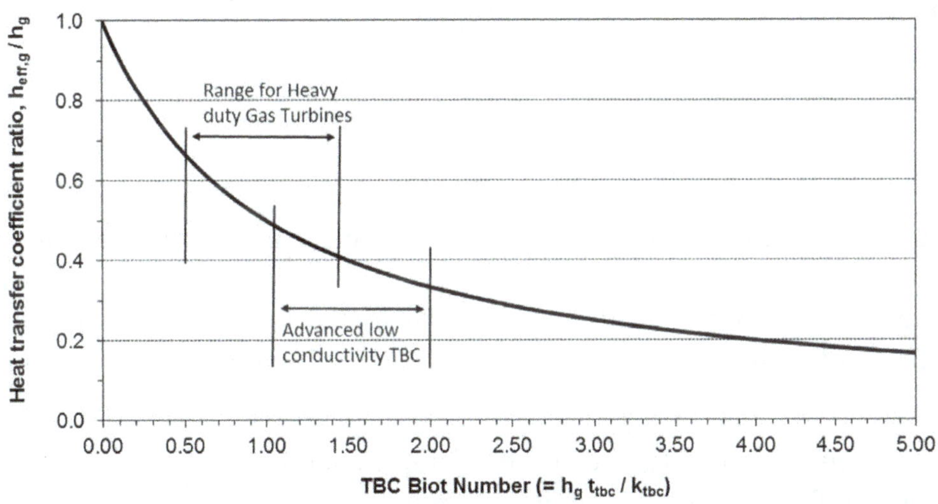

Figure 17. Effect of thermal barrier coatings (TBC) on heat transfer.

For typical heavy duty gas turbines, **Figure 17** shows that the thermal barrier coating reduces the effective heat transfer coefficient by almost 50% compared to no application of the TBC. Additionally, **Figure 17** shows that, for the new generation of advanced TBC's [6, 7], with lower thermal conductivity, the effective heat transfer coefficient can be further reduced. It is clear from **Figure 17**, that thermal barrier coatings are an integral and significant part of the overall blade thermal design system.

4.5. Film Cooling

Film cooling is generally applied at different locations along the perimeter of an airofoil by rows of discrete holes, through which coolant air is discharged into the airofoil external boundary layer. The coolant, which is several hundred degrees colder than the hot gas, then creates a film of air on the airofoil surface, whose temperature is significantly lower than the surrounding hot gas. Consequently, the incident hot gas temperature for heat transfer is reduced. **Figure 18** shows an example of the application of a film row at the blade trailing edge and the key parameters which define the performances of film cooling. **Figure 19** shows that as the average film cooling effectiveness on a turbine blade is increased, the film to hot gas temperature ratio reduces and the film temperature close to the wall can be reduced by several hundred degrees relative to the surrounding hot gas temperature.

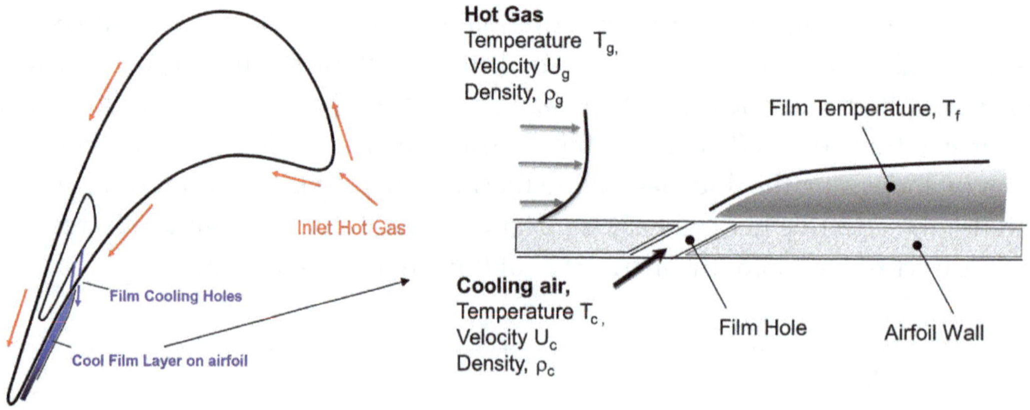

Figure 18. Film cooling of gas turbine blades.

Increasing the average film cooling effectiveness can be achieved by using many film rows, but this would be at the expense of high coolant consumption and reduced turbine efficiencies. Alternatively, increased film cooling effectiveness can also be achieved by using advanced film cooling hole designs without increasing the coolant consumption [4, 6, 11, 13, 20, 21].

The film cooling effectiveness depends on the complex aerothermal interaction between the high speed hot gas flow and the ejected film cooling jets in the external gas boundary layer. It is also dependent on several geometrical and operational parameters such as film cooling hole shape, hole angle, velocity and temperature of the ejected coolant, temperature and velocity of the surrounding hot gas, blade curvature, and local turbulence levels. As highlighted in **Figure 19**, increasing the film cooling effectiveness results in significant reduction in the film to hot gas temperature ratio, and hence there continues to be a significant research effort on developing film cooling technology due to its significant benefits in reducing local near wall

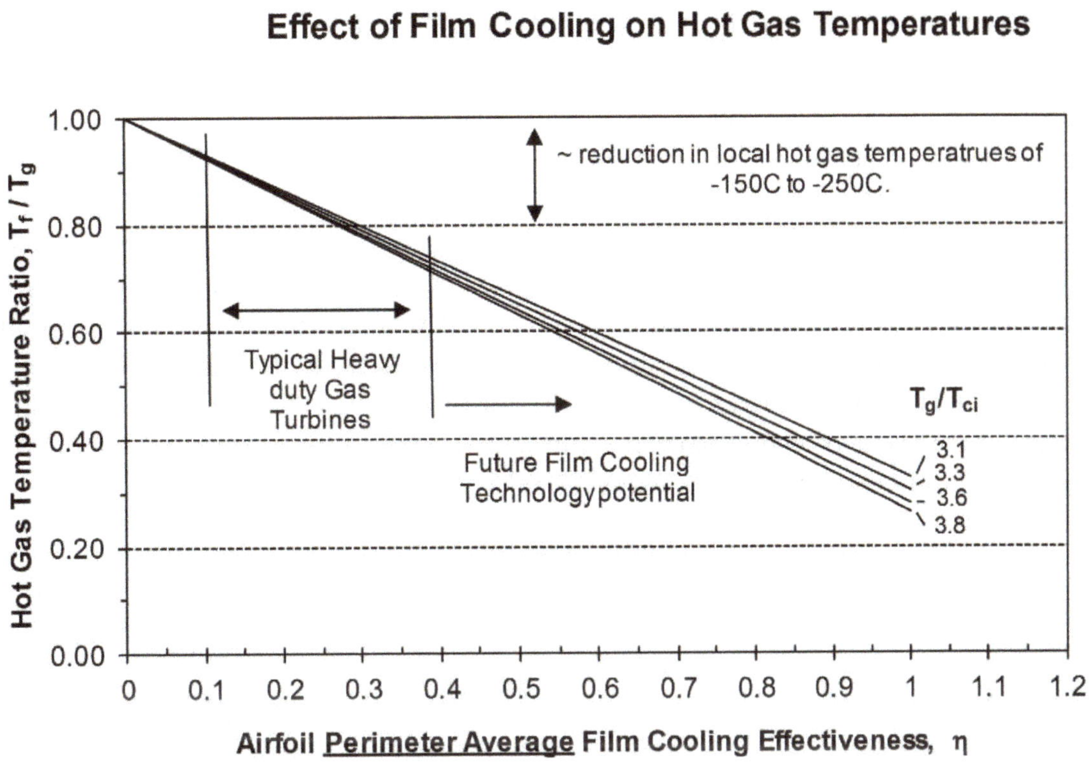

Figure 19. Effect of film cooling effectiveness on hot gas temperatures.

hot gas temperatures [4–6, 11, 16, 17, 20, 21]. Over the last decade, there has been a significant focus on airofoil, platform, and blade tip film cooling with more recent focus on advanced shapes of film cooling holes, such as three-dimensional shaped holes and trench holes. In a recent study [13], the multirow film cooling characteristic on a high lift vane and blade were demonstrated. **Figure 20** shows that the use of three-dimensional advanced fan shaped holes can provide high airofoil average film cooling effectiveness and the use of only one or two row of shaped holes located upstream of the suction side shoulder can provide high film cooling effectiveness until the trailing edge.

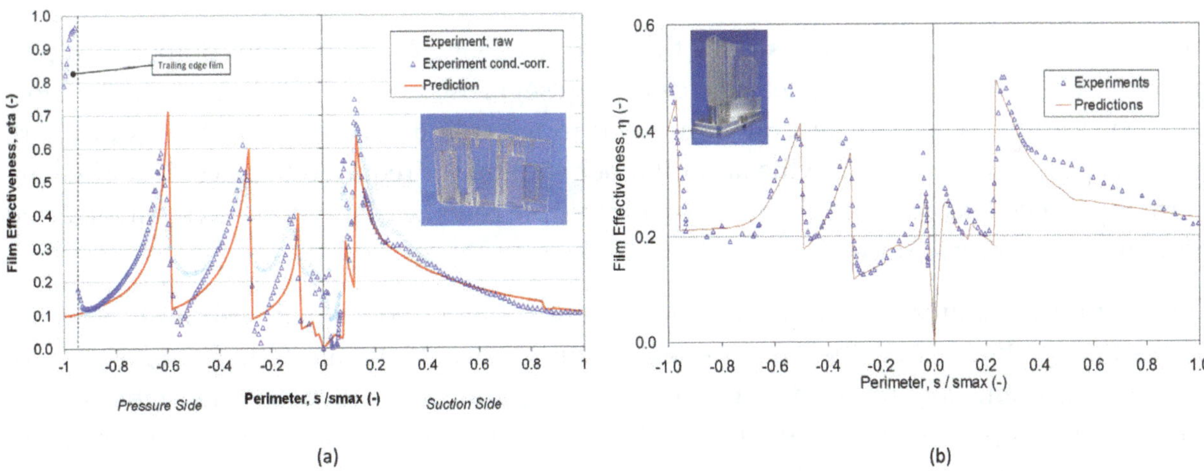

Figure 20. Multi row film cooling characteristics on a gas turbine (a) vane and (b) blade [13].

5. Internal heat transfer of cooled turbine airofoils

The need for the internal cooling of gas turbine blades is primarily defined by the magnitude of the incident heat load on the airofoils, which range from 0.5 to 5 MW/m^2, and the requirements of the component durability for long operating hours against thermomechanical fatigue (TMF), low cyclic fatigue (LCF), creep, oxidation, and high cyclic fatigue (HCF). While the external airofoil profile defines the airofoil aerodynamic performance, the internal cooling geometry is defined by the amount of coolant required to maintain the airofoil at a certain material temperature and the temperature gradients across critical wall sections of the airofoil. **Figure 21** shows some typical examples of turbine vane and blade cooling designs. The internal heat transfer technologies used in these vanes and blades include impingement cooling, turbulators or ribs, pin or pedestal banks, dimples, shaped internal passages, and combinations of the above cooling features.

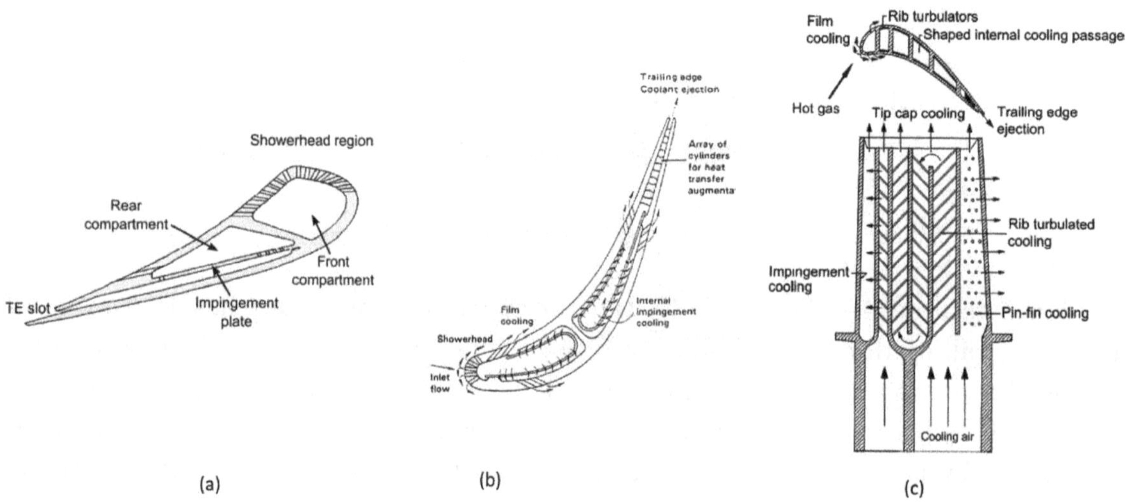

Figure 21. Typical blade cooling designs, (a) nozzle guide vane [22], (b) turbine vane [23], and (c) turbine blade [19].

5.1. Convective cooling with jet impingement

Impingement cooling is widely used for the internal cooling of gas turbine components, particularly static airofoils (vanes), heatshields (casing segments), combustor liners, and fuel nozzles. The impinging jets are generally formed through cylindrical holes in a thin wall insert, which is positioned adjacent to the airofoil inner wall that is required to be cooled. They are normally directed as a single row of jets or as multiple rows of jets, and are generally injected normal to the target surface.

For the midchord areas of airofoils, impingement cooling is designed with multiple rows of jets and is directed on the pressure and suction sides of the airofoil. The efficiency of the impingement cooling is defined by several parameters such as the standoff distance of the impingement jet relative to the target surface, the axial and radial pitch of the neighbouring impingement hole, the arrangement pattern (e.g., in-line, staggered, or other combinations), and the amount of cross-flow from the upstream impingement jets. An overview of recent research into impingement cooling is given in Ref. [24] with impingement cooling applications detailed in Refs. [4–6, 27]. For relatively flat surfaces, **Figure 22** shows the impingement hat transfer for a flat surface with multiple impingement holes, which is based on the correlations in Ref. [25]. This figure highlights, that although jet impingement cooling is highly effective, the design of the impingement system requires careful consideration of several influencing parameters, such as the standoff distance from the target surface, the axial and lateral pitch of the impingement holes and the amount of crossflow from upstream jets.

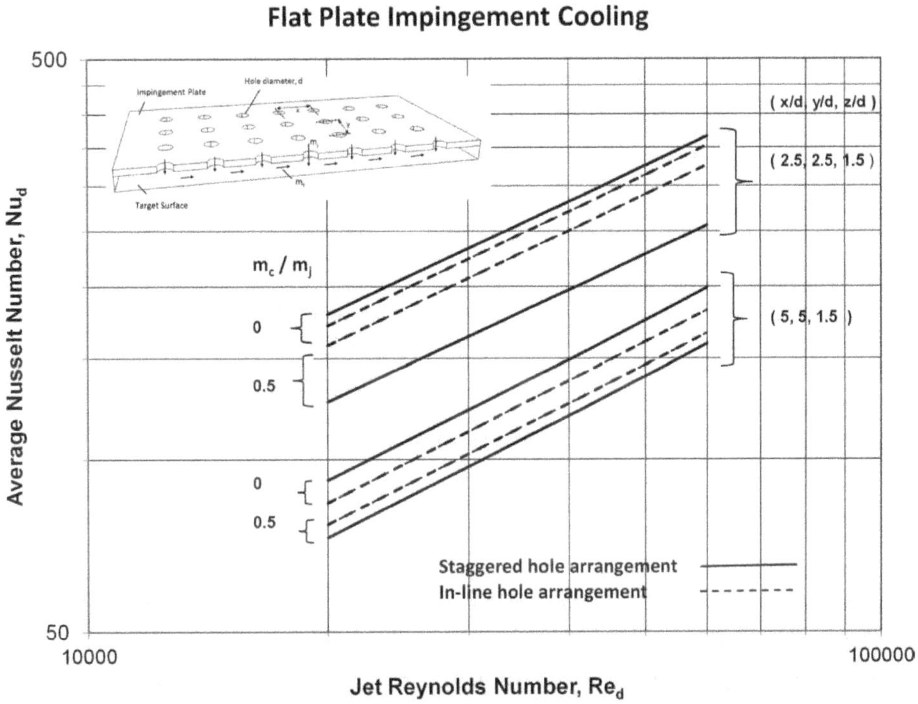

Figure 22. Impingement heat transfer with multiple rows on flat target surfaces.

Airofoil curved leading edges are normally subjected to very high heat loads, and at these locations, internal impingement cooling in combination with turbulators and film cooling is quite common. **Figure 23** shows the dependency of the standoff distance and the leading edge curvature on the coolant Nusselt numbers at varying Reynolds numbers, based on the correlation in Ref. [26]. Highest stagnation heat transfer can be achieved if the jets are arranged very close to the target surface and the highest average heat transfer are achieved for airofoils with small internal leading edge diameters. At the internal leading edges, there are also several other additional factors that influence the airofoil heat transfer, such as showerhead film cooling, turbulators, surface roughness, and the amount of impingement crossflows [4–6].

More recently, there have been several studies on the use of narrow channel impingement passages and inclined impingement jets in variable shaped passages. Such design configurations can provide higher internal heat transfers and have been mainly driven by the introduction of near wall cooling features in gas turbine blades. Such configurations can be manufactured with 3D printing technologies such as selective laser melting (SLM) and direct laser melting (DLM), which allows greater manufacturing flexibility with geometrically complex cooling passages.

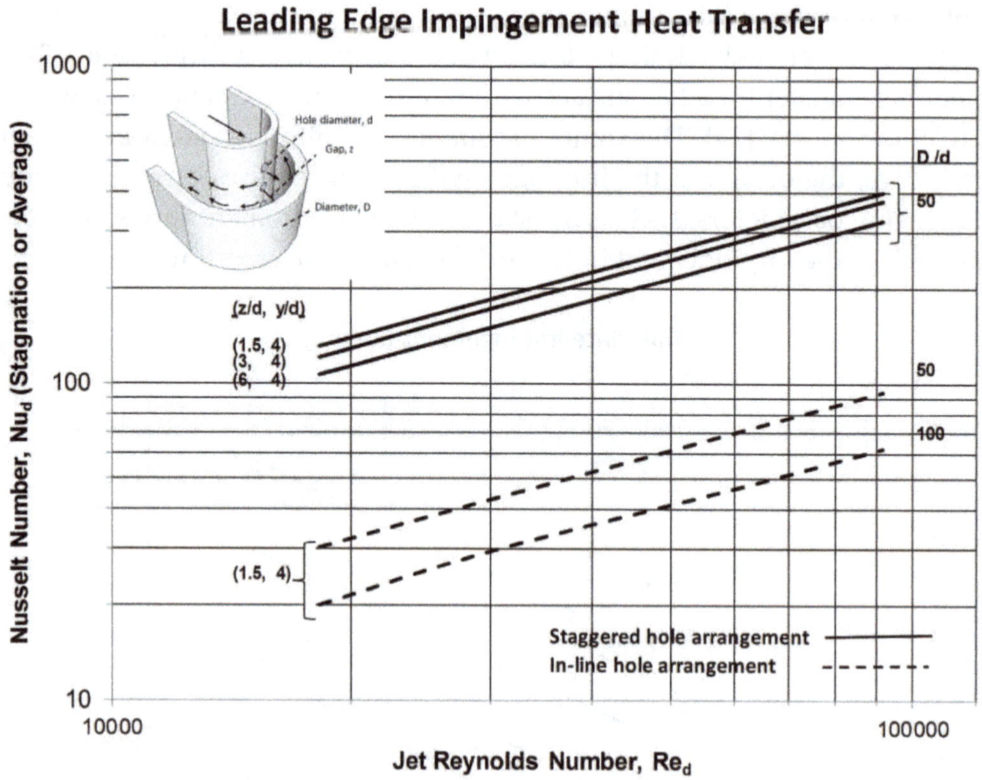

Figure 23. Impingement heat transfer on curved leading edge surface.

For narrow channel impingement, it was recently highlighted in Ref. [28] that in addition to the high heat transfer from the target surface, the heat transfer from the impingement cavity side walls can also be significant. **Figure 24** shows that for a narrow channel with in-line impingement holes, the heat transfer from the side walls can be up to 50% of that from the target plate.

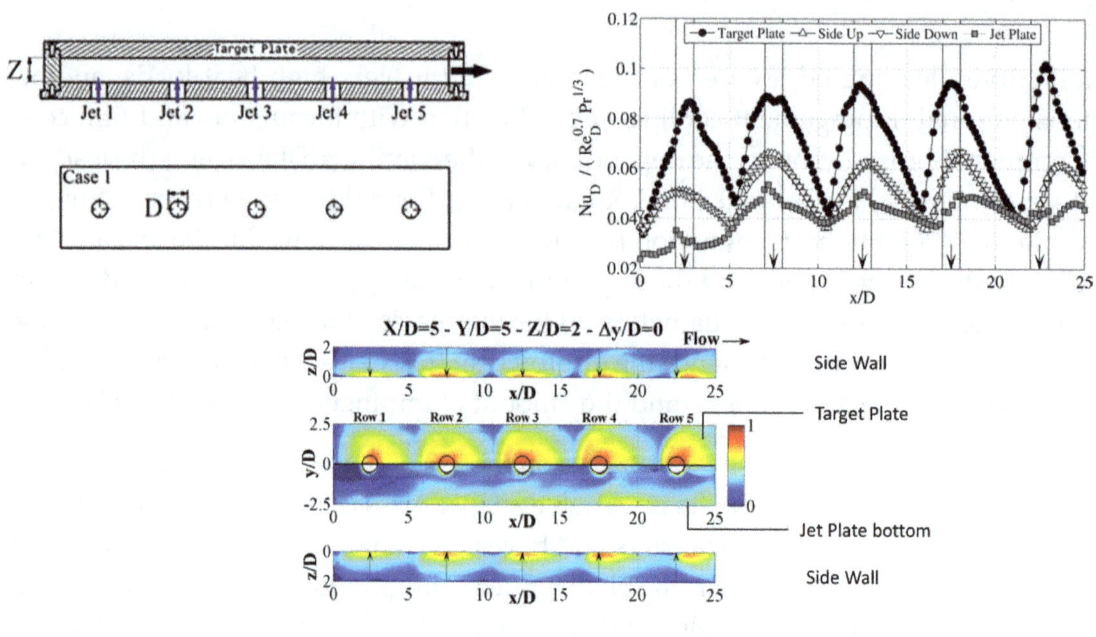

Figure 24. Impingement heat transfer in narrow channel passages [28].

The use of inclined impingement jets on shaped turbulators in irregular shaped passages can also result in very high heat transfer. In a recent study [29, 30], and as shown in **Figure 25**, it was highlighted that directed inclined impingement can result in relatively high heat transfers from the target walls and additionally produces intense convective fluid mixing within the passage. In a turbine blade, such a combination can result in greater total heat removal by the coolant from the hot airofoil walls. The use of directed impinging cooling jets in the leading edge passages was demonstrated in Ref. [31]. **Figure 26** shows that directing a double impingement jets on a curve leading edge with showerhead cooling results in high heat transfer coefficients at the pressure and suction surfaces, and additionally generates significant turbulent mixing within the leading edge passage.

Based on the many different design variations of impingement cooling, it is expected that impingement cooling systems will continue to play a significant role in gas turbine heat transfer technology.

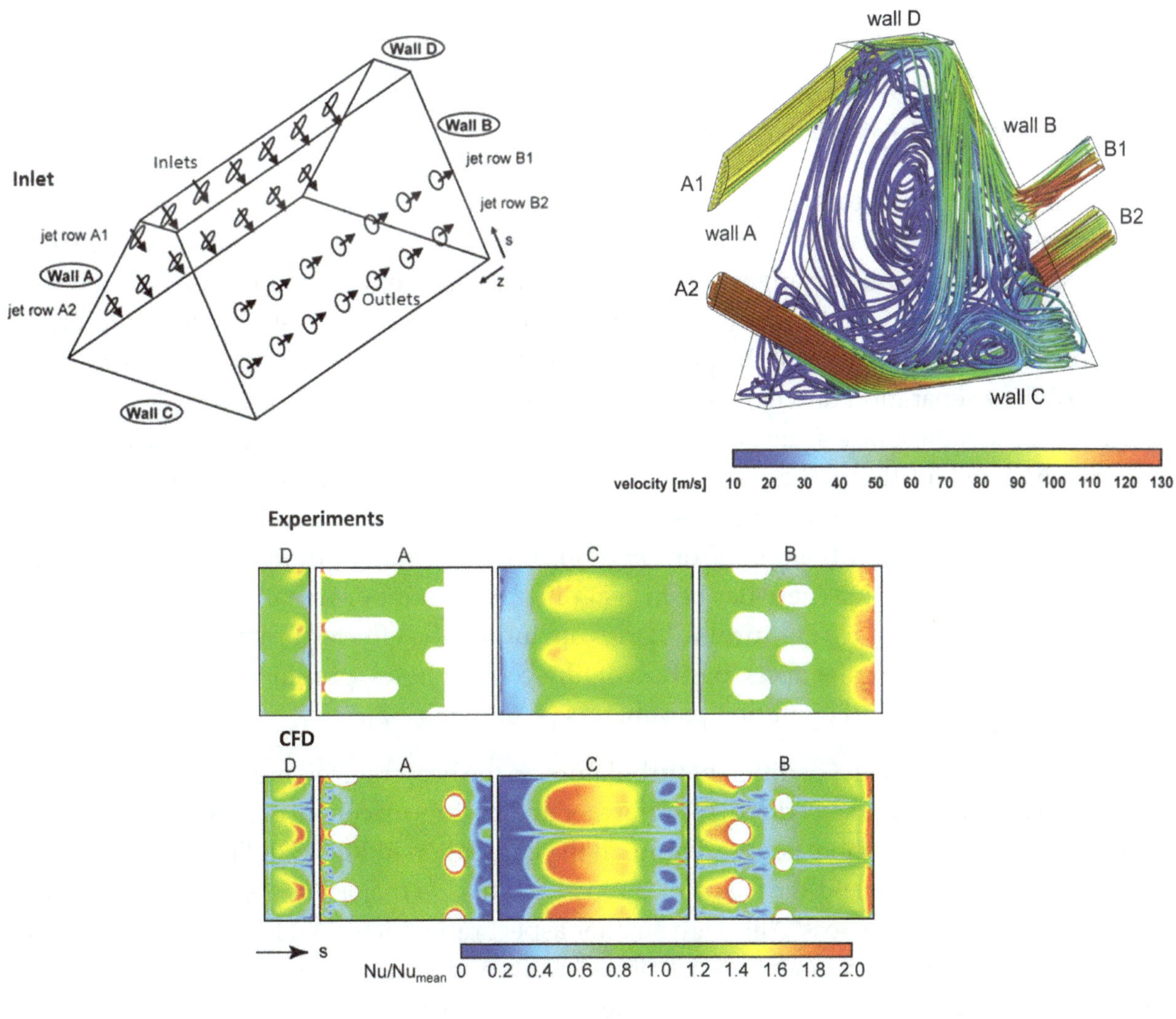

Figure 25. Impingement heat transfer in irregular passages [29, 30].

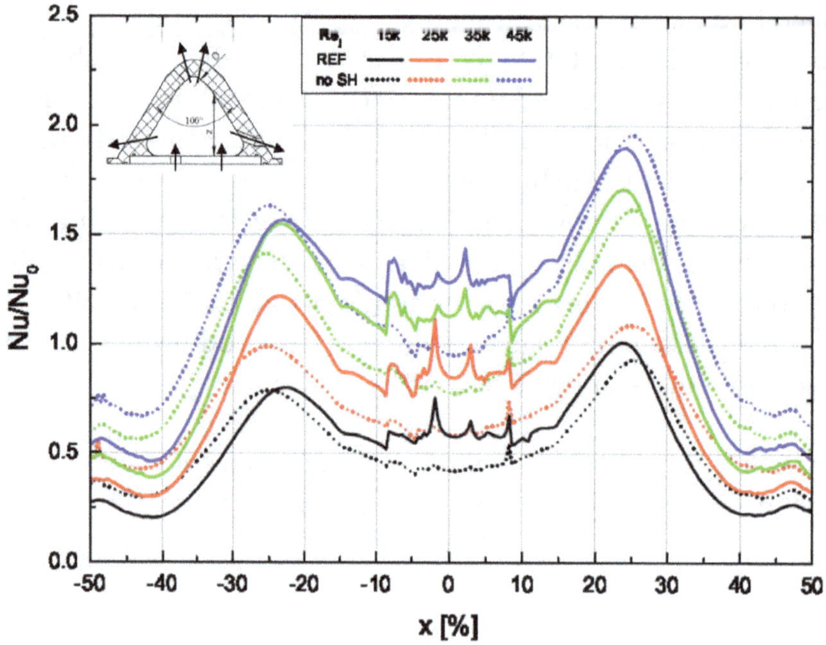

Figure 26. Impingement heat transfer in leading edge channels [31].

5.2. Convective cooling with turbulators

The use of ribs or turbulators for cooling gas turbine blades is a major heat transfer technology and has been employed largely in rotating blades with radial passages as shown in **Figure 27**. The passages are typically arranged as multiple radial passages and commonly referred to as serpentine passages or multipass systems, and the turbulators are generally designed on the pressure and suction surfaces of the passages. The key function of the turbulators is to create regions of flow separation downstream of the turbulators, which promotes intense regions of turbulence, secondary flows, and rapid mixing between the air warmed by the heated walls and the core coolant flow. The nature of the flow structure and the amount of heat transferred in the passages with turbulators is dependent significantly on the turbulator shape, configuration pitch, height, angle of orientation, flow Reynolds number, rotation, passage shape, and its aspect ratio. There has been a significant amount of research conducted on the application of turbulators in gas turbine blades, including the effects of rotation, shapes, sizes, orientation, entrance length effects, position of film cooling holes, presence of bends and other enhancement devices, and operating parameters [4–6, 19, 33, 52].

To assess the relative impact of different turbulators on the heat transfer and frictional characteristics, **Figure 27** shows the ribbed wall heat transfer and passage frictional enhancement of turbulators in various aspect ratio passages, based on the correlation in Ref. [32]. **Figure 27** highlights that the turbulator angle and the shape of the passage have a major effect on both the passage heat transfer and pressure loss. Although smaller aspect ratio ducts (W/H = 0.25) generally give higher heat transfer enhancement on the ribbed walls and lower pressure losses, the passage average heat transfer values can be lower due to the larger perimeter of the nonribbed walls, which have much lower heat transfer enhancement. Although the results shown in **Figure 27**

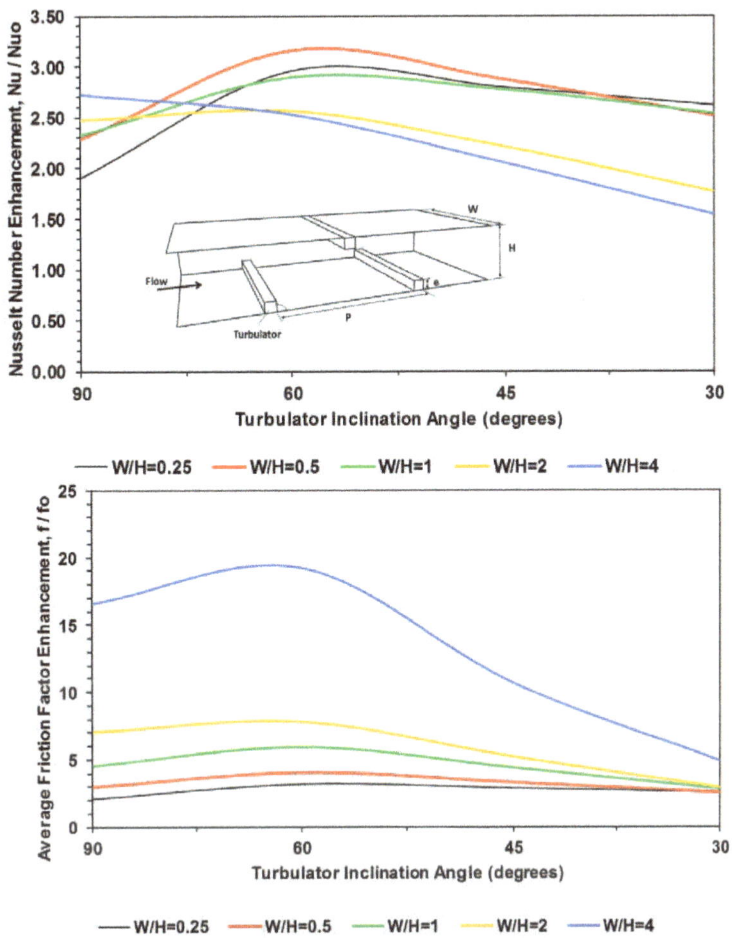

Figure 27. Passage heat transfer and frictional losses due to turbulator design and passage shape.

are for low Reynolds number, and with idealized geometry, care needs to be taken when implementing such results in real turbine blades with cast geometries, where the dimensions, shape, and position of the turbulators can be different from the predicted idealized geometries.

The impact of rotation on the heat transfer from gas turbine blades can be significant and is dependent on several additional parameters such as the rotational and buoyancy numbers [34, 35]. **Figure 28** shows a schematic overview of the impact of rotation on the flow field in a two-pass rotating passage of a gas turbine blade. Under rotating conditions, Coriolis and buoyancy effects can significantly alter the temperature and velocity profiles within the passages.

In a study from Ref. [35], **Figure 28** shows that with the coolant flowing radially upwards, the heat transfer with increasing rotation numbers, increases on the pressure side and reduces on the suction surfaces. Similarly, when the coolant flows radially inwards, the heat transfer increases on the suction side and reduces on the pressure side, especially for smooth passages. However, for passages with turbulators, the turbulators tend to dampen the effect of rotation and the heat transfer enhancement on the pressure and suction sides. This overall trend of rotation with different type of passages and turbulators has been observed by several studies [33–35], and these effects play an important role in the design of gas turbine blades.

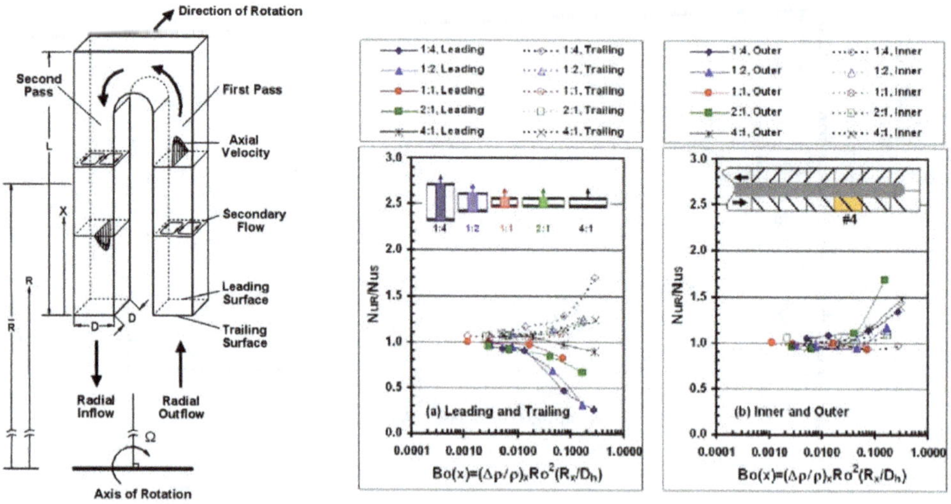

Figure 28. Impact of blade rotation on passage heat transfer [6, 19, 35].

The trailing edge regions of rotating blades are in general the most difficult to thermally design, largely due to the thin airofoil geometry, complex internal flow geometry, and the coolant flow conditions. The trailing edge region generally consists of coolant passages with very high aspect ratios, typically between 4 and 7, and which have cooling features such as turbulator and pedestals. Previous studies in such triangular and wedge-shaped passages with various turbulator shapes have been investigated by Refs. [36, 37, 38, 39]. They reported significant variation in the heat transfer distribution in both stationary and rotating cases. In a recent study [40], it has been shown that the impact of high Reynolds number typically found in heavy duty gas turbines can have a significant effect on the overall thermal performances of angled, broken, and chevron turbulators in a very large aspect ratio passage. **Figure 29**

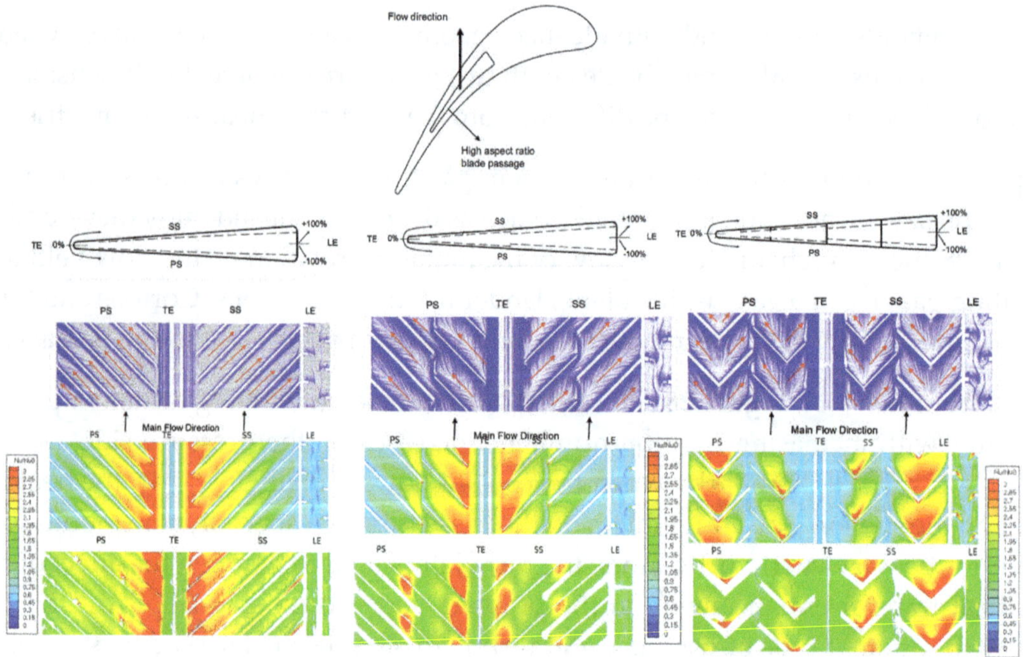

Figure 29. Impact of turbulator design in trailing edge passage [40].

shows the complex flow structures and the high three-dimensional heat transfer distribution which exist within the high aspect ratio triangular passages with different turbulator shapes. **Figure 30** shows the comparison of the average thermal performances and shows that at high Reynolds numbers, the differences between the various designs are very similar. When considering the investigated turbulator design for gas turbine cooling applications, all three configurations show comparable levels of heat transfer performances.

For leading edge passages of gas turbine blades, the application of turbulators is also widespread. However, due to the leading edge geometry, the heat transfer is significantly different to that in midchord or trailing edge passages. Several studies show the impact of the turbulator geometry on the overall heat transfer in gas turbine blade leading edges [33, 41–43]. In a recent study by Saxer-Felici et al. [44], several turbulator geometries were tested at engine representative Reynolds numbers. **Figure 31** shows that the flow structure is significantly modified due to the presence of the turbulators, and this dominates the strength and distribution of the local and average heat transfer coefficients.

The final selection and implementation of turbulator designs in a turbine blade are dependent on several additional complex requirements. These include blade metal temperature, metal temperature gradients, cooling flow pressure margins, and the amount of required cooling flow. A further key criterion is for the fulfilment of the blade mechanical integrity, which is determined by the blade low cycle fatigue and creep behaviour, both of which are driven by the local metal temperature gradients and the absolute metal temperatures. An optimal balance of these factors is therefore needed to select the best turbulator concept for a blade design system.

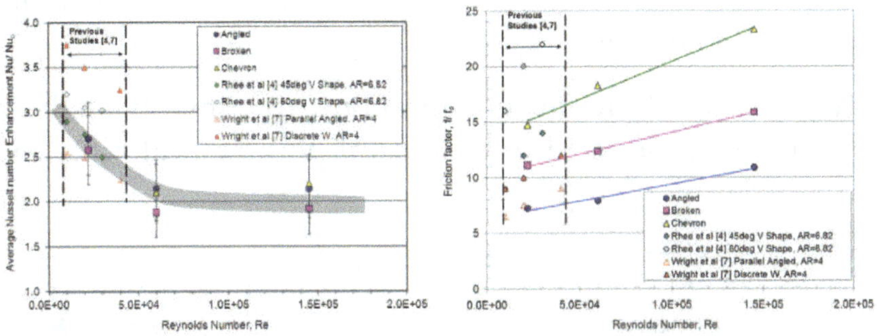

Figure 30. Average heat transfer and frictional loss in trailing edge passages [40].

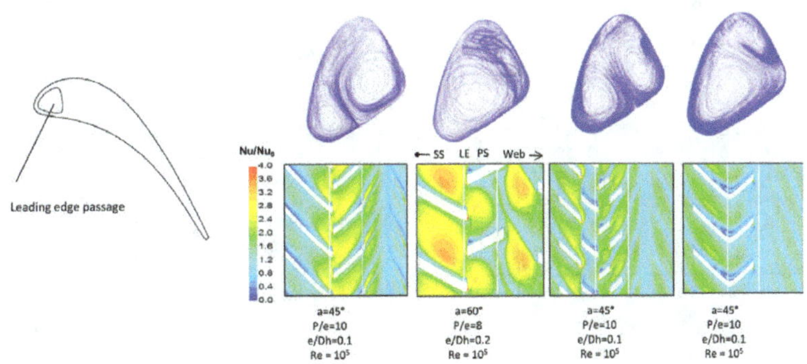

Figure 31. Heat transfer in leading edge passages [44].

5.3. Convective cooling with pins and pedestals

The use of pins and pedestals for enhancing the internal heat transfer in gas turbine blades and vanes is quite common particularly at the airofoil trailing edges, which generally demands aerodynamically small wedge angles and thin trailing edge diameters. Pin banks and pedestals are sometimes the only method of cooling in the space constrained narrow, converging trailing edges. They are also preferred from a manufacturing point as they tend to offer structural stability for casting. An additional advantage is that the pin banks also provide good mechanical integrity of the blades due to the robust structural support between the pressure and suction surfaces of the airofoil. In general, the pins are cylindrical in shape, tend to be thick relative to their height, has fillets imposed at the interface with blade walls, and are typically arranged in a staggered pattern. Although they provide high heat transfer due to a combination of high heat transfer coefficients from the base wall and the pins, they also entail high pressure losses. The latter disadvantage is generally not a major issue for convectively cooled blades, where there is availability of higher coolant pressure ratios relative to the surrounding hot gas pressure at the blade trailing edge.

There have been a large number of heat transfer studies on pin banks which have addressed the influences of pin geometry, channel shape, arrangements, pin shapes and combination of pins with dimples and turbulators [45–49]. For straight passages with pin banks, the local heat transfer generally increases from the first row of pins until the second to third row and then starts to decrease. For converging ducts, **Figure 32** shows that with a converging duct, the heat transfer increases in the downstream section. Additionally, by using a thicker pin in the rear portion of the passage, further increase in heat transfer can be achieved, which is driven by the increased Reynolds number.

In a recent study [50], several pin fin configurations were investigated in a trailing edge converging channel which consisted of cylindrical pins, conical pins, and a hybrid cyl-

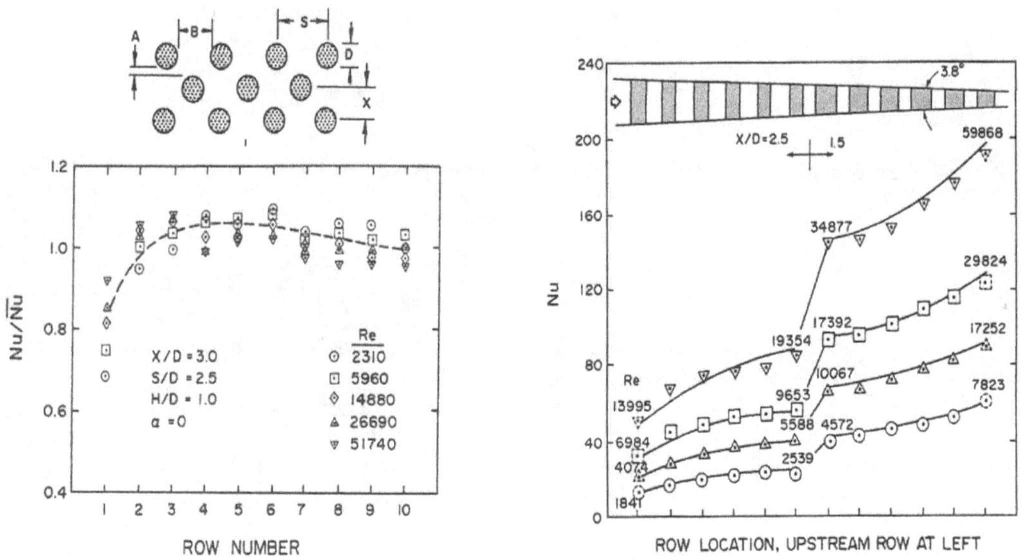

Figure 32. Heat transfer in pin banks and pedestals [45, 46].

inder pin/turbulator configuration. **Figure 33** highlights the investigated geometries. **Figure 34** shows from both predictions and measurements that the flow is highly turbulent downstream of the pins and that the complex heat transfer distribution exists on both the endwall and the pins. High levels of local heat transfer occur at the leading edge stagnation point of the pins and at the leading edge endwall. Lower heat transfer coefficients were predicted and measured in the wake region of the pin trailing edge. Laterally, averaged local distributions of the heat transfer enhancement are shown in **Figure 34** for the tested geometries, and the nonuniform nature of the heat transfer in the pin banks are further highlighted.

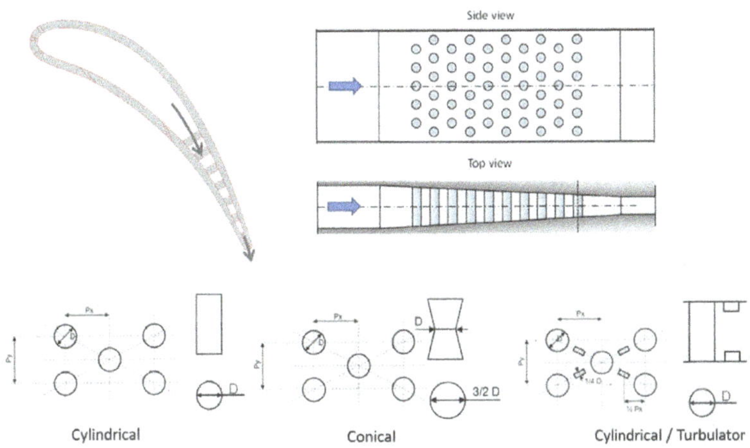

Figure 33. Trailing edge passages with different pin bank configurations [45, 46].

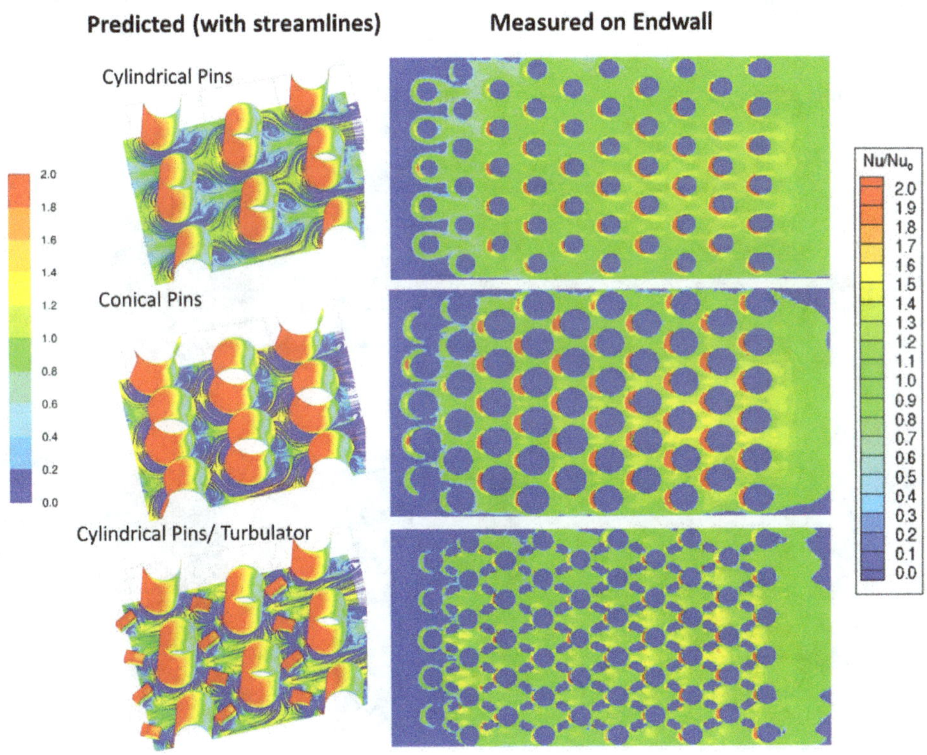

Figure 34. Heat transfer in trailing edge passages with different pin bank configurations [50].

6. Gas turbine heat transfer validation

One key aspect in the aerothermal design of gas turbine airofoils is the validation of the airofoil thermal performances under engine operating conditions. The design aspect outlined in the previous sections focused on individual design features such as film cooling, turbulators, pins, and impingement. For the overall validation of the cooling system of turbine blades and vanes, static perspex model testing is very common and is generally scaled to match engine operating Reynolds and Mach numbers. Typically, methods for the heat transfer testing include thermochromic liquid crystals (TLC) with embedded pressure and temperature sensors, which together provide a full map of the internal heat transfer and pressure drop characteristic of the blade cooling system for a range of flow conditions. **Figure 35** shows some examples of perspex models used for heat transfer testing.

For the testing and validation of the external film cooling, cascade test rigs are generally employed to validate the external aerodynamics and the film cooling performances. The blade or vane models are generally scaled to engine geometry and the cascade is operated at engine Mach and Reynolds number conditions. **Figure 36** shows an example of a first stage vane and

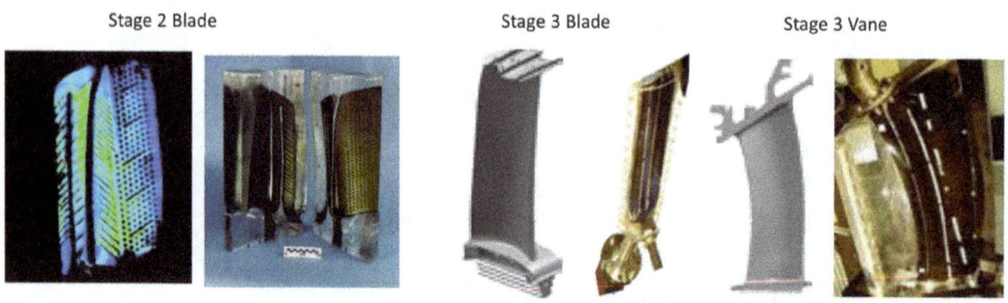

Figure 35. Perspex model testing of gas turbine blades and vanes [2].

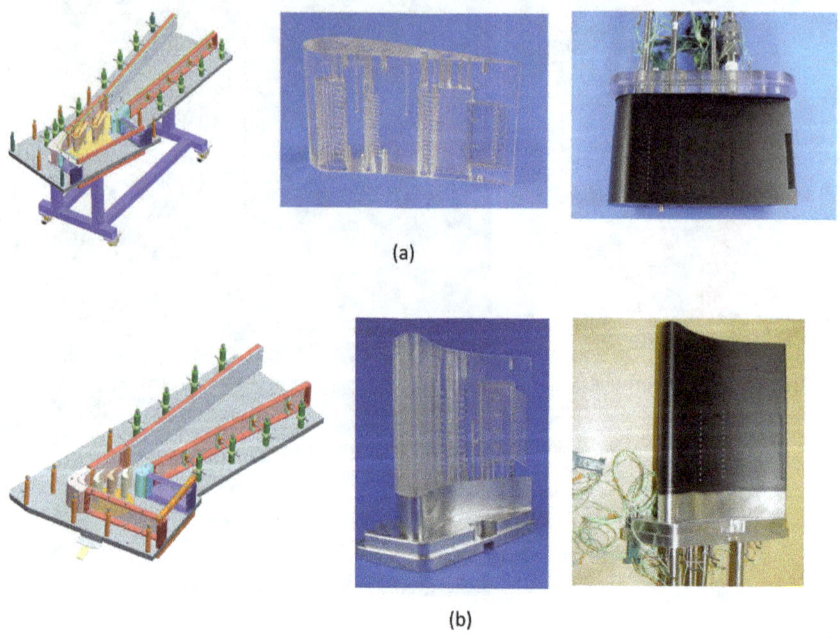

Figure 36. High speed cascade model testing of gas turbine components, (a) Turbine Vanes, (b) Turbine Blades, [13].

blade cascade with test models. These rigs enable measurements of heat transfer coefficients, film cooling effectiveness, airofoil pressure distributions, and oil flow visualization at engine representative operating conditions.

In addition to the above tests, the final validation test is conducted in a test gas turbine, which effectively represents the full operating boundary conditions that are prevalent for the entire operating range. For full engine testing, the measurement techniques employed include thermal paint, thermocouples, thermos-crystals, pyrometers, pressure taps, Kiel-Temperature and pressure probes and several other operating instrumentations. **Figure 37** shows a gas turbine with the airofoils painted with thermal paint and the additional instrumentation required to validate and monitor the airofoil and engine performances.

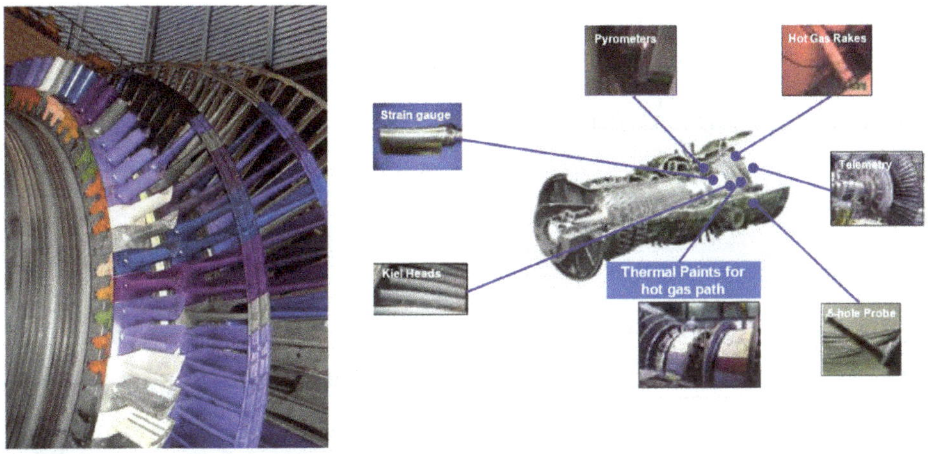

Figure 37. Gas turbine test engine and instrumentation [2].

7. Conclusion

The aerothermal design of gas turbine components has progressed at a rapid pace in the last decade with all gas turbine manufacturers, in order to obtain higher thermodynamic efficiencies. This has been achieved by using higher turbine inlet temperatures and pressures, advanced turbine aerodynamics, efficient cooling systems for turbine airofoils and advanced high temperature alloys, metallic coatings, and ceramic thermal barrier coatings.

In this chapter, some of the basic heat transfer phenomenon associated with both the external hot gas side and the coolant internal flows in turbine airofoils has been outlined. The external hot gas side heat transfer is largely driven by the unsteady and transonic high pressure and high temperature aerodynamic flows. By establishing the hot gas side external heat loads, which generally varies with different turbine design, it is possible to design for efficient airofoil internal cooling systems. The gas turbine airofoil internal cooling systems are however complex and varied in design, particularly with static and rotating airofoils. The internal cooling design of airofoils generally involves integration of several technology features such as film cooling, three-dimensional turbulators, pedestals, impingement cooling, and thermal barrier coatings.

Due to the complexity of the heat transfer phenomenon associated with gas turbine airofoils, it is also relatively common that final airofoil designs are thermally validated in several validation carriers which are representative of engine conditions. These validation carriers include high speed cascade rigs, scaled perspex models, airflow and qualitative heat transfer flow benches, and gas turbine test engines.

Author details

Shailendra Naik

Address all correspondence to: shailendra.naik@ansaldoenergia.com

Ansaldo Energia, Baden, Switzerland

References

[1] R.H. Kehlhofer, J. Warner, H. Nielsen, R. Bachmann, Combined Cycle Gas - Steam Turbine Power Plant, 2nd Edition, Pen Well Publishing Company, Oklahoma, ISBN 0.87814-736-5, 1998.

[2] S. Naik, T. Sommer, M. Schnieder, Aero-thermal design and validation of an advanced turbine, Proceedings of ASME Turbo-Expo Power for Land, Sea and Air, Paper GT2012-69761, June 5-7, Copenhagen, Denmark, 2012.

[3] S. Can Gülen. Étude on gas turbine combined cycle power plant—next 20 years. Journal of Engineering for Gas Turbines and Power, **138**(5) 051701, pp 1-10, 2015.

[4] J.C. Han, S. Dutta, S. Ekkad, Editors, Gas turbine heat transfer and cooling technology, Taylor and Francis, ISBN 1-565032-841-X, 2000.

[5] C. Lechner, J. Seume, Editors, Stationäre gasturbinen, Springer-Verlag, Heidelberg, ISBN 3 540-42831-3, 2003.

[6] The Gas Turbine Handbook, Ed. R. Dennis, published by National Energy Technology Laboratory, Department Of Energy, Morgantown, WV, USA, 2006.

[7] D.R. Clarke, M. Oechsner, N.P. Padture, Thermal-barrier coatings for more efficient gas-turbine engines, MRS Bulletin. 2012;**37**(10).

[8] W. Krebs, J. Hellat, A. Eroglu, Technische Verbrennungssystem, Ch. In C. Lechner and J. Seume, Editors, Stationäre Gasturbinen, Springer-Verlag, Heidelberg, ISBN 3-540-42831-3, 2003.

[9] T. R. Kingston, Rapid design, analysis and optimisation of cooled turbine blades, Paper

ISABE 2015-20275, 22nd International Symposium on Air Breathing Engines, October 23–25 2015, Phoenix, Arizon, USA.

[10] J. Krueckels, T. Arzel, T.R. Kingston, M. Schnieder, Turbine blade thermal design process enhancements for Increased firing temperatures and reduced coolant flow, Proceedings of ASME Turbo-Expo Power for Land, Sea and Air, Paper GT2007-27457, May 14–17, Montreal, Canada, 2007.

[11] M.G. Dunn, Convective heat transfer and aerodynamics in axial flow turbines, ASME Journal of Turbomachinery. 2001;**123**:637.

[12] J.A. Tallman, C.W. Haldeman, M.G. Dunn, A.K. Tolpadi, R.F. Bergholz, Heat transfer measurements and predictions for a modern, high pressure transonic turbine including endwalls, Transactions of the ASME, Journal of Turbomachinery. 2009;**131**.

[13] S. Naik, J. Krueckels, M.Gritsch, M. Schnieder, Multi-row film cooling perfomances of a high lift blade and vane, Transactions of ASME, Journal of Turbomachinery. 2014;**136**:051003-1.

[14] K. Takeishi, M. Matsuura, S. Aoki, T. Sato, An experimental study of heat transfer and film cooling on low aspect ratio turbine nozzles, Trans. ASME, J. Turbomach, 112(3), pp 488-496, 2008.

[15] J. Krueckels, W. Colban, M. Gritsch, M. Schnieder, Validation of a first vane platform cooling design, Proceedings of ASME Turbo Expo, GT2011-45252, June 6–10, Vancouver, Canada, 2011

[16] Barigozzi, G., Benzoni, G., Franchini, G., Perdichizzi, A., "Fan-shaped hole effects on the aero-thermal performance of a film cooled endwall," Transactions of the ASME, Journal of Turbomachinery. 2006;**128**:43–52.

[17] Colban, W., Thole, K.A., Haendler, M., A comparison of cylindrical and fan-shaped film-cooling holes on a vane endwall at low and high freestream turbulence levels, Trans. of ASME, Journal of Turbomachinery. **130**(3), 031007, pp 1–9, 2008.

[18] S. Naik, C. Georgakis, T. Hofer, D. Lengani, Heat transfer and film cooling of blade tips and endwalls, Transactions of ASME, Journal of Turbomachinery. 2012:**134**:041004-1.

[19] J.C. Han, Recent advances in turbine blade cooling, International Journal of Rotating Machinery. 2004;**10**(6):443–457.

[20] J.C. Han, A.P. Rallabandi, Turbine film cooling using PSP technique, Frontiers in Heat and Mass Transfer. 2010;**1**:013001.

[21] S. Baldauf, M. Scheurlen, A. Schulz, S. Wittig, Correlation of film-cooling effectiveness from thermographic measurements at enginelike conditions, ASME Journal of Turbomachinery. 2002;**124**:686.

[22] S. Luquel, J. Batstone, D.R.H. Gillespie,T. Povey, and E. Romer, Full thermal experimenta assessment of a dendritic turbine vane cooling scheme, Transactions of The

ASME, Journal of Turbomachinery. 2014;**136**:021011.

[23] J.R. Taylor, Heat Transfer Phenomena in Gas Turbines, Proceedings of ASME International Gas Turbine Conference, Paper 80-GT-172, March 10-13, New Orleans, USA, 1980.

[24] B. Weigand and S. Spring, Multiple jet impingement—a review, Heat Transfer Research. 2011;**42**(2):101–142.

[25] L.W. Florschuetz, D.E. Metzger, C.C. Su, Heat transfer characteristics for jet array impingement with initial crossflow, ASME Jorunal of Heat Transfer. 1984;**106**:34–41.

[26] R.E. Chupp, H.E. Helms, P.W. McFadden, T.R. Brown, Evaluation of internal heat transfer coefficients for impingement cooled turbine airfoils, Proceedings of AIAA, 4th Propulsion Joint Specialist Conference, Paper AIAA 68-564, 10–14 June, Ohio, USA,1968.

[27] M.E. Taslim, L. Setayeshgar, S.D. Spring, An Experimental Evaluation of Advanced Leading Edge Impingement Cooling Concepts, Trans of ASME, Journal of Turbomachinery, 123(1), pp 147-153, 2000.

[28] A. Terzis, P. Ott, J. Wolfersdorf, B. Weigand, M. Cochet, Detailed heat transfer distributions of narrow impingement channels for cast-in turbine airfoils, Transactions of ASME, Journal of Turbomachinery. 2014;**136**:091011-1.

[29] F. Hoefler, S. Schueren, J. Wolfersdorf, and S. Naik, Heat transfer characteristics of an oblique jet impingement configuration in a passage with ribbed surfaces, Transactions of ASME, Journal of Turbomachinery. 2011;**134**.

[30] S. Schueren, F. Hoefler, J. Wolfersdorf, S. Naik, Heat transfer in an oblique jet impingement configuration with varying jet geometries, Transactions of ASME, Journal of Turbomachinery. 2012;**135**(2).

[31] B. Facchini, F. Maiuolo, L. Tarchi, N. Ohlendorf, Experimental investigation on the heat transfer of a leading edge cooling system: effects of jet-to-jet spacing and showerhead extraction, Proceedings of ASME Turbo Expo 2013, GT2013-94759, June 3–7, San Antonio, Texas, USA, 2013.

[32] J.C. Han, J.S. Park, 1988. Developing heat transfer in rectangular channels with rib turbulators. International Journal of Heat and Mass Transfer. 1988;**31**(1):183–195.

[33] P. Ligrani, Heat transfer augmentation technologies for internal cooling of turbine components of gas turbine engines, International Journal of Rotating Machinery, v2013, Article ID 275653, 32 pages, 2013.

[34] B.V. Johnson, J.H. Wagner, G.D. Steuber, F.C. Yeh, "heat transfer in rotating serpentine passages with trips skewed to the flow," ASME Paper No. 92-GT-191, ASME Journal of Turbomachinery. 1992;**116**:113–123.

[35] W.L. Fu, L.M. Wright, J.C. Han, Rotational buoyancy effects on heat transfer in five different aspect-ratio rectangular channels with smooth walls and 45 degree ribbed walls, Transaction of ASME Journal of Heat Transfer. 2006;**128**(11):1130–1141.

[36] D.H. Rhee, D.H. Lee, H.H. Cho, Effects of duct aspect ratios on heat/mass transfer with discrete V-shaped ribs, Proceedings of ASME Turbo-Expo Power for Land,Sea and Air, Paper GT2003-38622, June 16-19, Atlanta, USA, 2003.

[37] R. Kiml, S. Mochizuki, A. Murata, J. Sulitka, Rib-induced secondary flow structures inside a high aspect ratio trapezoidal channel, Proceedings of the International Gas Turbine Congress, Paper IGTC2003Tokyo-078, Nov 2-7, Tokyo, Japan, 2003.

[38] L.M. Wright, Y.-H. Liu, J.-C. Han, S. Chopra, Heat transfer in trailing edge, wedge shaped cooling channels under high rotation numbers, ASME Journal of Heat Transfer. 2008;**130**.

[39] M.E. Taslim, T. Li, and S.D. Spring, "Experimental study of the effects of bleed holes on heat transfer and pressure drop in trapezoidal passages with tapered turbulator," ASME Journal of Turbomachinery. 1995;**117**:pp. 281–289.

[40] S. Naik, S. Retzko, and M. Gritsch, Impact of turbulator design on the heat transfer in a high aspect ratio passage of a turbine blade, Proceedings of ASME Turbo-Expo for Land, Sea and Air, Paper GT2014-25841, June 16-20, Duesseldorf, Germany, 2014.

[41] M.E. Taslim, T. Li, S.D. Spring, "Measurements of heat transfer coefficients and friction factors in rib-roughened channels simulating leading-edge cavities of a modern turbine blade". Transactions of The ASME, Journal of Turbomachinery. 1997;**119**:602–609.

[42] N. Domaschke, J. von Wolfersdorf, K. Semmel, Heat transfer and pressure drop measurements in a rib leading edge channel, Trans of ASME, Journal of Turbomachinery, 134(6), 061006, pp 1-9, 2012.

[43] C. LeBlanc, S.V. Ekkad, T. Lambert, V. Rajendran, Detailed heat transfer distributions in engine similar cooling channels for a turbine rotor blade with different rib orientations, Trans of ASME, Journal of Turbomachinery, 135(1), 011034, 2012.

[44] H. Saxer-Felici, S. Naik, M. Gritsch, Heat transfer enhancement for a turbine blade leading edge passage using various turbulator geometries, Proceedings of ASME Turbo-Expo for Land, Sea and Air, Paper GT2014-26130, June 16-20, Duesseldorf, Germany, 2014.

[45] J. Armstrong D. Winstanley, "A review of staggered array pin fin heat transfer for turbine cooling applications," Transaction of the ASME, Journal of Turbomachinery. 1988;**110**:94–103.

[46] D.E. Metzger, W.B. Shepard, S.W. Haley, Row resolved heat transfer variations in pin/fin arrays including effects of non-uniform arrays and flow convergence, Proceedings of ASME Turbo-Expo for Land, Sea and Air, Paper 86-GT-132, June 8-12, Dusseldorf, Germany, 1986.

[47] G.J. VanFossen, "Heat-transfer coefficients for staggered arrays of short pin fins", Transactions of the ASME, Journal of Heat Transfer. 1982;**104**:268–274.

[48] M.K. Chyu, "Heat transfer and pressure drop for short pin-fin arrays with pin-endwall

fillet", ASME Journal of Heat Transfer. 1990;**112**:926–932.

[49] T.K. Kumaran, J.C. Han, and S.C. Lau, "Augmented heat transfer in a pin fin channel with short or long ejection holes", International Journal of Heat Mass Transfer. 1991;**34**;(10):2617–2628.

[50] J. Krueckels, S. Naik, A. Lerch, Heat transfer in a vane trailing edge passage with conical pins and pin-turbulator integrated configurations, Proceedings of ASME Turbo-Expo for Land, Sea and Air, Paper GT2014-25522, June 16-20, Duesseldorf, Germany, 2014.

[51] R. Bunker, Gas turbine heat transfer: 10 remaining hot gas path challenges, GT2006-90002, Proceedings of ASME Turbo Expo 2006: Power for Land, Sea and Air, May 8–11, Barcelona, Spain, 2006.

[52] M.K. Chyu and S.C. Siw, Recent advances of internal cooling techniques for gas turbine airfoils, Journal of Thermal Science and Engineering applications, 2013;**5**:021008-1.

[53] M. Cochet, W. Colban, M. Gritsch, S. Naik, and M. Schnieder, Thermal validation of a heat shield surface for a high lift blade profile, Proceedings of ASME Turbo Expo, Paper GT2011-46294, June 6–10, 2011, Vancouver, Canada, 2011

Permissions

All chapters in this book were first published in HEDES, by InTech Open; hereby published with permission under the Creative Commons Attribution License or equivalent. Every chapter published in this book has been scrutinized by our experts. Their significance has been extensively debated. The topics covered herein carry significant findings which will fuel the growth of the discipline. They may even be implemented as practical applications or may be referred to as a beginning point for another development.

The contributors of this book come from diverse backgrounds, making this book a truly international effort. This book will bring forth new frontiers with its revolutionizing research information and detailed analysis of the nascent developments around the world.

We would like to thank all the contributing authors for lending their expertise to make the book truly unique. They have played a crucial role in the development of this book. Without their invaluable contributions this book wouldn't have been possible. They have made vital efforts to compile up to date information on the varied aspects of this subject to make this book a valuable addition to the collection of many professionals and students.

This book was conceptualized with the vision of imparting up-to-date information and advanced data in this field. To ensure the same, a matchless editorial board was set up. Every individual on the board went through rigorous rounds of assessment to prove their worth. After which they invested a large part of their time researching and compiling the most relevant data for our readers.

The editorial board has been involved in producing this book since its inception. They have spent rigorous hours researching and exploring the diverse topics which have resulted in the successful publishing of this book. They have passed on their knowledge of decades through this book. To expedite this challenging task, the publisher supported the team at every step. A small team of assistant editors was also appointed to further simplify the editing procedure and attain best results for the readers.

Apart from the editorial board, the designing team has also invested a significant amount of their time in understanding the subject and creating the most relevant covers. They scrutinized every image to scout for the most suitable representation of the subject and create an appropriate cover for the book.

The publishing team has been an ardent support to the editorial, designing and production team. Their endless efforts to recruit the best for this project, has resulted in the accomplishment of this book. They are a veteran in the field of academics and their pool of knowledge is as vast as their experience in printing. Their expertise and guidance has proved useful at every step. Their uncompromising quality standards have made this book an exceptional effort. Their encouragement from time to time has been an inspiration for everyone.

The publisher and the editorial board hope that this book will prove to be a valuable piece of knowledge for researchers, students, practitioners and scholars across the globe.

List of Contributors

Vitor da Silva Rosa and Deovaldo de Moraes Júnior
Chemical Engineering Department, Santa Cecília University, Santos, São Paulo, Brazil

Adeniyi Jide Isafiade
Department of Chemical Engineering, University of Cape Town, South Africa

Alireza Bahadori
School of Environment, Science and Engineering, Southern Cross University, Lismore, NSW, Australia

Erick Salcedo
Departamento de Termofluidos, Facultad de Ingeniería, UNAM México, Ciudad de México, Mexico

César Treviño
UMDI, Facultad de Ciencias, Universidad Nacional Autónoma de México, Sisal, Yucatán, Mexico

Juan C. Cajas
Barcelona Supercomputing Center (BSC), Barcelona, Spain

Lorenzo Martínez-Suástegui
ESIME Azcapotzalco, Instituto Politécnico Nacional, Ciudad de México, Mexico

Tianyi Gao, Bahgat Sammakia and James Geer
SUNY Binghamton University, Binghamton, NY, USA

Barbaros Çetin
Department of Mechanical Engineering, Bilkent University, Ankara, Turkey

Kadir G. Güler
YETSAN Auto Radiator Co. Inc., Çorum, Turkey
General Electric Aviation Marmara Technology Center, Gebze, Istanbul, Turkey

Mehmet Haluk Aksel
Department of Mechanical Engineering, Middle East Technical University, Ankara, Turkey

Mark Christian E. Manuel
School of Mechanical and Manufacturing Engineering, Mapua Institute of Technology, Manila, Philippines

Po Ting Lin
Mechanical Engineering Department, Chung Yuan Christian University, Chungli City, Taoyuan, Taiwan

Magdalena Jaremkiewicz
Cracow University of Technology, Institute of Thermal Power Engineering, Cracow, Poland

Hua Wang, Qingtai Xiao and Jianxin Xu
Kunming University of Science and Technology, Kunming, PR China

Shailendra Naik
Ansaldo Energia, Baden, Switzerland

Index

www.ingramcontent.com/pod-product-compliance
Lightning Source LLC
LaVergne TN
LVHW060556210926
848485LV00005B/1321